2008

2008
北京奥林匹克公园及五棵松文化体育中心
规划设计方案征集

International Competition for Conceptual Planning and
Design of Beijing Olympic Green & Wukesong Cultural and
Sports Center

Beijing Olympic Games

北京市规划委员会　北京水晶石数字传媒　编　Edit: Beijing Municipal City Planning Commission　Beijing Crystal Digital Media

中国建筑工业出版社　China Architecture & Building Press

图书在版编目(CIP)数据

2008北京奥林匹克公园及五棵松文化体育中心规划设计方案征集／北京市规划委员会等编．－北京：中国建筑工业出版社，2003
ISBN 7-112-05556-3

Ⅰ.2...Ⅱ.北...Ⅲ.夏季奥运会－体育建筑－设计方案－北京市－图集 Ⅳ.TU245.4-64

中国版本图书馆CIP数据核字(2002)第095247号

顾问主编：刘敬民　单霁翔
Consulting Editors: Liu Jingmin　Shan Jixiang
主　　编：平永泉　陈　刚
Chief Editors: Ping Yongquan　Chen Gang
副 主 编：姚　莹　魏成林　黄　艳　邱　跃　邢玉海
Associate Chief Editor: Yao Ying　Wei Chenglin　Huang Yan　Qiu Yue　Xing Yuhai
执行主编：黄　艳
Executive Chief Editor: Huang Yan
执行副主编：张立新　李晓鸿
Executive Associate Chief Editor: Zhang Lixin　Li Xiaohong
编　　委(按姓氏笔划为序)：冯琳　白晨曦　卢正刚　成砚　李晓鸿　张立新
　　　　　　　　　苏晶　何小坚　郑燕华　夏林茂　黄艳　黄凯陵
Editorial Board: Feng Lin　Bai Chenxi　Lu Zhenggang　Cheng Yan　Li Xiaohong　Zhang Lixin
　　　　　　　Su Jing　He Xiaojian　Zheng Yanhua　Xia Linmao　Huang Yan　Huang Kailing
责任编辑：何　楠　尹怀梅
Editor: He Nan　Yin Huaimei
装帧设计：邬永柳　蒙安童　郭丽伟　陈卓夫
Layout: Wu Yongliu　Meng Antong　Guo Liwei　Chen Zhuofu
封面设计：蒙安童
Cover: Meng Antong

2008北京奥林匹克公园及五棵松文化体育中心规划设计方案征集
北京市规划委员会　编
北京水晶石数字传媒
＊
中国建筑工业出版社出版、发行(北京西郊百万庄)
新华书店经销
北京水晶石数字传媒制版
利丰雅高印刷（深圳）有限公司印刷
＊
开本：965×1270毫米 1/12 印张：24 1/3
2003年1月第一版 2003年1月第一次印刷
印数：1—3000册 定价：298.00元
ISBN 7-112-05556-3
TU·4883 (11174)
版权所有　翻印必究
如有印装质量问题，可寄本社退换
（邮政编码：100037）
本社网址：http://www.china-abp.com.cn
网上书店：http://www.china-building.com.cn

序 言

百年奥运，中华圆梦。在党中央、国务院的正确领导下，在全国各族人民和海外华人、华侨及国际友人的热情支持下，经过不懈的努力，2001年7月13日，北京终于赢得了2008年奥运会的承办权。这为首都的现代化建设提供了千载难逢的历史机遇，增添了强大的发展动力。

党的十六大报告明确指出：新世纪头20年，对我国来说，是一个必须紧紧抓住并且可以大有作为的重要战略机遇期。从现在起到2008年，以至更长的一段时期，首都北京正在步入一个以筹办和举办奥运会为特征的加速发展时期，我们面临着前所未有的有利环境和条件。

为贯彻落实中央"努力办好2008年奥运会"的要求，我们一定要以"三个代表"重要思想为指导，以"新北京，新奥运"为主题，突出"绿色奥运、科技奥运、人文奥运"的理念，举办一届最出色的奥运会，始终坚持把发展作为第一要务，以奥运促发展，以发展助奥运，在全面建设小康社会的基础上，加快率先基本实现现代化的步伐。

目前，筹办奥运会各项工作正在紧张而有序地展开。2002年，北京市规划委员会对奥林匹克公园和五棵松文化体育中心的规划设计方案组织了大规模的国际征集活动，国内外上百家知名规划设计单位递交了众多优秀方案，为我们进一步深化完善和最终确定规划设计方案奠定了良好的基础。

今后，在其他奥运项目的规划设计和建设中，我们将始终坚持公平、公正、公开的原则，广泛采取向国内外企业公开招标等国际通行的做法。在此，我们真诚欢迎国内外知名规划设计单位积极参与北京奥运项目的规划设计，热切期盼广大中外企业家来京投资，我们将竭诚提供一流的环境和最优质的服务。

2008年转瞬即至，拥抱北京，拥抱奥运，机遇无限！

让我们传承历史，开创未来，共同发展，共铸奥运辉煌！

北京市市长
第29届奥林匹克运动会组织委员会主席

2002年11月

Preface

It has been our dream for hundred of years to hold the Olympic Games. Under the leadership of the Party and the State Council, the support of the people of all nationalities and overseas Chineses, as well as the friends from other countries, and through relentless effort, Beijing has eventually won the bid to hold the 2008 Olympic Games on July 13, 2001. This has offered Beijing an unprecedented opportunity for its construction of modernization and a momentum for its development.

As the 16th Party congress points out, for our country, the first 20 years in the new century is an important strategic chance that we must seize and do something worthwhile. From now on till 2008 and longer, Beijing has entered a period of rapid development characterized by preparing and holding the Olympic Games. The favorable environment and conditions we are encountering are unprecedented.

In order to carry out the requirement of " holding the Olympic Games successfully" put forth by the Party central committee, we must follow the important thought of the "Three Represents", use "New Beijing ,Great Olympic" as the theme and stress the concept of the "Green, High-tech and People's Olympic Games" to make the Games the most excellent one. We must also regard the development as the priority, boost development by hold the Games and aid the Games with development, quickening the pace of fundamentally realizing modernization on the basis of constructing a well-off society.

At the moment, the preparation for the Olympic Games is carried out in a tense and orderly way. In 2002, the Beijing Urban Planning Commission organized a large scale public bidding for the designs of the Olympic Green and the Wukesong Cultural and Sports Center, receiving more than a hundred excellent designs from well-known designing firms from China and abroad. This has laid down a solid foundation for us to further perfect and decided on the final designing plan.

In the future, for the designing and construction of other programs of Olympic Games, we will consistently abide by the principle of being just, fair and open, and adopt the international practice of open bidding to enterprises in China and other countries. Thus, we would like to welcome the famous designers both from China and abroad to actively take part in the designing work for the Olympic Games projects, and warmly expect Chinese and foreign entrepreneurs to make investment in Beijing. We will go all out to provide first-class environment and best quality services for them.

The year of 2008 will arrive soon, so let's embrace Beijing, the Olympic Games and seize the opportunity!

Let us carry forward the history, create our future, advance together and achieve a brilliant glory for the Olympic Games!

Liu Qi
Mayor of Beijing
Chairman of the Beijing Organizing Committee for the Games of the XXIX Olympiad
November 2002

前言

奥运会，一个我们期盼了多年的梦想，终于在新世纪的第一年实现了。自2001年7月13日北京赢得2008年夏季奥运会主办权以来，北京奥运会设施的建设筹备工作就开始成为社会各界关注的焦点。面临这样一次大型的世界体育盛会，机遇与挑战并存，北京正在以前所未有的发展速度，创新的姿态，迎接这样一次世纪盛会。

千里之行，始于足下。目前，距北京举办奥运会只有不到六年的时间，相关的规划设计工作正在按计划紧锣密鼓地展开。根据场馆总体规划，北京奥运设施的布局基本可描述为"一个中心加三个区域"。其中的"一个中心"即奥林匹克公园，这里集中了10多个比赛场馆，奥运会一半以上的金牌将产生于此；"三个区域"包括"西部社区"、"大学区"和"北部风景旅游区"，而五棵松文化体育中心即是"西部社区"中最主要的一个场馆建设区。

2002年3月至7月，在北京市政府和第29届奥运会组委会的授权下，由北京市规划委员会组织，面向国内外公开征集奥林匹克公园和五棵松文化体育中心的规划设计方案。这次活动作为北京申办奥运成功以来第一次大型的规划设计方案征集活动，受到了世界各国规划设计团体的热切关注，前后共有21个国家和地区的100多家设计单位参与，最终收到了90多个方案，使之成为目前为止，北京地区乃至全国范围内最大规模的一次方案征集活动。所征集到的方案创作思想活跃、表现手法多样、文化内涵丰富、文件制作精良，充分显示了国内外规划设计单位对北京举办2008年奥运会的热情支持。在方案评审活动结束后，主办单位组织了公开展览，也得到了各社会团体、广大市民和国内外友人的热烈反响，收到了许多有益的建议和意见。

为了答谢所有的竞赛参与单位，完成对这一历史事件的纪录，现由北京市规划委员会组织，将此次征集活动的方案编辑出版，希望能够进一步促进我国规划设计市场的开放和中外规划设计单位的交流，推动城市设计和奥运会相关规划设计的研究。

由于编者水平所限，有挂一漏万，取舍不当之处，敬请见谅，并欢迎提出宝贵意见。

编者
2002年11月

Preface

The first year in the new century witnessed the realization of our long-waited dream-----to host the Olympic Games. People of all walks of life have been following with interest the preparation work for the games ever since Beijing won the bid on July 3, 2001 to host the 29th Games in 2008. We will encounter both opportunities and challenges in such a large-scale sports gala; therefore, Beijing is welcoming this grand event with an unpresidented speed of development and innovative spirit.

A thousand-li journey is started with taking the first step. As there are only 6 years to go before the Games is held in Beijing, all planning work is under way in an intense way. According to the general planning, sports facilities in Beijing can be referred to as "one center plus three areas". The center refers to the Olympic Park which comprises over 10 sports venues, and where more than half of the gold medals will be competed for. The three areas refer to the western area, college student area, and northern tourist area. The Wukesong Culture and Sports Center is in fact the most important venue in the western area.

From March to July 2002, authorized by the Beijing municipal government and the Organizing Committee of the 29th Olympic Games, the Beijing City Planning Commission organized an open design competition for the Olympic Park and the Wukesong Culture and Sports Center. As the first large event after the successful bid to host the Games, and open to designers both at home and abroad, this activity has aroused enthusiastic responses all over the world. All told, over 100 designers from 21 countries and regions were involved with more than 90 designs submitted, making it the largest design colleting event not only in Beijing, but also in China. The design entries were all creative in concept, varied and rich in cultural contents, and executed excellently, showing the positive support for Beijing to host the Games in 2008 from domestic and foreign designers. An exhibition was held right after the appraisal work was finished, which got favorable responses and constructive suggestions from people of China and other countries.

In order to thank all the participants and record this historic event, the Beijing City Planning Commission has organized the publication of the participating designs, with a view of further opening the designing market to the outside world, and promoting exchanges in this field and the research on city and Olympic-related planning.

As the selection is far from perfect, any suggestion is welcome for future revision.

Editor
November 2002

General 总目录 contents

2008 北京奥林匹克公园及五棵松文化体育中心规划设计方案征集
Conceptual Planning and Design of Olympic Green and Wukesong Cultural and Sports Center

北京奥林匹克公园和五棵松文化体育中心规划设计方案征集活动新闻发布稿 6
Press Release for Conceptual Planning and Design of Beijing Olympic Green and Beijing Wukesong Cultural and Sports Center

北京奥林匹克公园规划设计方案征集书 8
Document of International Competition for Conceptual Planning and Design of Beijing Olympic Green

北京五棵松文化体育中心规划设计方案征集书 16
Document of International Competition for Conceptual Planning and Design of Beijing Wukesong Cultural and Sports Center

北京奥林匹克公园现场踏勘提问及答疑 22
Faq on the Field Survey of Beijing Olympic Green

北京五棵松文化体育中心现场踏勘提问及答疑 26
Faq on the Field Survey of Beijing wukesong cultural and Sports Center

评委简历 28
Background of the Jury Members

北京奥林匹克公园和五棵松文化体育中心规划设计方案评审总结 30
Assessment Report of International Competition for Conceptual Planning and Design of Beijing Olympic Green and Wukesong Cultural and Sports Center

北京奥林匹克公园规划设计方案 34
Conceptual Planning and Design of Beijing Olympic Green

北京五棵松文化体育中心规划设计方案 208
Conceptual Planning and Design of Beijing Wukesong Cultural and Sports Center

后记 292
Postscript

北京奥林匹克公园
和五棵松文化体育中心
规划设计方案征集活动
新闻发布稿

经北京市政府和第29届奥林匹克运动会组委会授权，由北京市规划委员会组织进行北京奥林匹克公园和五棵松文化体育中心规划设计方案的征集活动。

北京奥林匹克公园和五棵松文化体育中心是北京为举办第29届奥运会确定的两处大型奥运会设施集中建设场地。为充分体现"绿色奥运、科技奥运、人文奥运"的宗旨，将项目所在地规划建设成为提供最佳赛事组织条件，并有利于城市长远发展的多功能公共活动中心，北京市规划委员会诚邀国内外关心奥林匹克运动、富有创新精神、具有丰富规划设计经验的设计单位参与这两个项目的规划设计。

北京奥林匹克公园位于北京市区北部、城市中轴线的北端。将规划建设成为集体育、文化、展览、休闲、观光旅游为一体，并有配套商业、酒店、会议等服务设施的多功能区域和充满活力的、市民喜爱的城市公共活动中心。

规划总用地约1135ha，其中森林公园约680ha，奥运中心区用地约405ha，中华民族园及部分北中轴路用地约50ha。

在奥运中心区四环路以北的建设用地中安排总建筑规模约216万m²的设施，其中包括体育场馆(80 000座的国家体育场、18 000座的国家体育馆、15 000座的国家游泳中心)及其附属商业运营设施共约40万m²；文化设施约20万m²；会展博览设施约40万m²；奥运村运动员公寓约36万m²；商业服务设施(包括商业、酒店、商务办公、娱乐等内容)约80万m²。

规划设计内容包括景观规划、总体布局及分期建设概念规划、交通规划、市政概念规划方案、地下空间规划等。

北京五棵松文化体育中心位于城市西部，长安街西延长线的北侧，既是北京为举办第29届奥运会规划安排的一处主要的比赛场地，同时，也将规划成为北京市西部地区市民进行体育活动和文化休闲活动的城市公共活动中心。

规划总用地约50ha，包括城市公共绿化带用地约11ha、公园绿地10ha、城市建设用地约29ha。总建筑规模约20万m²，包括奥运会比赛用篮球馆(18 000座，永久设施)、棒球场(25 000座，临时设施)和垒球场(8500座，临时设施)、群众性文化体育设施、商业服务设施等。规划设计内容包括用地布局规划、交通规划、绿地及景观规划、篮球馆单体建筑概念设计等。

本次方案征集采用公开方式。两个项目的方案征集和评审分别进行，设计单位可以选择参加其中的一个，也可以同时参加两个。

报名表可从网站www.bjghw.gov.cn免费下载，也可专人到北京市规划委员会领取。

Press Release

征集书通过购买方式获得。发售方式有现场发售和邮寄两种。

报名时间从2002年4月2日9时(北京时间，下同)起，截止为2002年4月15日17时。提交方案的截止时间为2002年7月2日17时。2002年7月由专家评审委员会进行评审。

北京奥林匹克公园规划设计方案征集设一等奖1名，二等奖2名，优秀奖5名。

北京五棵松文化体育中心规划设计方案征集设一等奖1名，二等奖1名，优秀奖3名。

本次活动由北京市公证处进行公证。

有关两个项目方案征集的详细信息可查询第29届奥林匹克运动会组委会网站和北京市规划委员会网站。

我们期待着您的参与。

北京市规划委员会
2002年3月31日

Press Release for Conceptual Planning and Design of Beijing Olympic Green and Beijing Wukesong Cultural and Sports Center

Press Release

and commercial affairs, hotels, entertainment, etc.).

The contents of the planning and design of Olympic Green include landscape planning and design, overall layout planning, phasing conceptual planning, transportation and traffic planning, utility conceptual planning, underground space conceptual planning, etc.

Located in the west of the city, Beijing Wukesong Cultural and Sports Center is in the north of the extension of the West Chang'an Avenue. It will not only serve as one of the major areas for the 29th Olympic Games, but also be designed as a public center of sports and entertainment for residents living in the west of the city.

International Competition for Conceptual Planning and Design of Beijing Olympic Green and Beijing Wukesong Cultural and Sports Center

Authorized by Beijing Municipal Government and Beijing Organizing Committee for the Games of the XXIX Olympiad, Beijing Municipal Planning Commission will solicit the conceptual planning and design schemes for Beijing Olympic Green and Beijing Wukesong Cultural and Sports Center.

The Beijing Olympic Green and Beijing Wukesong Cultural and Sports Center are two large-scale sports areas designated for holding the 29th Olympic Games in Beijing. In order to achieve the aim of "Green Olympics, High-tech Olympics and People's Olympics" and make the two sites the best places not only for sports competition, but also multi-functional centers for the long-term development of the city, Beijing Municipal Planning Commission would like to have designers both from home and abroad, who are enthusiastic in the Olympic Games, full of innovation and have rich experiences in planning and design, to compete for the planning and design of these two projects.

Located in the north tip of the central axis, the Beijing Olympic Green will be built into a multi-functional area and a lively public center with sports, cultural, exhibition, entertaining and sightseeing facilities complimented with business, hotel and conference services.

The project will cover an area of approximately 1,135 hectares, out of which about 680 hectares will be the Forest Area, 405 hectares.

for the Olympic Central Area, and around 50 hectares for the Chinese Ethnic Culture Park and part for the Beizhongzhou Road.

Facilities of about 2.16 million square meters will be constructed to the north of the Fourth Ring Road in the Olympic Central Area. About 400,000 square meters will be utilized to build stadiums (a National Stadium with 80,000 seats; a National Gymnasium with 18,000 seats and a National Swimming Center with 15,000 seats) and auxiliary commercial facilities; 200,000 square meters for cultural facilities; another 400,000 square meters for convention and exhibition facilities; 360,000 square meters for athlete apartments and 800,000 square meters for business and service facilities (business

The planned area of Wukesong is approximately 50 hectares, in which 11 hectares are green belt along urban roads; 10 hectares open green space in parks; 29 hectares for urban construction. The total construction area is 200,000 square meters, including a basketball hall (a permanent facility with a capacity of 18,000 seats), a baseball field (a temporary facility with a capacity of 25,000 seats) and a softball field (a temporary facility with a capacity of 8,500 seats) for the 29th Olympic Games. The center will also include cultural, sports and business facilities for the public. The content of the planning and design includes site planning, transportation and traffic planning, green area and landscape planning, conceptual architecture design of the basketball hall, etc.

The competition is opene to all qualified design firms; the two proposals will be assessed separately. Designers can select and design one of the two projects or both projects.

Applications can be free of charge downloaded from: www.bjghw.gov.cn, or picked up in person at the Beijing Municipal Planning Commission.

The competition documents will be released either on spot or through mail.

The registration starts from 09:00, April 02nd, 2002 (Beijing time, same below) and ends at 17:00, April 15th, 2002. The deadline for submission will be at 17:00, July 02nd, 2002 and the assessment will be conducted on July.

1 first prize winner, 2 second prize winners and 5 honorable prize winners will be selected from the proposals for the Beijing Olympic Green.

1 first prize winner, 1 second prize winner and 3 honorable prize winners will be chosen from the proposals for the Beijing Wukesong Cultural and Sports Center.

The event will be supervised by the Beijing Notary Public Office.

For more detailed information, please visit the following websites listed below:

www.beijing-olympic.org.cn; www.bjghw.gov.cn

We are looking forward to your participation.

Beijing Municipal Planning Commission
March 31st, 2002

北京奥林匹克公园规划设计方案征集书

北京市规划委员会 2002年3月

第一部分 应征者须知（略）
第二部分 规划设计任务书

1 名称

北京奥林匹克公园

2 位置及用地范围

北京奥林匹克公园是为举办第29届奥运会而规划的一处大型体育设施集中建设场地。其选址位于北京市区北部、城市中轴线的北端（详见附图1）。

奥林匹克公园规划用地共1135ha（不包括穿过规划用地中部的五环路、四环路和辛店路），其中森林公园680ha（包括现状碧玉公园别墅区用地7.7ha），奥运中心区总用地405ha（包括四环路以北用地291ha、现状国家奥林匹克体育中心及其南部预留地共114ha），中华民族园及部分北中轴路用地50ha。

奥林匹克公园用地南部为元大都遗址公园中轴段。

3 规划设计目的和任务
3.1 目的

（1）规划建设国际先进水平的体育设施，为成功举办第29届奥运会创造条件；

（2）规划建设满足市民需要的多功能的城市公共活动中心，有利于城市长远发展。

3.2 任务

（1）城市景观规划研究范围：包括奥林匹克公园及其南部的元大都遗址公园中轴段；

（2）重点规划研究范围——奥运中心区四环路以北用地（B区，291ha）：总体布局规划、交通规划、市政概念规划、地下空间规划等，上述规划分2008年奥运会赛时和城市长期发展两个阶段进行表述。不进行单体建筑设计。

规划方案应具有可操作性并有利于实施，成为单体建筑设计的规划依据。

4 现状条件
4.1 用地条件
4.1.1 用地内现状

奥林匹克公园规划总用地约1135ha，分为A、B、C、D四个区，其中体育设施主要分布在A区和B区，这两个区组成奥运中心区（详见附图2）。

A区约114ha，A-1区为现状国家奥林匹克体育中心，约56ha，需保留；A-2区为其南部预留地，约50ha，为城市总体规划预留的体育用地；A-3区为现状清洁车辆四厂，约8ha，暂作为规划预留地。

B区约291ha，位于四环路以北、中轴线两侧，包括城市建设用地和道路、绿化用地等。其中B-1区约261ha，B-2区约30ha。B区现状主要为农田、村庄、仓库等，用地内的建筑大部分将拆除，包括位于四环路以北的观天下别墅区。但位于B-1区内的沿北辰东路西侧的现状凯迪克大酒店需保留（用地约1ha）；

C区约680ha，为规划城市森林公园用地，位于市区中心区与边缘集团之间的绿化隔离地区内。其中C-1区373ha，C-2区300ha，目前已形成的部分林地和水面原则上予以保留；C-3为现状碧玉公园别墅区，用地7.7ha，需保留。用地内现有22万V高压供电走廊，需保留；

D区约27ha，为中华民族园，需保留；

E区约23ha，为已实现的北中轴路部分路段，红线宽度200m，道路中线两侧已建成以草坪为主的绿化带；

F区位于奥林匹克公园外，约28ha，为元大都遗址公园中轴段。

4.1.2 用地地质条件

从宏观地质条件分析，奥林匹克公园内的城市建设用地位于地形平坦的河洪冲积沉积区，地质构造条件中等，地质环境条件属简单类型，适于奥运场馆及其他各项设施的建设。

4.1.3 周边地区情况

1990年第11届亚洲运动会的召开，带动了中轴线北部地区的建设，除建成现状国家奥林匹克体育中心、亚运村外，还建设了国际会议中心、酒店、博物馆等公共设施，相关交通设施、市政设施比较完善。亚运会后，经过近十几年的发展，该地区已经成为北京最具吸引力的区域之一。

奥林匹克公园周围土地开发强度较高，东部有亚运村、慧忠里、安慧里和安慧北里、小关北里等居住区，西部有华严里和安翔里等居住区，总建筑量超过1000万m²。城市土地使用情况见下表：

用地性质	用地面积（ha）	所占比例（%）	建筑面积（万m²）	所占比例（%）
行政办公用地	49.99	3.32	96.20	4.56
商业金融用地	124.35	8.26	339.07	16.07
文化娱乐用地	15.74	1.05	31.62	1.50
医疗卫生用地	29.84	1.98	42.94	2.04
教育科研设计用地	226.57	15.05	264.52	12.54
宗教社会福利用地	3.24	0.22	3.71	0.18
水域	15.86	1.05	0.00	0.00
公共绿地	179.01	11.89	0.00	0.00
城市工业用地	40.92	2.72	42.26	2.00
乡镇企业	15.63	1.04	15.63	0.74
居住用地	547.75	36.37	1145.32	54.29
中学、小学、托幼用地	74.45	4.94	44.00	2.09
停车场库用地	1.57	0.10	0.62	0.03
市政公用设施用地	177.49	11.79	81.73	3.87
仓储用地	3.49	0.23	2.02	0.10
总计	1505.90	100.00	2109.64	100.00

DOCUMENT OF INTERNATIONAL COMPETITION FOR CONCEPTUAL PLANNING & DESIGN OF BEIJING OLYMPIC GREEN

Part I Procedure and Rules (Omitted)
Part II Planning Brief

1. Project Name
Beijing Olympic Green

2. Location and Site
The Beijing Olympic Green is a large-scale cluster of sports facilities purposely built for the holding of the 29th Olympic Games. The site is located in the northern part of the Beijing City, and at the northern end of the City Central Axis (refer to Figure 1 for details).

The total site area in the planning for the Olympic Green is 1,135ha (excluding the Fifth Ring Road, Fourth Ring Road and Xindiancun Road which dissect through the site). Part of the Olympic Green accommodates the Forest Area of 680 ha (including the existing Biyu Villa of 7.7ha), the Olympic Central Area of 405 ha (including the land of 291 ha north of the Yuan Dynasty City Wall Relics Park, the existing National Olympic Sports Center, and the land to its south, which add up to a total of 114 ha), Chinese Ethnic Culture Park and part of the northern axis which add up to a total of 50ha.

To the south of the Olympic Green is the Yuan Dynasty City Wall Relics Park.

3. Objectives and Missions
3.1 Objectives
(1) To plan and build world-class sports facilities, for successfully holding the 29th Olympic Games;
(2) To plan and build a multi-functional city center with public spaces for activities, in benefiting the long-term development of the city.

3.2 Missions
(1) A master layout and landscape planning for a wider context, including the Olympic Green and the central part of the Yuan Dynasty City Wall Relics Park.
(2) Focused analysis of key area which is to the north of the Fourth Ring Road in the Olympic Central Area, that is the master planning, the conceptual phasing planning, transport and traffic planning, utility conceptual planning, and underground space planning, etc. No design of specific building is required.

The planning and design proposal should be feasible in implementation, and form the basis for architecture design of individual buildings.

4. Existing Conditions
4.1 Land Use Conditions
4.1.1 Existing land use conditions
The Olympic Green occupies a total land of approximately 1,135 ha, and is divided into four districts: A,B,C and D. The sports facilities are mainly distributed in Site A and B, which form the Olympic Central Area (see Figure 2).

Site A is about 114 ha, Site A-1 accommodates the existing National Olympic Sports Center of approximately 56 ha, which needs to be retained;
To the south is Site A-2 of approximately 50 ha, which is the land reserve for sports facilities in the city master planning;
Site A-3 of approximately 8 ha accommodates the existing depot for the No.4 sanitary vehicles, which is the land reserve in this master planning.

Site B is about 291 ha, situated to the north of the Fourth Ring Road and to the two sides of the Central Axis, including the infrastructure land, road and green spaces, etc;
Site B-1 of about 261 ha is for the Olympic sports facilities, city public facilities and services, and public open spaces;
Site B-2 of about 30 ha is reserved for the athlete apartments in the Olympic Village;
Existing land use of Site B are predominantly agricultural land, villages and warehouses etc., which will mostly be demolished upon development, including the Guan Tian Xia Villa to the north of the Fourth Ring Road. However, the CATIC Hotel (B-3) at the western side of the Beichendong Road should be retained. (Land accommodating the hotel is about 1 ha.)

Site C is about 680 ha, and is planned for the Forest Area. This Site is situated between the fringe of the urban Center and the reserved green buffer. Site C-1 is 373 ha, existing forest and water body should be retained in principle;
Site C-2 is 300 ha, existing forest and water body should be retained in principle;
Site C-3 is 7.7 ha, which accommodates the Biyu Villa, should be retained. A 220,000V corridor running through the District should also be retained;
Site D is 27 ha, which accommodates the Chinese Ethnic Culture Park, should be retained;
Site E is 23 ha, it is part of Beizhongzhou Road. The width of the planned property line is 200m. Green area of lawns are located along two sides of the central-line of the road, this can be retained or enhanced with landscaping;
Site F is 28 ha, which is outside the boundary of the Olympic Green. It cuts across the Yuan Dynasty City Wall Relics Park. It needs to be improved with visual and landscaping treatments.

4.1.2 Geological conditions of the site
Having analyzed the geological conditions on a macro scale, it is found that the building area of the Olympic Green is located on the terrain of flat topography. Geological conditions of ground structure are to medium standard. No complications are found on the geological conditions. It is suitable for the construction of the Olympic sports facilities, other ancillary facilities and infrastructure.

4.1.3 Land uses of the surrounding
The XXI Asia Games in 1990 brought about development projects along the area north to the Central Axis. Apart from the existing National Olympics Center, the Asian Games Village, there are also public facilities such as the international convention Center, hotels, museums, etc. Provision of ancillary transport facilities and infrastructure are also satisfactory. After the Asian Games, through recent developments in this decade, the district has become one of the most attractive districts in Beijing.

Development intensity of land surrounding the Olympic Green has been high. To its east are residential areas like the Asian Games Village, Huizhongli, Anhuili, North Anhuli and North Xiaoguan, etc. To its west are residential areas like Huayanli and Anxiangli. Total Floor Area of the above covers more than 10,000,000 sqm. The following table shows the usage of land:

Land use	Size (Ha)	Ratio (%)	Total Floor Area (in 10,000 sqm)	Ratio (%)
Administrative Uses	49.99	3.32	96.20	4.56
Commercial & Financial Uses	124.35	8.26	339.07	16.07
Culture & Leisure Uses	15.74	1.05	31.62	1.50
Medical & Health units	29.84	1.98	42.94	2.04
Education & Research Uses	226.57	15.05	264.52	12.54
Religion & Social Service	3.24	0.22	3.71	0.18
Water Area	15.86	1.05	0.00	0.00
Green Space	179.01	11.89	0.00	0.00
Industry	40.92	2.72	42.26	2.00
Township Enterprises	15.63	1.04	15.63	0.74
Residential Area	547.75	36.37	1145.32	54.29
School and Kindergarten	74.45	4.94	44.00	2.09
Parking	1.57	0.10	0.62	0.03
Utilities	177.49	11.79	81.73	3.87
Warehouse	3.49	0.23	2.02	0.10
Total	1505.90	100.00	2109.64	100.00

The above land uses are located within the city planning and development area. In addition, the green buffer to the north-east occupies an area of 715 ha.

4.2 Transport and Traffic
The road network of the Beijing City is made up of expressways of ring roads and radical roads. The Beijing Old City is surrounded by the Second Ring Road. Major part of the urbanized Beijing City is within the Fourth Ring Road. Green belt is planned between the Fourth Ring Road and Fifth Ring Road.

上述用地均位于规划城市建设区内,另外,用地东北方向邻近的绿化隔离地区绿化用地约715ha。

4.2 交通条件

北京城市中心区骨干路网主要由环路加放射路构成(详见附图3),二环路以内为旧城,四环路以内为主要的城市建设区,四环路与五环路之间规划有绿化隔离地区,奥林匹克公园即位于北四环路与北五环路之间。

目前,奥林匹克公园周边尚无轨道交通线路可达。根据城市总体规划,未来将有五条规划轨道交通线路在奥林匹克公园周边通过,分别为4号线、5号线、8号线、10号线和13号线(北京城市铁路),其中13号线正在建设,预计2002年底竣工,5号线正在进行试验段建设。规划设想主要利用8号线和10号线部分线位修建奥运支线到达奥林匹克公园中心区,奥运支线沿地铁8号线从奥林匹克公园向南修建到10号线,再沿10号线向东与5号线和13号线构成2个交叉换乘站(详见附图4、5)。

奥林匹克公园周边有多条高速公路和城市快速路、主次干道等,与其他地区交通联系方便。

从奥林匹克公园所处位置和交通流向来看,主导交通流向为市中心方向,主要的集散道路为八达岭高速路、四环路、五环路、中轴路、安立路等,大屯路、成府路和北土城路为贯穿城市北部地区的东西向城市主次干道,也将承担部分集散交通量(详见附图5)。

4.3 市政条件

奥林匹克公园四环路以南部分现状市政条件较好,土城北路、北四环路、安立路、中轴路等道路上敷设有各种市政管线。规划新建项目的外部市政条件基本可由上述现状市政管线提供。

奥林匹克公园四环路以北部分,供水、雨水、污水、燃气等具备一定基础条件,有若干现状管线。但需要根据规划新增的建设项目,在供水、供电、供热、供气、排水、信息网络等方面规划建设大量新的设施和管线。如新建220kV变电站一座和110kV变电站两座;敷设综合信息管道;新建独立电信局等。

5 功能定位和规划设计指导思想

5.1 功能定位

奥林匹克公园的功能定位是充满活力的、市民喜爱的城市公共活动中心,是集体育、文化、展览、休闲、观光旅游为一体,并有配套商业、酒店、会议等服务设施的多功能区域。应具有十分鲜明的纪念性,成为奥林匹克运动遗产的标志。建设形成空间开敞、绿地环绕、环境优美的城市形象,并体现整体性、民族性和时代性。

5.2 规划设计指导思想

5.2.1 考虑奥运会的建设和城市长远发展相结合

要考虑举办奥运会和残奥会期间约两个月的使用要求,更要注重奥运会前后、尤其是奥运会后的长期发展和使用要求,与城市整体规划紧密衔接,平衡发展,保持该地区长久的活力。

其中,要特别着重研究奥运会设施的赛后利用问题。并且要坚持勤俭节约,力戒奢华浪费。

5.2.2 延续和发展城市中轴线,体现文化内涵

北京的城市中轴线既是传统的轴线,更是不断发展的轴线。1993年《北京城市总体规划》(1991年~2010年)中明确提出:"中轴北延长线要保留宽阔的绿带,在其两侧和北端的公共建筑群作为城市轴线的高潮与终结,突出体现21世纪首都的新风貌。"

如何处理好中轴北端的城市空间形象十分重要。规划方案应反映并强化中轴线在北京城市规划建设中的地位,既保护和延续历史,又反映出新的时代精神和北京发展建设的新成就,形成新的城市形象。

International

5.2.3 体现"绿色奥运、科技奥运、人文奥运"的宗旨

规划设计方案应体现"绿色奥运、科技奥运、人文奥运"的宗旨。要有利于环境保护和可持续发展,采用当今国际上最新的规划设计理念和先进的技术手段,提高生态环境质量;广泛普及奥林匹克精神,弘扬中华民族优秀传统文化,展现北京历史文化名城风貌,建立与举办奥运会相适应的人文环境;充分考虑各类人员的需要,尤其要关注残疾人和有行动障碍人员的需求等。

5.2.4 近远期规划建设结合,预留发展空间

规划布局应能满足分期建设实施的要求。用于奥运会的体育设施、会展博览设施、广场以及必要的配套设施等应适当集中布局,有利于在2008年奥运会举办前建成,形成相对完整的城市形象。同时要考虑为奥运会赛后的城市建设预留发展空间。

应充分考虑奥运会期间需要的临时设施空间和组织人流集散的广场空间。

5.2.5 规划具有可操作性,有利于开发运营

奥林匹克公园内安排的建设项目中,有的能够带来收益,有的是非盈利性的,要研究如何规划有利于建设实施,鼓励提出有关商业开发和市场化运作方面的建议。

6 规划设计内容

6.1 中轴线

从北土城路向北到四环路段的中轴线已基本形成,四环路以北的中轴线将随着奥林匹克公园的建设而形成。

在规划中,要着重研究中轴线新建部分与已有部分的衔接,考虑北京城市整体空间环境规划和中轴线整体城市设计的背景,研究如何延续旧城中轴线的传统,整合已建成的和规划的各项景观要素。

6.2 森林公园

森林公园应具有改善生态、美化环境、提供游憩等功能,以大面积林木为主,目前已经建成的树林原则保留,并与规划有机结合。

可考虑人工建设形成100ha左右的水面,并可利用挖掘扩大水面所取土方堆筑山体,用山水合一的自然景观作为北中轴的对景或背景。

规划北辰东路和北辰西路可考虑穿过公园,也可考虑终止于辛店村路。在四周主要道路上应设出入口,并在出入口附近适当安排停车场和公交车站。

6.3 元大都遗址公园

在奥林匹克公园南部有一条东西向为主的城垣遗址,是

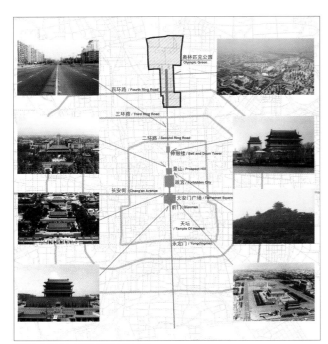

附图1:北京城市中轴线主要景观示意图
LANDMARKS ALONG THE CENTRAL AXIS

Currently, there is no railway reaching the Olympic Green. According to the master planning of the city, there will be five planned railways reaching the Olympic Green. They are the No.4 Rail, No.5 Rail, No.8 Rail and No.10 Rail. No. 13 Rail is under construction and will be accomplished at the end of 2002, No.5 Rail is under trail construction. It is suggested that Olympic Subway will use No.8 Rail and part of No. 10 Rail. The Subway will use No.8 Rail to south and meet the No.10 Rail, then the Lateral will go east to meet No.5 Rail and No.13 Rail, two exchange stations will be built (see Figure 4.5).

The Olympic Green is served by several expressways, city highways, primary and secondary distributor roads. Accessibility and connection to other districts are good.

Based on the location of the Olympic Green and the traffic generation and capacity, the traffic should flow towards the city Center. Primary distributor roads should include the Badaling Expressway, Fourth Ring Road, Fifth Ring Road, Axis Road, Anli Road, etc. Datun Road, Chengfu Road and Beitucheng Road intersect with the major distributor that connects the north and north-east of the city. They will also absorb some of the traffic flow (see Figure 5).

4.3 Utilities

Existing utility conditions to the south of the Fourth Ring Road of the Olympic Green is satisfactory. Beitucheng Road, North Fourth Ring Road, Anli Road and Zhongzhou Road(the central axis road) are laid with utility pipes and lines. Newly planned development projects can be sustained by the existing utilities.

Water supply, drainage, sewage and gas supply facilities are accommodated to the north of the Fourth Ring Road of the Olympic Green. However, these facilities should be upgraded according to the additional need by the newly planned development projects, in aspects such as water supply, electricity supply, heating supply, gas supply, sewage system, information network, etc. For example, a

long-term utilization and operational needs before and after the Olympic Games. There should be a mix of land use, integration with the overall planning of the City, a balanced and sustainable development that ensures the long-term vibrancy of the area.

Specifically, the utilization of the sports facilities after the Olympic Games should be studied with focus. With a major aim to meet the technical requirements of the Olympic Games, all facilities should be designed to the maximum flexibility to achieve the maximum socio-economical effectiveness and efficiency, such that these facilities could be catered for the daily-needs of the general public. The design should also be cost effective. All unnecessary and inefficient proposals should be avoided.

5.2.2 Enhancement of the Central Axis Reflects the Cultural Heritage

It has been stated clearly in the 'Beijing City Master Plan from 1991 to 2010' in 1993, that 'wide green belts should be preserved along the northern extension of the central axis, and the public architectural complexes there and at its northern end represent the climax and ending of the central axis, which shall mirror the new image of the capital in the 21st Century.''

The City Central Axis is where the cultural heritage concentrates, and is also a continuously expanding axis. Treatment of the relationship between the North Central Axis and the Traditional Central Axis, and the urban spatial relationship with the north of the Central Axis is an important issue. The design proposal should reflect and strengthen the prominence of the Central Axis in the planning of Beijing. It should on one hand preserve and conserve the history, and on the other, reflect the contemporary spirit and the great achievement of the new developments in Beijing. It should create a new image for the city, and to imprint the unique cultural heritage of Olympic Beijing.

The need of temporary facilities space and the square space for people's circulation during the Olympic Games should be fully considered.

5.2.5 Feasibility of the Planning

The planning and development of the Olympic Green includes large-scale plazas with green open spaces, large-scale sports facilities and the Olympic Village, convention, exhibition, commercial and business facilities, etc. Some facilities will generate revenue, while some will not. The proposal should study on the feasibility and implementation of the development, to encourage commercial development and market-oriented operation.

6. Components of the Project
6.1 The Central Axis (see Attachment 2)

The Central Axis, from Beitucheng Road running north towards Fourth Ring Road has basically been formed. The Central Axis is comprised mainly of roads and lawns. Along the Axis is the Chinese Ethnic Culture Park, the existing National Olympic Sports Center etc. The Central Axis to the north of Fourth Ring Road will be formed upon the development of the Olympic Green.

The planning proposal should focus on the connection of the new and existing section of the Central Axis, to consider the overall spatial planning of the Beijing city and the background of design relationship between the Central Axis and the city, to study on how to extend the traditional Central Axis of the ancient Beijing city. The proposal should complement all visual elements of the existing and planned developments.

competition

new 220,000V electrical transformer sub-station, and two 110,000V electrical transformer sub-stations, integrated information cables, and information towers should be built.

5. Functional Requirements and Guiding Principals of the Master Planning
5.1 Functional Requirements

The Olympic Green should be a Center of vibrant and popular public spaces for the enjoyment of the citizens. The function should be a combination of sports, cultures, exhibition, leisure, and tourism. It should be served with multi-functional districts of supporting facilities such as commercial, hotel, convention, etc. The Olympic Green should be spacious, with abundance of green open spaces, and should project an image of an elegant city with a high quality of environment. The design should be comprehensive, integrated, representative of the Chinese culture, and contemporary.

5.2 Guiding Principles of the Master Planning
5.2.1 Integration of the Olympics Constructions with the long term development of the city

The planning and development of the Olympic Green caters for the need of hosting the 29th Olympic Games, and also the need of the long-term development of the city. On one hand, it should address the practical needs during the 2-month period of the Olympic Games and the Paralympic. On the other, it should focus on the

5.2.3 Vision of 'Green Olympics, High-Tech Olympics and People's Olympics'

The proposal should highlights characteristics of this Olympic Games in relation to the history of the Olympics, and to realize the vision of 'Green Olympics, High-Tech Olympics and People's Olympics'. The proposal should facilitate environmental conservation and sustainable development of the city. It should adopt the most advanced planning concept and technology, to improve the quality of the ecological environment. The proposal should aim at spreading the Olympic spirit, in praising the excellence of the traditional Chinese culture, to show the landmarks and scenery of the Beijing city, and to build a humane environment that is integrated with the hosting of the Olympics. The proposal should fully consider the needs of different groups, especially the physically impaired.

5.2.4 Combination of the short and long term planning development of the Olympic Green

The master planning of the Olympic Green should consider the need of the 2008 Olympic Games, and to cater for the need of the long-term development of the city. Full consideration should be given to the land and spatial requirements, and the pedestrian flow during the Olympic Games. Olympic sports facilities, the Olympic Village, the convention and exhibition facilities, plazas, and other supporting facilities should be appropriately concentrated. Construction should be competed prior to the 2008 Olympic Games, exhibiting a comprehensive development within the city.

（2）会展博览设施：建筑面积约 40 万 m²，包括约 20 万 m² 的展览面积和约 20 万 m² 附属配套设施。要求设计为多功能、大空间建筑，可兼容会议、展览、商业、娱乐等其他功能。

在奥运会期间，有五个项目的比赛将临时使用其中的设施：乒乓球、击剑、现代五项（击剑、射击）、摔跤、羽毛球，所需总建筑面积约 8 万 m²。

主新闻中心（MPC）和国际广播电视中心（IBC）也将临时使用其设施，其中主新闻中心所需建筑面积约 5.84 万 m²，国际广播电视中心所需建筑面积约 8.35 万 m²。

结合资金筹措情况、奥运会的需求等可考虑进行分期建设，但近期应保证约 22 万 m² 的建设规模。这些设施在奥运会赛后应可改作其他用途，灵活使用。

（3）文化设施：建筑面积约 20 万 m²，包括：
● 首都青少年宫：建筑面积 15 万 m² 左右，是青少年儿童学习历史、文化、自然、科技等方面的知识，开展课外活动及娱乐活动的场所，宜结合园林绿地进行规划。
● 中国杂技马戏馆：结合国家体育馆进行设计，如需要，可增加适当规模的附属设施。
● 北京城市规划展览馆：建筑面积 5 万 m² 左右，展示北京城市历史、现状及未来。

以上文化设施的建筑面积为建议规模，应征者可根据各自的规划理念和经验提出建议。

赛场 5000 座。

7 规划设计成果要求

7.1 城市景观规划研究范围
7.1.1 规划研究内容
（1）空间形态规划和景观立意、景观结构布局、功能组织；
（2）道路交通系统规划；
（3）中轴线、森林公园、奥运中心区、元大都遗址公园中轴段等主要景观构成区内的景观规划设想；
（4）重要景观节点的规划设计构想。可考虑在奥林匹克公园内设置重要的景观构筑物形成地标，增强识别性和景观特性。

7.1.2 规划成果要求
7.1.2.1 方案说明
7.1.2.2 图纸
　　景观规划平面图（赛后，比例 1∶4000）
　　交通系统规划图（赛后，比例 1∶4000）
　　景观规划分析图及重要节点的景观设计图（比例自定）
7.1.2.3 模型（赛后，比例 1∶4000）

7.2 重点研究范围：奥运中心区四环路以北用地（B区，291ha）
7.2.1 总体布局和分期建设概念规划

Beijing Olympic

元代大都城北城墙的一部分，为北京市文物保护单位。

元大都规划建设在中国都城建设史中占有重要地位，但因历史久远已基本无存，仅存留北部分城墙遗址，在规划中被确定为遗址公园。

元大都遗址是体现北京城历史文化风貌的一处主要古迹，遗址公园应以园林绿化为主，通过规划整治，与奥林匹克公园共同形成良好的城市景观环境，同时也为城市居民和旅游者提供一处凭吊古迹、休憩交往的公共活动场所。

6.4 奥运中心区
6.4.1 奥运中心区四环路以北用地（B区）内的设施及规模用地面积：291ha
（1）体育设施及配套商业服务设施：建筑面积约 40 万 m²，包括：
● 国家体育场，在奥运会期间将进行田径比赛、足球决赛，并举行开、闭幕式，80 000 座，建筑面积约 14.5 万 m²。另外还有两块训练场地。
● 国家体育馆（赛后可兼作中国杂技马戏馆），在奥运会期间将进行体操比赛、手球和排球决赛，18 000 座，建筑面积约 6.5 万 m²。
● 国家游泳中心（赛后可兼作戏水乐园），在奥运会期间将进行游泳、跳水、水球、花样游泳比赛，15 000 座，建筑面积约 9 万 m²。

为利于赛后运营，每个场馆可适当增加相应的商业设施，总建筑面积约 10 万 m²。
● 奥林匹克公园射箭场：临时性比赛设施，座位数 5000 个。

（4）商业服务设施：建议建筑面积 80 万 m² 左右，包括商业、酒店、商务办公、娱乐等内容。其中部分面积可安排在地下。

（5）奥运村：其中运动员公寓建筑面积约 36 万 m²，应安排在 B-2 区；另外，还设置临时性的国际区设施，可安排在 B-1 区。

（6）地下停车设施：应根据建设项目的实际需求确定，与地面建筑、地铁及地下空间规划相衔接，集中与分散结合设置，建筑面积约 40 万 m²。

（7）集中公共绿地和广场：集中公共绿地要考虑和中轴线北部森林公园衔接，将绿化以城市从城市外围引入建设区，突出中轴线的形态，创造良好的城市环境。

公共广场是重要的市民公共活动空间，奥运会期间同时为人流集散服务。要处理好与周围开放绿地、城市交通和建筑的衔接问题，创造高品质的室外环境，并研究利用地下空间。

以上各类设施的总建筑面积约 216 万 m²，不含地下停车面积和奥运村国际区临时设施及其他临时设施的建筑面积。

6.4.2 奥运中心区四环路以南用地（A区）内的设施及规模
用地面积为 114ha，保留 A-1 区原有体育设施，在 A-2 区新建国家网球中心和曲棍球场。其中网球中心设 12 000 座的主赛场和分别为 3000 座和 2000 座的两个副赛场；曲棍球场包括主、副两个比赛场和两个训练场，其中主赛场 15 000 座，副

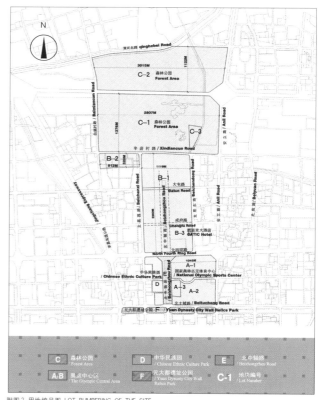

附图 2：用地编号图 LOT BUMBERING OF THE SITE

6.2 The Forest Area

The Forest Area is an organic part of the Beijing city green buffer, and is also the green backdrop of the Olympic Central Area District. The existing woods should be retained. New vegetation should be mainly of trees.

It is suggested that water area of 100 ha should be built, the dirt coming from the excavation can form a hill, and the landscape will become the background of the North Central Axis.

The planned Beichendong Road and the Beichenxi Road could be proposed to cut through the Park, or to terminate at Xindiancun Road. Accesses should be designed for the main roads, Parking spaces and public transport facilities should be provided according to the needs.

6.3 Yuan Dynasty City Wall Relics Park

To the south of the Olympic Green lies a ruin (running from east to west) of the Yuan Dynasty. It is part of the city wall of ancient Yuan Dynasty and is under the protection of the Beijing City Heritage Conservation List.

The building of the capital in Yuan Dynasty plays a significant role in the history of planning and construction of Chinese capitals. Unfortunately, the only remains nowadays are the northern section of the city wall. It is designated as the heritage park in the master planning.

The remains of the capital of Yuan Dynasty are an important site demonstrating the rich cultural heritage of Beijing. The Relics Park is mainly planned with landscaping and green treatment. Through planning, the Relics Park and the Olympic Green should form an environment with good visual and landscape treatment. It should at the same time provide citizens and archaeologists with a place for investigating on the ancient heritage, learning history, leisure and recreational public open spaces.

6.4 Olympic Central Area

6.4.1 Key area: North part of the Olympic Central Area (Site B)

(1) Sports and auxiliary facilities: Total Floor Area of about 400,000 sqm, including:

National Stadium, which will be used for track and field events and the final football events, holding of the opening and closing ceremonies. Seating capacity: 80,000 seats. Total Floor Area is about 145,000 sqm.

National Gymnasium (can also be used as the Chinese Acrobatic and Circus Hall), during the Olympic Games, will be used for gymnastics, handball finals, and volleyball finals. Seating capacity: 18,000 seats. Total Floor Area is about 65,000 sqm.

National Swimming Center (can also be used as the Paddling Paradise), during the Olympic Games, will be used for events of swimming, diving, synchronized swimming, water polo. Seating capacity:15,000 seats. Total Floor Area is about 90,000 sqm.

As for management during the post-Olympic period, auxiliary commercial facilities should be built for the expansion; the total construction area is approximately 100,000 sqm.

Archery Ranges: Temporary Competition Facility with 5,000 seats.

(2) Convention and Exhibition facilities: Total Floor Area of 400,000 sqm. including 200,000 sqm. of exhibition space and about 200,000 sqm. of ancillary supporting facilities. The design should cater for large multi-functional spaces that can duly accommodate convention, exhibition, business and entertainment functions.

During the Olympic period, five contest groups will temporary need the facilities provided, the five groups are: table-tennis, fencing, Modern Pentathlon (fencing, shooting), wrestling, and badminton. The Total Floor Area required is 80,000 sqm.

The Main Press Center (MPC) and International Broadcasting Center (IBC) will also temporarily need the facilities provided, where MPC needs a Total Floor Area of 58,400 sqm., and IBC needs a Total Floor Area of 83,500 sqm.

To raise enough building capital, the required facilities in the Olympics need to be carried out by phase. In any case, Total Floor Area of 220,000 sqm. should be completed in the short-run. Flexibility should be retained for these facilities to be converted to other uses after the Olympics.

(3) Cultural Facilities: Total Floor Area of about 200,000 sqm, including:

Capital Teenage Palace: Total Floor Area around 150,000 sqm, should be planned together with green parks;

The Chinese Acrobatic and Circus Hall: should be integrated in the design and development of the National Stadium. If necessary, certain scale of ancillary facilities could be provided.

The Exhibition Hall of Beijing City Planning: Total Floor Area of about 50,000 sqm.

Above mentioned building area is the suggestion scale, the competition units can make flexible advice on their own experience.

(4) Business Services Facilities: Total Floor Area of about 800,000 sqm., including commercial, hotel, business office, entertainment components, Some Total Floor Area can be provided underground.

(5) The Olympic Village: Total Floor Area for accommodation for the athletes should be about 360,000 sqm in Site B-2. In addition, an 'International Area' should be temporarily built to serve needs during the Games and may be removed upon completion of the Games in Site B-1.

(6) Underground parking facilities: Should be designed in accordance with the practical needs of the development, should have good integration with the buildings at-grade, the underground rail, and the design of underground space. Some should be concentrated, while some should be dispersed. Total Floor Area should be about 400,000 sqm.

(7) Central Green spaces and Plazas: to maintain an ecological environment that is of good quality, and to create an elegant urban space, central green spaces and plazas should be provided appropriately within the Olympics Green.

The planning of central green spaces should consider the integration with the north of Forest Park in the Central Axis, so as to extend the green spaces from the suburban green buffer into the city central areas. This would create a series of urban spaces, enhancing the legibility of the built form, strengthening the Central Axis in the planning of the city, improving the city image, and providing a good living environmental for the citizens.

Public spaces that encourage and allow activities are most im-

Green

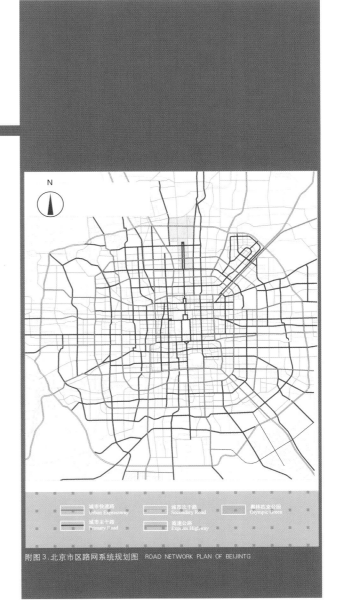

附图3.北京市区路网系统规划图 ROAD NETWORK PLAN OF BEIJINTG

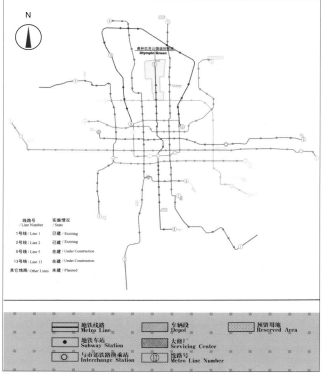

附图4：北京市区轨道交通线网规划图　PLANNED METRO LINE SYSTEM

各类建设项目、绿地和开放空间、道路交通设施、景观设施等的布局要统筹考虑，功能布局合理、结构清晰、有利于实施。其中与奥运会直接相关的项目可考虑集中建设，在2008年前形成相对完整的城市形象。赛后对预留用地、临时设施用地等可进行远期开发。

7.2.2 交通规划
7.2.2.1 规划目标

交通规划应遵循顺捷、快速、安全、经济的原则，满足该地区在非奥运会期间、没有特别交通管制的情况下交通良好运转的要求，同时也应考虑必要的临时性交通设施，制定相应的交通组织方案，满足奥运会期间的特殊需求，以保证各类人员的顺利集散。

7.2.2.2 规划内容
　道路及交通场站规划
　各类车流、人流的出入口位置
　各类车流、人流的集散方案和组织方案

7.2.2.3 规划要求

交通规划应分别考虑奥运会期间以及该地区建设达到终期规模时平日运营(非奥运会期间)两个阶段。

在奥林匹克公园内可以考虑设置联系各项设施和景点的轻型轨道交通工具，既为地区内部交通提供一种工具，又能作为观光旅游的设施。

(1) 非奥运会期间

规划要求以公共交通为主(含轨道交通)、小汽车和自行车为辅，解决该地区日常客运交通问题。

内部道路规划应维持地区道路系统的完整性和连通性，并按照给定道路性质、红线宽度进行设计，具体横断面和纵断面布置可结合规划进行研究。

其中中轴路在四环路以北区段没有划定道路红线，可结合规划研究端点位置、红线宽度、断面形式等。

原规划东西向的大屯路为城市主干道，红线宽度50m，原

附图5：奥林匹克公园周边道路、地铁规划示意图
DIAGRAM OF THE ROAD AND METRO NETWORK FOR THE OLYMPIC GREEN AND ITS SURROUNDING

规划东西向的成府路为城市次干道，红线宽度50m；这两条路可从地面或地下穿过，结合规划需进一步研究横、纵断面形式等。

应处理好内部道路与周围外部路网的衔接问题。
各种交通组织流线宜分别布设，减少相互干扰。

(2) 奥运会期间

交通规划除应满足非奥运会期间的规划要求外，还应满足奥运会期间的特殊要求。

应充分满足运动员与教练、工作人员、新闻记者、贵宾、赞助商、观众等不同出行要求,高峰日和开、闭幕式时观众疏散时间(从场馆到乘上交通工具时间)不超过120分钟。

奥运中心区实行封闭管理，只允许持证车辆驶入(奥林匹克官员及贵宾等特殊人员)。

奥运会期间可考虑增加临时性设施及特别交通组织方案以满足特殊交通需求。

交通组织规划应充分考虑安全检查的需要。

7.2.3 市政概念规划

鼓励使用新技术、新手段，贯彻环保、节能、资源综合利用的概念。如使用太阳能、建设高效节能的照明系统、回收利用可再生资源、利用地热供暖等。

以总平面为基础，研究建设用地内市政管线与外部市政管线的衔接问题；在市政专业说明及总平面图中应反映所需要的市政设施的位置、用地等。

7.2.4 地下空间规划

应充分利用地下空间，建设内容可为：商业、娱乐、停车场等，要处理好与北中轴路、大屯路、成府路等的交通连接和与地面公共建筑等的衔接、与地铁的关系等，组织好地下与地面层的竖向交通，方便大量车流、人流的集散。

7.2.5 规划成果要求
7.2.5.1 方案说明
　规划方案构思说明
　功能布局说明
　分期建设规划说明
　交通规划和交通组织说明
　市政概念规划说明
　地下空间规划说明
　用地平衡表和经济技术指标

7.2.5.2 图纸
　总体规划布局平面图(赛时和赛后比例1:2000)
　道路、轨道交通和交通场站规划平面图(赛时和赛后比例1:2000)
　交通组织规划(含车流、人流出入口)图(赛时和赛后比例1:2000)
　市政概念规划平面图(赛后比例1:4000)
　地下空间规划图(比例1:4000)
　规划范围整体鸟瞰图(赛时和赛后)

7.2.5.3 模型(能识别出奥运会赛时和赛后的不同比例1:1500)

7.3 提交文件的规格及格式

所有正本原图须裱在1200mm×1800mm的轻质板上，统一规格。图板总数不超过15张。

提供A3(297mm×420mm)图册21本。

提供电子文件2套(光盘)，其中文本文件使用doc格式(Office97或Office2000)，平面、立面、剖面等图纸使用dwg格式，透视图或鸟瞰图使用jpg格式。

所有规划成果的计量单位均应采用国际标准计量单位。长度单位：总平面图标注尺寸以米(m)为单位；面积单位以平方米(m^2)为单位。

第三部分 附件(略)
第四部分 附图(部分略)

OIYMPIC GREEN

portant in public plazas. During the Olympics, these spaces will also accommodate the crowd and pedestrian flow. The integration of these public plazas with surrounding green spaces, city transportation and individual building blocks should be considered. The utilization of underground spaces as daily cultural and leisure activities for the citizens should also be studied. The scale should be user-friendly, and the usability of the spaces should be considered, by introducing green spaces, water features, sculptures, etc.. High quality outdoor spaces that are humane should be created to accommodate the need of large-scale public activities.

The above facilities should account for a Total Floor Area of about 2,160,000 sqm. (excluding the underground parking spaces and the temporary 'International Zone' of the Olympic Village and Total Floor Area of other temporary facilities).

6.4.2 South Part for the Olympic Central Area (Site A)

The existing sports facilities in Site A-1 should be retained, National Tennis Center and Hockey ground should be built in the Site A-2. The Tennis Center should have a main court of 12,000 seats, and two side courts of 3,000 seats and 2,000 seats respectively. The Hockey ground should include one main court, two side courts, and two training courts. The main court should have 15,000 seats, and side courts should have 5,000 seats.

7. Design Contents and Deliverables
7.1 Overall landscape Planning and Design
7.1.1 Design Contents
(1) Planning of urban spaces and landscape treatments, visual treatment, functional clusters;

(2) Traffic and road network planning;

(3) The concept landscape planning of the main key areas including the Central Axis, the Forest Area, the Olympic Central Area, the central part of Yuan Dynasty City Wall Relics Park.

(4) The planning and design of important visual nodes. Landmark could be considered for the Olympic Green, to enhance identify and legibility of the place.

7.1.2 Deliverables
7.1.2.1 Description
7.1.2.2 Drawings
  Master Layout and Landscape Plan (Post-Games),Scale1:4000
  Transport Network Planning Diagram (Post-Games),Scale1:4000
  Landscape planning analysis and Landscape Design for Key Areas(to scale which is more appropriate)
7.1.3.3 Model (Post-Games)Scale 1:4000

7.2 Planning and Design of the Key area (Site B)
7.2.1 Layout planning both for during and after the Games
All kinds of development projects (including sports facilities, the Olympic Village, convention and exhibition facilities, cultural facilities, commercial facilities, etc.), green and open spaces, road network, visual and landscape treatments should be taken into consideration. It should create a beautiful and lively urban space, and demonstrate feasible implementation.

The land requirements and implementation phases should be based on the requirements of the organization of the Olympic events, the implementation of master planning, and the capital operation. A complete and comprehensive urban form should be formed before the 2008 Olympic Games, where development projects that are directly related to the Olympic Games should be planned in concentration. The development phasing should benefit the long-term development of the Post-Games. Master Layout Planning of short-term development (before 2008) and long-term development (after 2008) should be submitted.

7.2.2 Transport and Traffic Planning
7.2.2.1 Goal

The transport system shall be smooth, fast, safe and low-cost. The planning shall meet the requirements that under the circumstance of non-Olympic period, which is no traffic control, also the planning shall meet the special needs during the Olympic Games as to ensure the hosting of the Olympics.

7.2.2.2 Contents of transportation planning
Planning of road network and transportation station.
Accesses of the vehicle flow and pedestrian flow.
Traffic planning for various vehicles and pedestrian flow.

7.2.2.3 Planning Requirements
The transportation planning shall consider two phases: Within Olympic Games and normal operation (Non-Olympic Games Period)

Light rail may be suggested as a transportation tool to connect all the facilities as well as for tourists' sightseeing.

(1) Non-Olympic Games Period
The planning shall settle the daily transportation problem mainly by public transportation (including rail transportation); cars and bicycles will be considered as auxiliary tools.

The internal roads shall maintain the integrity and connectivity of the road system. The road planning should be made according to the given nature and the width of the red line of the road. The transect and vertical section can be disposed according to the planning.

The central axis road (Zhongzhou Road) at the north part of the Fourth Ring Road has not lined out the red line, the port location, width of red line, section format should be determined according to the planning.

Datun Road as arterial road going from east to west, the red line is 50m in width, Chengfu Road as secondary main road, the red line is 50m in width, these two roads can go through on the surface or below the ground, the width of the red line and the section format should be determined according to the plan.

The connection between the internal road and external road system should be treated carefully.

The transportation flow should be disposed separately to reduce jams.

(2) Olympic Games Period
The transportation planning should not only meet the requirements of the non-Olympic Games Period, but also for the specific requirements of Olympic Games Period.

The planning should consider the different demands of the athletes and coaches, working staffs, medias, VIP, sponsors, spectators. In the rush hour and the evacuate time on the opening and closing ceremony, the waiting time for personnel to get in the transportation tools from the playfield shall not exceed 120 minutes.

The Olympic Central Area shall be close managed, only the vehicles that have special credentials are allowed to access (i.e. The Olympic Officials and Vips), other social vehicles are not permitted to access, their transportation shall be settled by mass transportation-Buses and Railway.

The temporary facilities shall be considered to fulfill the specific demands during the Olympic Games period.

The requirement for security should be considered in the transportation plan.

7.2.3 Utility conceptual planning
Based on the needs of the Olympic Games and the long-term development of the city, new technology, and new approach, together with concepts of environmental conservation, energy savings, integrated utilization of resources should be adopted in the utility conceptual planning. This includes the use of solar energy and geological heat, high efficiency lighting system, and recycling etc.

Base on the master planning, the integration of all-utility conceptual planning within the built development, and integration with the infrastructures of the city should be studied. The proposal, project scale, estimated amount of investment, location of infrastructure required, land uses and Total Floor Area should be indicated for each aspect of infrastructure.

7.2.4 Underground space planning
Underground space should be utilized. Uses can include: commercial, entertainment, car parking spaces etc. They should be integrated with transportation of North Central Axis, Datun Road, Chengfu Road, and with the underground rail system. Special attention should be paid to the vertical integration between the underground space and the at-grade transport and buildings. It should accommodate and facilitate the large amount of pedestrian flow.

7.2.5 Deliverables
7.2.5.1 Descriptions of proposal
  Planning and design concept
  Function layout
  Landscape planning
  Phasing in the master planning (during and after the Olympics)
  Transport and traffic planning
  Utility conceptual planning
  Underground space planning
  Land use index and technical indicators

7.2.5.2 Drawings
  Master Layout Plan (during and after the Games Scale 1:2000)
  Road Network, Metro Line Network and Transportation Stations Plan (during and after the Games Scale 1:2000)
  (If necessary, other more appropriate scale can be used)
  Traffic Plan (including accesses for the vehicles and pedestrians Scale 1:2000)
  Utility Conceptual Plan?(after the Games Scale 1:4000)
  Underground Space Plan(Scale 1:4000)
  Birds-eye views of the overall Master Planning (during and after the Games)
7.2.5.3 Model (The images during and after the Games should be distinguished Scale 1:1500)

7.3 Document Format
All originals of the proposals should be mounted on lightweight panels of size 1200mm x 1800mm. There should be a maximum of 15 panels.

21 copies of A3 editions (297mm x 420mm)

2 sets of digital files (in CD). Words should be stored as doc. format (Microsoft Office 97 or 2000 version). Plans, elevations, and sections drawings should be in dwg. format. Perspectives and birds-eye views should be in jpg. format.

All the units used in the production should be International Standard Measure. The measurement of length in general drawing is Meter (m), the measurement of area is Square Meter (m^2).

Part III Attachments (Omitted)
Part IV Figures (Omitted Partially)

Wukesong

北京五棵松文化体育中心规划设计方案征集书

北京市规划委员会 2002 年 3 月

第一部分　应征者须知（略）
第二部分　规划设计任务书

1 名称
北京五棵松文化体育中心

2 位置及用地范围
北京五棵松文化体育中心规划建设用地位于北京市区西部，复兴路（西长安街延长线）以北，西四环路以东，西翠路以西，总用地约50ha（详见附图1）。

3 规划设计目的和任务
3.1 目的
（1）规划建设国际先进水平的体育设施，为成功举办2008年北京奥运会创造条件。
（2）规划建设符合城市发展需要的群众性城市公共活动中心，为城市的长远发展服务。

3.2 任务
（1）用地总平面布局、交通规划、市政规划、空间形态和景观环境规划设计等。上述规划内容分为2008年奥运会赛时和赛后远期发展两个阶段分别表述。
（2）重要建筑单体的概念设计。

4 项目概况
4.1 五棵松文化体育中心在奥运会设施总体规划中的位置
根据第29届奥运会奥运会设施总体规划，奥运会比赛场馆主要集中在城市的北部和西部，分为1个中心区和3个分区。中心区位于奥林匹克公园内，3个分区分别是大学区、西部社区和北部风景旅游区。其中五棵松文化体育中心是位于西部社区内的主要体育设施集中建设场地，安排了用于篮球、棒球、垒球比赛的3个比赛场馆。

4.2 功能定位
北京五棵松文化体育中心既是北京为举办第29届奥运会规划安排的一处主要的比赛场地，同时，也将规划成为北京市西部地区市民进行体育活动和文化休闲活动的重要场所，要建成充满活力的、市民喜爱的城市公共活动中心，形成建筑与开放空间结合、绿地开敞、环境优美的城市形象。

5 现状条件
5.1 用地条件
5.1.1 用地内现状
用地共分为两部分（详见附图2），包括城市公共绿化带用地和城市建设用地。
根据北京市城市总体规划，用地内沿西四环路东侧和复兴路北侧分别预留了100m和50m宽的城市公共绿化带用地，共约11ha，目前该绿化带已基本形成。
其余为城市建设用地和公共绿地，共约39ha，目前用地内的原有建筑已大部分拆除，没有需要保留的现状建筑。

5.1.2 用地周边情况
用地周边城市建设区已基本建设形成，东侧主要为现状居住用地，北侧为规划住宅开发用地，南侧隔复兴路相对为解放军总医院，西侧紧邻城市快速路四环路（详见附图3）。

5.2 交通条件
五棵松文化体育中心位于西三环路和西四环路之间，用地南侧紧邻东西向贯穿北京市区的地铁1号线，交通极为方便。其周边共涉及10条城市快速路和主、次干道，从西向东依次为：永定路、西四环路、西翠路、东翠路、西三环路；从北向南依次为：阜石路、金沟路、玉渊潭南街、万寿路西街、复兴路（详见附图4）。

5.3 市政条件（详见附图5）
雨水：沿西四环路、复兴路分别有一条Φ1000～Φ1800mm、Φ1000mm～□1600mm×1000mm现状雨水管道，规划沿玉渊潭南街新建管径Φ800mm～Φ1000mm雨水管道，分别接入上述两条现状管道。
污水：现状沿西四环路东侧有一条Φ1750mm的现状污水管道，

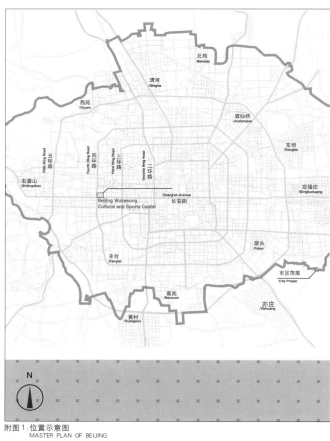

附图1：位置示意图
MASTER PLAN OF BEIJING

Document of International Competition for Conceptual Planning and Design of Beijing Wukesong Cultural and Sports Center

Part I Procedures and Rules(Omitted)
Part II Planning Brief(Omitted)

1. Project Title

Beijing Wukesong Cultural and Sports Center

2. Location and Site

Beijing Wukesong Cultural and Sports Center is located at the western part of the Beijing City, to the north of the West Chang'an Avenue, to the East of the West Fourth Ring Road, and to the west of the Xicui Road. It occupies a total land of 50 ha. (refer to Figure 1 for detail).

3. Objectives and Missions
3.1 Objectives

(1) To plan and build world-class sports facilities, for the successful holding of the 29th Olympic Games;
(2) To plan and build a multi-functional area with public spaces for activities, in benefiting the long-term development of the city.

3.2 Missions

(1) Site planning, transport and traffic planning, utility conceptual planning, landscape planning and design, etc. The above-mentioned contents shall be formulated in two phases: during and after the Games .
(2) Architectural conceptual design for the Basketball Hall.

4. Introduction
4.1 Olympic Contexts

According to the master planning of the Olympic facilities for the 29th Olympics, sports facilities for the Games are mainly concentrated in the northern and western part of the city, and are divided into one core district and three sub-districts. The core district is located within the Olympic Green. The three sub-districts are the University District, the Western Community District, and the Northern Scenic District. The Wukesong Cultural and Sports Center is located within the Western Community District, in the main sports facilities cluster, accommodating courts for basketball, baseball and softball.

4.2 Urban Contexts

The Beijing Wukesong Cultural and Sports Center is a main sports venue for the 29th Olympics, and will also be planned as an important venue for the sports, cultural and recreational activities in the Western Community District. It will be built into a vibrant, and popular public activity center and help to create the image of a city where the architectural form and open spaces are integrated, green spaces are spacious, and the environment is of high quality.

5. Existing Conditions
5.1 Site Conditions

5.1.1 Existing site conditions

Development land is separated into two portions (see Figure 2), including the urban public open space and the urban development land.

In accordance with the Beijing City Master Plan, two urban public green spaces reserves are allocated on the eastern side of the West Fourth Ring Road and the northern side of the Fuxing Road, with a width of 50 meters and 100 meters respectively. The total area of the urban green spaces reserves is 11 ha. The green buffer has basically been formed.

The portion of the land for urban development is about 39 ha. Original buildings have mostly been demolished and none of the existing buildings need to be retained.

5.1.2 Land uses of the surrounding areas

The urban development zones surrounding the subject site have mostly been completed. To the east is an existing residential area; to the north is planned residential development land. To the south facing Fuxing road is the General Hospital of the Liberation Army. The Fourth Ring Road, the expressway of the city, abuts west of the site. (See Figure 3)

5.2 Traffic

The Wukesong Cultural and Sports Center is situated between the West Third Ring Road and the West Fourth Ring Road. The site is abutted by the Line One of the subway that runs east-west through the Beijing City. So it enjoys convenient transport and is easily accessible. The site is nearby to 10 major expressways, main roads and secondary roads including successively the following ones from west to east: Yongding Road, West Fourth Ring Road, Xicui Road, Dongcui Road and the West Third Ring Road, South Yuyuantan Street, West Wanshou Street and Fuxing Street (Figure 4 for detail).

5.3 Utilities(Figure 5 for detail)

Drainage: There is an existing drainage pipeline of $\Phi1000\sim\Phi1800mm$, $\Phi1000\sim1600mm \times 1000mm$ along the Fourth Ring Road and Fuxing Road. A new drainage pipeline of $\Phi800\sim\Phi1000mm$ is planned along South Yuyuantan Street to connect the above two existing pipelines.

Sewage: There is an existing sewage pipeline of $\Phi1750mm$ along West Fourth Ring Road. A new sewage pipeline of diameter $\Phi500mm$ is planned along South Yuyuantan Street to connect the existing sewage pipeline at the west.

Potable water supply: New potable water pipelines of $DN600mm$, $DN400mm$ are planned along the South Yuyuantan Street and Xicui Road respectively. They are to connect to the existing pipelines along the West Fourth Ring Road ($DN600mm$) and Fuxing Road ($DN400mm$).

Non-potable recycled water supply: Source of the non-potable recycled water is from the Wujiacun Waste Treatment Plant. A new non-potable recycled water pipeline of $DN400mm$ is planned along the West Fourth Ring Road.

Gas supply: Existing $DN500mm$ natural gas supply pipelines with medium pressure are positioned along the West Fourth Ring Road and Fuxing Road respectively. Two new pipelines of $DN500mm$ are planned along South Yuyuantan Street and Xicui Road respectively.

Heat supply: One heat generation station is planned. One heat pipeline of $DN300mm$ along the West Fourth Ring Road is planned to be connected with the existing pipeline along Fushi Road. Other alternatives of environmental friendly energy resources can be considered.

Power supply: It is planned that the Balizhuang Transformer Substation will be responsible for electricity supply to the area. Two sets of switching stations are planned within the subject site.

Integrated information network: Existing telecommunication cables and TV cables are presented along the West Fourth Ring Road and Fuxing Road. A new 24-36 holes integrated information network is planned along Yuyuantan Street and Xicui Road.

6. Planning and Design Requirements
6.1 Guiding Principles

6.1.1 Vision of 'Green Olympics, High-tech Olympics and People's Olympics'

The proposal should highlight the guiding principles of the 2008 Olympics, namely 'Green Olympics, High-tech Olympics and People's Olympics'. The proposal should facilitate environmental protection and sustainable development of the city. It should adopt the most advanced planning concept and technology, and be conducive to improve the quality of the ecological environment. The proposal should aim at: disseminating the Olympic spirit; carrying forward the traditional Chinese culture; presenting the landmarks and sceneries of the Beijing city; building a humane environment that is integrated with the hosting of the Olympics. The proposal should fully consider the needs of different groups, especially the physically disabled.

6.1.2 Post-Olympic uses and developments

In the planning and building of sports and other ancillary facilities to meet the needs of the Olympics, the post-Olympic utilization of the mix of land uses and multi-functional sports facilities should

规划沿玉渊潭南街新建管径Φ500mm雨水管道，向西接入四环路现状管道。

供水：在规划中沿玉渊潭南街、西翠路分别新建DN600mm、DN400mm给水管道，与西四环路、复兴路上的现状DN600mm、DN400mm管相连。

中水：中水水源来自吴家村污水处理厂，规划沿西四环路新建一条DN400mm中水管道。

燃气：现状在西四环路、复兴路上分别有DN500mm天然气中压管道，规划沿玉渊潭南街、西翠路分别新建DN500mm天然气中压管道。

供热：规划新建1座热力站，沿西四环路新建DN300mm热力管道，与阜石路现状热力管道相连，也可考虑采用其他清洁能源供热。

供电：规划由八里庄变电站为该地区供电，在用地内需新建2座开闭站。

综合信息网络：西四环路、复兴路有现状电信管道、有线电视管道，规划在玉渊坛南街、西翠路新建24～36孔综合信息管道。

6 规划设计要求

6.1 规划设计指导思想

6.1.1 体现"绿色奥运、科技奥运、人文奥运"的主办宗旨

规划设计方案应突出"绿色奥运、科技奥运、人文奥运"的主办宗旨。要有利于环境保护和城市可持续发展，采用当今国际上最新的规划设计理念和先进的技术手段，注意节能，鼓励使用可再生资源和清洁能源，参照国际先进的设计标准，保护环境、改善局部地区环境质量。要充分考虑各类人员的需要，尤其要关注残疾人和有行动障碍人员的需求等。

6.1.2 考虑城市长远发展的需要，有利于赛后使用

规划建设体育场馆及其他设施的同时，要研究土地的多功能混合使用和体育设施的赛后利用，为地区的平衡发展和将来的长期运营提供条件。

6.1.3 合理节约使用土地，创造优美的城市空间

在规划设计中，要注意节约土地，保证城市公共绿化用地和开放空间的建设，形成绿地开敞、环境优美的城市空间。建设区和绿化带应作为一个整体进行景观环境规划设计。

6.1.4 考虑近远期结合，在奥运会举办前形成相对完整的城市形象

从赛事组织、规划实施和资金运作的需要出发，用地内设施布局和分期实施计划应考虑在2008年奥运会举办前形成相对完整的城市形象，并有利于奥运会后的远期开发。

6.2 设施及规模要求

总建筑规模控制在20万m²左右。建设内容包括篮球馆、临时性的棒球比赛场、临时性的垒球比赛场等奥运会场馆设施，群众性文化体育设施（包括五棵松游泳馆、群众文化活动中心、室外田径场和练习场）和商业服务设施。

6.2.1 奥运会场馆设施

① 五棵松篮球馆

五棵松篮球馆在奥运会期间将用于举办篮球比赛。包括一个比赛馆和一个热身馆，比赛馆座位为18 000个，建筑面积约3.5万m²。热身馆安排两块练习场地。为方便赛后利用和长期运营，篮球馆设计中可安排适量商务、商业娱乐等附属设施，面积约2～3万m²。

② 五棵松棒球场

五棵松棒球场为临时性设施，在奥运会期间将用于举办部分棒球比赛，包括一个比赛场地和两个热身场地。其中比赛场座位数为25 000个，两个热身场地应安排在比赛场旁边。

③ 五棵松垒球场

五棵松垒球场为临时性设施，在奥运会期间将用于举办部分垒球比赛，座位8500个。包括一个比赛场地，两个练习场地。

6.2.2 群众性文化体育设施

① 五棵松游泳馆

是主要用于市民平时进行体育锻炼的室内游泳场地，要求设有国际标准游泳池（50m长水道）和练习池，建筑面积约3万m²。

② 群众文化活动中心

6.3 近远期规划安排

奥运会场馆设施、公共绿地和广场、酒店及部分商业娱乐设施应作为近期建设项目；群众性文化体育设施等可作为远期建设项目。

在近期建设项目中，棒球场和垒球场是临时性的奥运会比赛设施，赛后可拆除部分或全部临时设施，进行改造利用或再开发。计划远期开发的群众性文化体育设施可考虑结合上述设施进行建设。

6.4 规划控制指标

6.4.1 建筑高度控制要求

根据市区中心地区控制性详细规划要求、周边地区规划建设情况和飞行净空要求，酒店建筑的控高为45m，体育场馆根据使用需要决定建筑高度。

6.4.2 建筑退用地红线要求

沿用地东、南、西、北要求分别退红线10m、50m、100m、10m。

6.4.3 绿化指标要求

除城市公共绿地外，场馆建设用地内的绿化率应不小于30%。在满足体育比赛、交通、市政等要求外，应尽量扩大绿地面积。并与公共绿地有机衔接，形成统一的绿化效果。

6.5 交通规划要求

6.5.1 奥运会期间交通组织方案

现状沿复兴路有地铁1号线。除地铁外，地面公共交通是奥运会期间大量观众进出该地区的主要交通方式，应根据需求评估奥运会期间外部交通设施负荷水平，必要时设置临时公交场站设施。

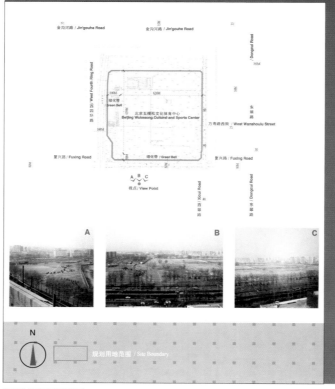

附图2：规划用地范围图
SITE BOUNDARY OF THE WUKESONG CULTURAL AND SPORTS CENTER

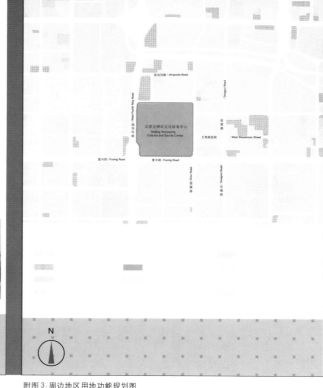

附图3：周边地区用地功能规划图
LAND USE PLAN OF THE SURROUNDING AREAS

be studied. There should be a balanced development of land and usage of land. The planning should cater for the long-term operational criteria.

6.1.3 Efficient land use and viable urban space

In the site plan, particular attention should be paid to the reasonable use of land, to secure the provision of green spaces and public open spaces, to create a green, spacious and high quality urban space. An integrated landscape and visual treatment should be introduced in the development area and the green amenity area.

6.1.4 Phasing of development

Based on the needs of the Olympic events, implementation of planning and operation of capital, phasing and implementation of the development should aim at creating a completed and comprehensive urban built form before the 2008 Olympics, and aim to facilitate long-term development after the Olympics.

6.2 Venues and Services

The total floor area restriction for the site is about 200,000 m^2. Development should include a Basketball Hall with a seating capacity of 18,000 seats, site area of 35,000 m^2 (Additional commercial facilities of 20,000-30,000 m^2. can be implemented after the Olympics.), a temporary Baseball Court, and a temporary Softball Court for the Olympics events. Possible sports facilities (including indoor swimming pool, outdoor track field and training field) of about 30,000 m^2; public cultural activities center of about 20,000 m^2; and commercial services facilities of about 100,000 m^2 should be planned.

6.2.1 Venues for the Olympics

(1) Wukesong Basketball Hall

The Wukesong Basketball Hall will be used to hold basketball events during the Games. It should include an event center and a training center. The event center should accommodate 18,000 seats, with a total floor area of 35,000 m^2. The training center should accommodate two training courts. To facilitate post-Olympic utilization and operation, appropriate scale of ancillary business and entertainment facilities could be introduced. The total floor area should be about 20,000 - 30,000 m^2.

(2) Wukesong Baseball Field

The Wukesong Baseball Field should be a temporary facility, used to hold part of the baseball events during the Games. It should include one event center and two training centers. The event center should accommodate 25,000 seats. The two training centers should be planned beside the event center.

(3) Wukesong Softball Field

The Wukesong Softball Field should be a temporary facility, used to hold part of the softball event during the Games. It should accommodate 8,500 seats.

6.2.2 Cultural and sports facilities

(1) Wukesong Indoor Swimming Complex

Mainly used for daily sports training for the citizens. It should include one swimming pool of international standard (50m lapse) and one training pool. The total floor area should be about 30,000 m^2.

(2) Public Cultural and Sports Center

The total floor area of a venue for public activities, performances, exchange and training should be about 20,000 m^2. It should accommodate one theater of 800 - 1,000 seats.

(3) Outdoor sports ground

It will be mainly used for daily sports training for the citizens, including track field, football court, basketball court, etc. The track field should accommodate one football court of international standard, including a viewing deck of 2,000 seats and a training ground.

6.2.3 Commercial services facilities

It should mainly provide hotel and business entertainment facilities, with some office floor area, which all adds up to a total floor area of 100,000 m^2. The hotel should be four-star standard, with 400 rooms.

6.2.4 Parking

Car parks should be provided at a rate of 65 spaces/ 10,000 m^2. A total of 1,300 spaces are needed. The parking provision can be over ground or underground. In addition, the car parks should consider the needs for special vehicles (media vehicles, vehicles for Olympic officials, mass transportation vehicles).

6.2.5 Open spaces

The City Master Plan requires green areas along the West Fourth Ring Road and Fuxing Road. A large area of public green and open spaces should also be provided within the developed area at appropriate scale. The provision of public green spaces should not be less than 10 ha. The design of open space should take into consideration the pedestrian flow during the Games and the daily leisure activities after the Games. Certain scale of paved plaza should be planned.

6.3 Project Management

The Olympic sports venues, green spaces and plazas, hotel and part of the commercial and entertaining facilities should be included as short-term projects, to be completed by the 2008 Olympics. The public cultural and sports facilities can be designated as long-term development projects.

In the short-term development projects, the baseball and softball fields are temporary facilities; part or whole of the facilities could be removed, redesigned or redeveloped after the Games. The above program and flexibility should be taken into consideration in the design of the long-term development of the public cultural and sports facilities.

6.4 Planning Guidelines

6.4.1 Building height limits

In accordance with the regulatory plan of the downtown district, the planning and development conditions of the surrounding areas, and the aviation height restriction requirements, the building height limit of the hotel is 45 meters whereas that of the Sports Complex should be determined by its functional requirements.

Green Olympics High-tech Olympics and People's Olympics

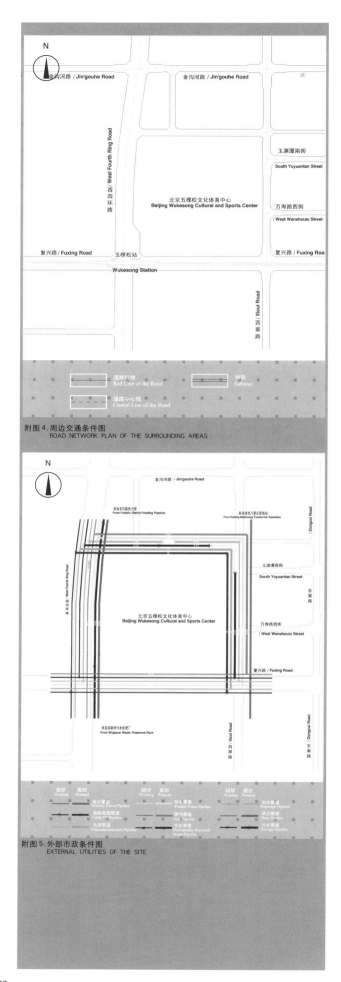

附图4：周边交通条件图
ROAD NETWORK PLAN OF THE SURROUNDING AREAS

附图5：外部市政条件图
EXTERNAL UTILITIES OF THE SITE

奥运会期间，应充分满足运动员与教练、工作人员（裁判员、技术人员等）、新闻记者、贵宾、赞助商、观众等不同出行要求（包括方式、路线、出行时间、停车泊位等）；奥运交通组织应符合安全检查的需要。

奥运会期间，在用地内部结合永久停车设施安排奥委会官员（国际、国内）、国际单项组织官员、贵宾、新闻媒体和赞助商等特殊人员的停车，并根据需要安排货运停车。其他人员如运动员、裁判员、普通观众、组委会成员、服务人员和志愿者等乘大巴或公交车辆出入，需安排临时停车场。

6.5.2 平日交通组织方案
规划要求利用地下大运量轨道交通和地面公共交通解决该地区日常交通运输问题；

应处理好用地范围内的人流、物流集散与外部交通的关系，人、车分流，出入口的设置不得影响周边外部道路的交通；满足高峰时段最大集散量的各种交通方式出入口的集散要求；处理好体育、文化娱乐、酒店、商务办公等各种活动人流、物流之间的关系；

平日停车设施安排，应按照不同建筑的停车需求在用地内解决。

6.6 市政规划要求
以总平面规划为基础，研究用地内的市政管线综合以及与外部市政管线的衔接关系；在市政规划说明及总平面图中反映各专业的方案、所需要的市政设施的位置、用地、建筑面积等指标。

6.7 篮球馆单体建筑概念设计要求

7 规划设计成果要求
7.1 总体布局规划
7.1.1 规划设计说明（奥运会赛时和赛后）
规划方案构思说明
功能布局说明
交通组织说明
景观环境规划设计说明
环境保护规划说明
市政设施规划说明
用地平衡表
总体规划经济技术指标

7.1.2 图纸
总平面规划图（赛时和赛后；比例1：1000）
交通系统规划图（赛时和赛后；比例1：1000）
绿化系统规划图（赛时和赛后；比例1：1000）
市政设施系统规划图（赛后；比例1：1000）
景观环境规划分析图（赛后；比例自定）
主要节点的景观设计构想图（比例自定）
西四环路和复兴路沿街立面图（比例1：1000赛时和赛后）
规划范围整体鸟瞰图（赛时和赛后）

7.1.3 总体规划模型（比例1：1000 应能识别奥运会赛时和赛后的不同）

7.2 篮球馆单体建筑概念设计
7.2.1 单体建筑设计说明
篮球馆功能布局、结构选型、人流、环保设计说明等经济技术指标。

7.2.2 图纸
总平面布局图（比例1：500）
主要层平面图（比例1：500）
交通流线分析图（比例1：500）
立面、剖面图（比例1：500）
透视图1~2张

7.2.3 篮球馆模型（比例1：500）

7.3 提交文件的规格及格式
所有正本原图须裱在1200mm×1800mm的轻质板上，统一规格。图板不超过10张。

提供A3图册（297mm×420mm）21本。

提供电子文件2套（光盘），其中文本文件使用doc格式（Office97或Office2000），平面、立面、剖面等图纸使用dwg格式，透视图或鸟瞰图使用jpg格式。

所有规划成果的计量单位均应采用国际标准计量单位。长度单位：总平面图标注尺寸以米（m）为单位；建筑设计图中平面图标注尺寸以毫米（mm）为单位；立面、剖面图标注尺寸以米（m）为单位；面积单位以平方米（m²）为单位。

第三部分 附件（略）
第四部分 附图（部分略）

6.4.2 Building line
The setback requirements are 10m, 50m, 100m and 10m for site boundary on the east, south, west and north respectively.

6.4.3 Green coverage
Excluding the public green area, the green coverage in the venues should exceed 30%, and the green area in the venues should integrate with the public green area.

6.5 Transport and Traffic Planning
6.5.1 During the Games
The existing Fuxing Road is served with No.1 Subway. Apart from subways, at-grade public transport will be the major means for the spectators during the Games. Temporary public transport facilities should be provided according to the estimated additional traffic demand during the Games.

During the Games, the traffic demands of the athletes, team officials, juries, technicians, media, VIPs, sponsors and spectators, including means of transport, routes, time, and parking, etc. should be met. The operation for transport should meet the requirements of the security check.

During the Games, the permanent parking facilities should be used for IOC members, the IF members, VIPs, media and sponsors. The others like athletes, juries, spectators, BOCOG staffs and volunteers will use buses and shuttles. The temporary parking facilities should be arranged for them.

6.5.2 During the ordinary time
Subway and other public transport should be the major tools for the daily transportation.

Special attentions should be paid in managing the pedestrian flow and the logistics flow, integrating with outside transportation. The pedestrian flow and vehicle flow should be separated. The location of accesses should not affect the external traffic.

The capacity of the access points for various traffic flows should meet the peak hour demand.

The pedestrian flow and logistic flow of various functions in sports, cultural and entertainment, hotel and business/office should be well considered.

The parking facilities during non-Games time should be arranged according to the official standard for different buildings.

6.6 Utility Requirements
The utility planning and its integration with the external utilities should be studied based on the site plan. Location, land use and floor area of each utility should be reflected in the site plan.

6.7 Conceptual Architectural Design for Wukesong Basketball Hall

7. Design Contents and Deliverables
7.1 Site Plan
7.1.1 Design contents (During and after the Games)
    Description of the planning concept
    Function layout
    Transport and traffic planning
    Landscape planning and design
    Environmental protection considerations
    Utility planning
    Land use index
    Technical indicators

7.1.2 Drawings
    Site Plan (during and after the Games Scale 1:1000)
    Transport and Traffic Plan (Scale 1:1000 during and after the Games)
    Green Network Plan (during and after the Games Scale 1:1000)
    Utility Network Plan (during and after the Games Scale 1:1000)
    Landscape Planning and Design Free scale (during and after the Games)
    Landscape Design of Important Visual Nodes (Free scale)
    Elevation Along the West Fourth Ring Road (Scale 1:1000 and Fuxing Road during and after the Games)
    Birds-eye Views of the Site Plan (during and after the Games)

7.1.3 Model (Scale 1:1000 The images during and after the Games can be distinguished clearly.)

7.2 Conceptual Architectural Design for Wukesong Basketball Hall
7.2.1 Design contents
ILLustration on the scheme conception of the Basketball Hall, structures, pedestrian flows and environmental protection considerations
    Technical indicators

7.2.2 Drawings
    Site Plan (Scale 1:500)
    Floor Plans (for main floors Scale 1:500)
    Circulation Analysis Figure (Scale 1:500)
    Elevations and Sections (Scale 1:500)
    Perspectives (1-2 pieces)

7.2.3 Model (Scale 1:500)

7.3 Document Formats
All originals of the plans should be mounted on lightweight panels of size 1200mm x 1800mm. There should be a maximum of 10 panels.
    21 copies of A3 editions (297mm x 420mm)
    2 sets of digital files (in CD). Words should be stored as doc. format (Office 97 or Office 2000). Plans, elevations, and sections drawings should be in dwg format. Perspectives and birds-eye views should be in jpg format.
    All the units used in the drawings should be in International Standard Measurement. The dimension units showing in the site plan should be in meter (m). The floor plans should be in millimeter (mm). The elevations and sections should be in meter (m). The measurement of the area should be in square meter (m^2).

Part III Attachments(Omitted)
Part IV Figures(Omitted Partially)

北京奥林匹克公园现场踏勘
提问及答疑

FAQ on the field survey of Beijing Olympic Green

问1：中轴线的北端是到规划的B区为止，还是一直延伸到森林公园里面？

答1：中轴线是城市规划一条重要的城市轴线，这次奥林匹克公园的规划当中，对中轴线是否规划北端点没有明确要求，希望设计单位结合城市总体规划和用地周边的发展情况综合考虑这个问题。

Q1: Shall The north terminal of the central axis road be ended at Barea or be across in the Forest Park?

A: The central axis is an important city axis in the city planning, we did not specify the terminal of the central axis in the plan of Olympic Green. We hope all design firms can give comprehensive consideration regarding the city master plan and its surrounding situation.

问2：A－2区的网球场和曲棍球场是露天的还是有顶的？

答2：设计要求只要满足比赛要求就可以了，不做特别规定。

Q2: Should the tennis ground and the hockey ground be opened or be covered ? Do you need the roof?

A2: All you have to do is to design the ground to meet the requirements for the match. No specified requirements to follow.

问3：对用地内建筑物的限高有没有明确规定？

答3：高度没有特别限制，根据建筑使用功能的要求确定。用地内可以设置标志性景观建筑物或构筑物，高度根据方案研究确定。

Q3: Are there any building height restrictions in the Olympics Green?

A3: No.The building height is decided by its function requirements. You can design some landmark buildings in site, and the height of the landmark buildings can be determined by the design firms.

问4：C区中可以考虑安排100ha左右的水面，是否可以在这个基础上再加大？

答4：征集书中提到的水面面积是一个参考值，可以结合规划方案灵活安排。

Q4: In C area, we have seen over 100 ha water area, can we expand the water area?

A4: The water area listed in the documents is just a reference. You can arrange the water area at your own decision.

问5：网球场和曲棍球场赛后是被移走，还是永久性保留？

答5：三个网球场里一个是永久性设施，另外两个是临时设施。曲棍球场地里一个是永久性设施，另一个是临时的。但都是在A－2区。

Q5: The tennis ground and hockey ground will be removed after the Games or will be remained permanently?

A5: One of the tennis ground will be remained, the other two will be removed, one hockey ground will be remained, the other one will be removed, all above mentioned grounds are located in A2 area.

问6：奥运村在赛后，一般都用作住宅，可否考虑其他用途？

答6：对奥运村的赛后利用可以提出其他的建议，一般的国际惯例是作为居住建筑比较好，因为容易改造，容易回收资金。

Q6: The Olympic village will be used as residential building after the Games, what about other uses, will it be acceptable?

A6: For the usage of the Olympic Village after the Games, we suggest you treat it as residential buildings. According to the international practice, the residential building will do good on the funds reclaim and will be easy for renovation.

问7：征集书"附件3 奥林匹克公园交通条件"中提及的奥林匹克大家庭的含义？

答7：在国际奥林匹克运动这个领域当中，奥林匹克大家庭是一个特殊的概念，在表3中"奥林匹克大家庭"所指的主要是国际奥委会的官员、国家体育协会的官员、国际单项体育协会的官员和国家奥委会的观察员等。

Q7: In the "Attachments 3 Transportation Condition in Olympics Green", what is the meaning of " Olympics Family"?

A7: In the Olympic field, "Olympics Family" is a special conception. In the Chart 3, it refers to the officials of IOC, the National Sports Association, the International Sports Association and the observers of IOC, etc.

问8：用地周边的安慧桥、健翔桥等有没有进一步的规划改造的设想？

答8：健翔桥改造方案已经确定，可以提供改造方案。安慧桥的改造还没有考虑。目前，和奥林匹克公园有关的道路如五环路、北苑路等正在建设。

Q8: Have you got any plans on further renovations on the traffic condition of the surrounding area, like Anhui flyover and Jianxiang flyover.

A8: We have made plans for Jianxiang flyover. Anhui flyover is the nearest flyover to the Olympic Green, but we have not made any plan for the reconstruction of that flyover. For road construction, the Fifth Ring Road and Beiyuan Road are under construction.

问9：征集书"附件4 奥林匹克公园市政条件"中北小河污水处理厂的工作范围？

答9：奥林匹克公园内的污水近期和远期都由北小河处理厂处理，包括中水考虑也由北小河污水处理厂水生物设施处理。

Q9: In the Appendix 4 " the infrastructure of Olympics Green" of the competition documents, what is the working scope of the Beixiaohe Water Treatment Farm?

A9: The drainage of the Olympics Green, in the short or long run, is treated by the Beixiaohe Water Treatment Plant. The non-potable recycled water is also treated by this plant.

问10：现状高压线是否保留？如果保留，会不会对周边建筑有特殊要求？

答10：现状高压线将改造入地，不会对建筑布局产生重大影响。

Q10: The high voltage cables composed a High voltage corridor, will these cables be removed or not? If the cables have to be remained, will the cables influence the surround buildings?

A10: The high voltage cables will be gone through the underground tunnels, but it will not influence the surrounding buildings.

问11：我们考虑是否可以把奥运会体育设施从其他竞赛场地挪到奥林匹克公园？

答11：不能。

Q11: We want to know, if it is possible to transfer sport facilities from other competition sites into the Olympic Green site?

A11: No.

问12：可否提供用地北侧的周边用地性质？

答12：奥林匹克公园北侧为绿化隔离地区的绿化用地，西北和东北方向有规划城市建设用地。

Q12: Could you provide the quality of the surrounding land to the north of the site?

A12: The north part tothe Olympic Green is green belt area, the northwest and northeast to it is planned construction land use.

问13：在附6-2页第5条要求奥运会工程项目的绿化面积不低于40%～50%，请问这些绿化面积中是否含森林公园？绿化覆土深度是否需大于3m？

答13：这些绿化面积中不含森林公园，绿化覆土深度须按《北京市绿化管理条例》执行。

Q13: In attach 6-2,clause 5,the Green Coverage should be guaranteed not below 40-50%, the question is does this green coverage include "Forest Park"? Does the height of soil for the green coverage needs to be over 3 meters?

A13: The green coverage area doesn't include "Forest Park", the height of soil for the greenland shall follow "Administrative Rules on Green Coverage of Beijing".

问14：会展博览中心：规划为400 000m²，其中大约有200 000m²的展览区和200 000m²的配套设施区。在比赛之前，为了满足各种功能方面的需求，大约需要210 000m²的设施。配套设施可在以后相继完成。请解释一下"配套设施"的具体含义及它们的分期建设计划，如果只有展览大厅而没有配套设施是不实际的。

答14：奥运会前应保证比赛场地，主新闻中心，国际广播中心所需的面积。为便于市场运作和使用、运营之要求，可以同时提出建设辅助设施的建议，如会议、住宿、娱乐等等。

Q14: Convention Exhibition Center: The program calls for 400,000 square meters with approximately 200,000 square meters for exhibition spaces and 200,000 square meters for the supporting facilities. Roughly 210,000 square meters of exhibition spaces will be needed for various Olympic functions, which will be constructed before the Olympic? The remaining supporting facilities will be added in future phases. Please clarify the nature of these "supporting facilities" and the phasing ideas for this facility. It will be impractical to build only the exhibition halls without any supporting facilities, such as lobbies, concourse, meeting rooms and concessions spaces.

A14: The space for Competition Courts, MPC, IBC should be completed before the Games. In order to meet the need of the market operation and running, you can pose some suggestions about the supporting facilities ,for example, conference hall, hotels and entertainment facilities,etc to be constructed in short-term.

问15：各个地块有没有指定的后退红线及不可建设范围的规定？

答15：参照北京市现行的规划管理有关规定，建筑工程与城市主、次干道红线之间的距离一般不小于10～15m。目前为概念性总体规划，具体的建设控制指标将在下一步详细规划阶段确定，上述数据仅供参考。

Q15: Is there any set-back line control of each site and any area which cannot be developed?

A15: According to the relevant regulations, the distance between the set-back lines and the red lines of urban roads primary roads or secondary roads is not less than 10-15m. Now it is a conceptual master plan, the specific development controlling index will fix in the next stage of detailed plan.

问16：体育设施按6.4.1项由(1)体育设施中的国家体育场，建筑面积14.5万m²是否包括两块训练场地，若不包括，请明确训练场地之面积。

答16：不包括。训练场地之面积包括一个标准的田径场和一个标准的投掷场。

Q16: Sports Facilities: (1)Does the National Stadium(145,000 sqm) include two training sites? If not, can you offer the area of these two training sites?

A16: No. The area of these two training sites include a standard ground track field and a standard tossing field.

问17：会展博览设施于6.4.1项中，第(4)之会展博览设施，总建筑面积为40万m²，对此附件5中的面积表提供的会展博览设施有分歧，表中没有列出展厅E击剑练习的面积，但所需建设的总面积亦已超过40万m²。

答17：40万m²是一个大约数，展厅E考虑展览的要求面积大约5000～10 000m²，完全可以满足击剑练习所需面积要求。

Q17: Convention and Exhibition Facilities On the Page 2-11, in 6.4.1 (2) , total floor area of 400,000sqm. This number conflicts with the number provided on the page Attach5-10, you don't provide the area of Fencing area in the Exhibition Hall E, however, the requiring area is over 400,000sqm.

A17: 400,000sqm is a approximate number, concerning the need of exhibition, the area of Exhibition Hall E is about 5000-10,000sqm, which can meet the need of the area of Fencing Field.

问18：请明确所有必须遵守的规划设计规范，如除标书之外，北京市有关规范。请列示之以便参考。

答18：主要包括建筑高度控制、居住建筑间距（奥运村运动员公寓部分）、建筑退红线、市政站点配套要求、停车位要求、绿化率要求、消防要求、环保要求、人防工程要求等，可查北京市相关规范。目前概念性规划阶段，不要求提出各项具体建设控制指标，可从用地规划合理性出发进行设计。

Q18: Please provide us with the relevant rules and regulations which we must follow besides the Bidding Documents. Hopefully, please give us a clear list.

A18: They include the regulations of height control of buildings; the space distance control between residential buildings(Athletes' Apartments in the Olympic Village); set back line control; parking space; green ratio; fire fighting; environmental protection; people's air defence,etc. We don't require any specific construction-controlling index in the conceptual period. Welcome reasonable suggestions.

问19：绿化率是指实际的植被面积还是指包括树冠在内的面积？

答19：是指实际的植被面积。

Q19: Does the green coverage rate show the area of the actually planted ground, or show area including the projection area of a tree crown?

A19: It shows the area of the actually planted ground.

问20：A-1区需完全保留，还是可以根据规划统一进行考虑？

答20：除了现有的建筑物保留外，其他场地可结合规划统一考虑。

Q20: Will Site A-1 be fully retained or give some integral consideration in favor of the Plan?

A20: The existing buildings have to be retained; the other sites will be planned concerning the master plan.

问21：设施总建筑面积

按6.4.1末段提及，总建筑面积与B区内之奥运中心为216万m²，不含地下停车面积。但地下设施，如商场，部分场馆等，地下之建筑面积是否需包括在216万m²之内？

答21：总建筑面积216万m²不含地下停车场，地铁线路及车站和相关附属设施的面积，但包括地下商场、场馆地下部分等。

Q21: In 6.4.1, the total total floor area is 2,160,000sqm (excluding the underground parking spaces) Does the total floor area of 2,160,000sqm include the underground market place and the underground area of some stadiums?

A21: The total total floor area is 2,160,000sqm doesn't include the underground parking spaces and subway stations and other relevant facilities, but it includes the underground market place and the underground area of some stadiums.

问22：奥林匹克公园内C1区、C2区的未来土地用途是什么？它(们)将变为森林公园吗？部分土地可否用作建筑用地？

答22：奥林匹克公园内C1区、C2区未来用于森林公园用地，位于城市绿化隔离地区(Green Belt)不能用作建筑用地。

Q22: What is the planning land use of Site C1, C2? Will they become the Forest Area? Can the part of it become the construction land?

A22: Site C1,C2 are planned for the Forest Area. They are in the Green Belt. They cannot be used for the construction land.

问23：第2-2页3.2项：城市景观规划对元大都遗址是否有具体要求？元大都遗址公园是否作为奥林匹克公园一部分？

答23：目前没有具体要求，但以文物保护和绿化为主要考虑，元大都遗址公园不是作为奥林匹克公园一部分。

Q23: On the Page 2-2, 3.2 Concerning the master layout and landscape planning, are there any special requirements on the Yuan Dynasty City Relics Park? Is it a part of the Olympic Green?

A23: Presently, no but main concern will be the preservation of historical relics and construction of green space. The Yuan Dynasty City Relics Park is not a part of the Olympic Green,

问24：C区之规划城市森林公园用地，是否完全不允许建设任何建筑设施？是否允许若干比例的绿色产业建筑，如度假酒店，康乐活动中心等等？

答24：可以结合绿地、景观规划适当安排少量小型景观、休闲设施，但不允许建设度假酒店、康乐中心等大型盈利性设施。

Q24: In Site C, is it possible to set up some buildings and facilities in this area? Is it allowed that , in certain part, some Green Industry Facilities are constrcted, for instance, holiday hotel, entertainments and activities center, etc.?

A24: You can arrange properly some mini-sights, entertainment facilities. But the large profit-making facilities (holiday hotel, health centers)are not allowed to be constructed.

问25：根据征集书所说，在C区中有100ha的水域，在比赛期间，这些水域是否可以服务于水上比赛项目，如赛艇或皮划艇？

答25：水上比赛项目，如赛艇或皮划艇已被安排在其他地区。

Q25: According to the documents, a 100 ha water body is to be planned in C area, therefore, Is it quite possible to use this water body for the competition during the Olympic Games (aquatic sport events like rowing, canoe etc.)?

A25: Aquatic sport events like rowing, canoe, etc have already been arranged in other areas.

Q26: In the "Document of International Competition for Conceptual Planning and Design of Beijing Olympic Green", Page 2-12, Clause(4), Business Service Facilities, do these facilities have to be completed by the end of the Olympic Games? Can these facilities become the temporary works?

A26: These facilities can be completed step by step. Total Construction Area of about 800,000sqm is the area of the permanent works. If the design firms want to arrange some temporary Business Service Facilities, it's area can not be included in the above-mentioned 800,000sqm.

Q27: Please confirm : Is the Exhibition Center of Beijing City Planning just a current plan? Will it be constructed after the Games?

A27: It is not required to be completed before the Games.

Q28: Please confirm: Will China Acrobatics & Circus Hall be removed into the National Gymnasium after the Games?

A28: Concerning the multi-functional needs, we primarily planned it in the National Gymnasium.

Q29: Please confirm : the Capital Teenage Palace(150000sqm) is just a current plan or it will be carried out after the Games?

A29: It is a long-term project, which doesn't have to be completed before the Games.

Q30: Please confirm: The Paddling Paradise is just a current plan or it will be carried out before the Games?

A30: We suggest that National Swimming Complex should be designed integrally. The main concern of the Paddling Paradise is the multi-functional need. It doesn't have to be completed before the Games.

Q31: Please confirm: Environmental Education Center(3000m^2) needs to constructed before the Games?

A31: Yes, it is required.

Q32: Please confirm: the one half (200,000m^2) of Exhibition Center will be used during the Olympic Games, the other half will be constructed after the Games.

A32: Exhibition Center will be carried out by phase. But in the short run, it will provide the area of Competition, IBC,MPC during the Games.

Q33: On the Page 2-9, 6.4.1 The required facilities need to be carried out by phase. How many years does the "short-run" mean? Will these facilities be used at once after the construction?

A33: The "short-run" means the period before 2008. These facilities will be used at once after the construction?

Q34:On the Page 2-15, 7.2.2.3, Is there any requirement on the terminating construction scale?

A34:The area of 2,160,000 sqm. is for permanent facilities.

Q35: The national swimming complex: The program calls for a 15,000 seat 90,000 square meter building. Sydney Olympic Swimming Complex has some similar building function requirements. However, the total building area is about 25,000 square meter. Could you please confirm that 90,000 square meter is really the program intention. If so, please confirm the program contents.

A35: The national swimming complex includes the Paddling Paradise.Besides the area for competition, it will meet the other needs.

Q36: On Attach5-6, 5.2, the total construction area of residential zone is around 360,000sqm,single rooms and double rooms will take 50% each. This percentage(50%) refer to the 50% of area or residential number? Is it true that 1/3 of 17,600 is in single rooms and 2/3 of it is in double rooms? If so, can you confirm the requirements on residential number and the area of each unit?

A36: Single rooms and double rooms will take 50% each, which means the number of the rooms.1/3 of 17,600 is in single rooms and 2/3 of it is in double rooms. In the Application Report, there are altogether 2200 apartments, the construction space per capita is about 22.5m^2,which is a suggestion.

FAQ *on the field survey of Beijing Olympic Green*

Q37: Is there any requirement on the layers of the athletes' apartments?
A37: Because of the limitation of the land area, you can arrange some high-rises. Most of the athletes' apartments are multi-storey buildings.

问38：请问奥林匹克公园附近有无清洁水源？
答38：附近无较清洁的地表水。
Q38: Is there any clean water resources near the Olympic Green?
A38: No.

问39．规划中地铁线路的运送能力在奥运会期间估计是多少？
答39．地铁奥运支线的设计单向运输能力为2.9万人次／小时，每辆列车为6节车厢编组，最小列车间隔为3分钟，每辆列车定员1440人，车身长度约120m。
Q39: What is your estimation, in the plan, of the transportation capability of metro during the Olympic Games How long is one subway lane(the train itself)?
A39: The transportation capability of metro is 29,000 person-time/hour, 6 compartments of a subway train. The sequence interval is 3 seconds, the passenger capability of each train is 1440. The length of each train is 120 meters.

问40：在我们的建议中包括地铁站的位置吗？
答40：是的，包括在内。
Q40: Is the position of a subway station included in the range of our proposal?
A40: Yes, it is.

问41：奥林匹克公园四环快速路与其他道路是平面交叉还是立体交叉？对于交叉形式您有何要求？
答41：奥林匹克公园四环快速路与中轴路、京昌快速路、安立路等均为立交。立交形式根据交通规划确定，要保证交通顺畅、安全。其中与京昌高速路交叉的健翔桥立交已有改造方案(补充材料中已提供)。
Q41: In Olympic Green, the crossing way between the 4th Ring Road and the other roads is garde-cross or vertical cross? Is there any requirement on the crossing way?
A41: The crossing way between the 4th Ring Road and the Central Axis, Jingchang Expressway, Anli Road is vertical cross. According to the traffic plan, the cross way should guarantee the sound traffic. We have already got the renovation plan on the cross-section between the 4th Ring Road and Jingchang Bridge. (Provided in the supplementary materials).

问42：考虑到综合规划，北四环路是否可以改造？
答42：目前北四环路路面已经是四上四下共八车道，道路两侧基本为建成区，无法进行改造。四环路上健翔桥立交桥正在进行改造，方案示意图已在4月30日邮寄的补充材料（光盘）中提供。四环路与中轴路的立交桥不能进行改造。
Q42: Could the North Fourth Ring Road be reconstructed in favor of Master plan?
A42: The North Fourth Ring Road presently has 8 lanes. Its surroundings are all completed residential area. It cannot be renovated. Jianxiang Bridge is under renovation. The plan sketch map has been mailed with the supplementary document(CD) on April 30.The cross-sections between the Fourth Ring Road and the Central Axis cannot be renovated.

问43：北四环路是否可以和奥林匹克公园内的地下车库相连？
答43：可结合规划提出建议方案。
Q43: Could the North Fourth Ring Road be routed in a subterranean garage in the Olympic region?
A43: Your suggestions are welcome in favor of the master plan.

问44：成府路和大屯路是否必须穿过B区或与北辰西路和北辰东路连接？
答44：是。但可根据道路横、纵断面形式和结合规划提出建议方案。如设在地面、地下还是架空等。
Q44: Must the Chengfu Road and Datong Road cross the Site B or could they be rerouted through Beichenxi Road and Beichendong Road?
A44:The Chengfu Road and Datun Road must cross the Site B, you can give suggestions concerning the transection and vertical section.(set up on the ground or underground or overhead).

问45：关于城市中轴线的问题，本项目基本上以森林公园作为轴线北端处理手法，不知现有或已规划的城市中轴线南端是什么处理方法，请明确。
答45：城市中轴线南端延伸到南五环路附近，目前正在进行整个中轴线的城市设计研究，尚无具体方案。1993年《北京城市总体规划》中提出："旧城中轴线的南、北延长线两侧，主要作为大型公共建筑用地"；"中轴南延长线要体现城市'南大门'形象"。
Q45: City Central Axis Roads
What are the arrangements of the existing or planned south end of Central Axis? Please confirm.
A45: South part of Central Axis is extending to the 5th Ring Road. Presently, we haven't any specific plan for it. In 1993, "Beijing City Master Plan" said, the two sides of prolonging line of South and North Central Axis will be used for the large scale of public buildings. The south prolonging line of the Central Axis will embody the City's Image of " South Gate".

问46：中轴线是否可以穿过森林公园？
答46：中轴线线位未定，在总体规划中综合考虑用地布局、交通、景观等各方面的要求，可提出建议方案。
Q46: Can the Central Axis get through the Forest Area?
A46: You can put forward the suggestions in favor of the land overall arrangement traffic conditions and sightseeing.

问47：中轴路现状可否改造？
答47：位于奥林匹克公园内北土城路和北四环路之间的现状中轴路道路断面不能改变，道路中部的绿化带可结合总体规划提出景观规划建议。
Q47: Can we change the existing conditions of the Central Axis?
A47: Please give suggestions in favor of the Plan. You cannot change the existing transection of the Central Axis Road which is between Beitucheng Road and North 4th Ring Road in the Olympic Green.

问48：现状北辰西路可否拓宽？
答48：可以结合总体规划提出建议方案。
Q48: Can we widen the existing Beichengxi Road ?
A48: Yes, you can give us suggestions.

问49：标书中提到：物流、新闻采访车从北辰西路进出；观众从地铁、北中轴路进出……请问这些要求是否硬性规定？可否根据设计做一些调整？
答49：可根据设计作部分调整。
Q49: According to the requirements in the Bidding Document, Material flow and news coverage vehicles: passing through from Beichengxi Road; Audience: passing through from subway and Northern Axis Road and separating with the material flows. Are these requirements changeable? Can we make some adjustments in favor of the Plan
A49: You can make some adjustments in favor of the reasonable arrangements.

问50：规划设计中水域面积为100ha，这部分水域是在森林公园（C区）吗？在B区和A区有没有水域面积的限制？水源从哪儿来？
答50：目前规划考虑水源来自清河或北小河污水处理厂中水，必要时可补清水（如北京城市西北方向的京密引水渠的水）。A区和B区对水面面积未作要求。欢迎合理建议。
Q50: The area of the water body is about 100 ha in the planning scheme.Is the water body located in the forest area?Are there any constraints for the water bodies in Site A and Site B? Where is the water resource?
A50: The water resource of the water bodies is mainly from the non-portable recycled water of Qinhe River or Beixiao River Sewage Treatment Plants. It also could be supplied by the fresh water , such as from the Jingmi Canal which lies in the northeast of urban Beijing).There is no requirement on water area in Site B and Site A. We welcome reasonable suggestions.

问51：用地内22万V高压供电走廊对建筑物和环境有否要求？
答51：没有要求，供电线路沿主要城市道路入地。
Q51: In the construction site, is there any requirement on the 220kv power supply corridor against the buildings and its surrounding?
A51: No, it will be buried underground along urban roads.

问52：现存的建筑搬迁后，原有水渠的水路是否应该保留。
答52：C区的水路要保留。
Q52: Should the existing water courses of the canals be retained after removal of the existing buildings ?
A52: The existing water courses of the canal in Site C will be retained.

问53：A–3区现有清洁车辆四厂，作为预留用地，可否考虑进行统一规划？
答53：清洁车辆四厂用地作为规划预留地，尚无具体规划意向，在2008年前不能安排奥运设施。远期发展可结合规划进行统一考虑。
Q53: Can we move away the existing depot for the No.4 sanitary vehicles in site A-3,or must it be reserved?
A53: Before the Games it will be the reserved land, which needs to be considered as a part of the whole design for the long-term development.

FAQ on the field survey

北京五棵松文化体育中心现场踏勘提问及答疑

FAQ on the field survey of Beijing Wukesong Cultural and Sports Center

问1：篮球馆可增加经营性面积约2～3万m², 是不是包含在篮球馆3.5万m²的建筑面积之内？

答1：不包括在内。另外，篮球馆要求的3.5万m²建筑面积是北京申办奥运期间，根据满足篮球比赛的基本要求和附属功能用房基本要求的情况下得出来的。但各国的设计标准不完全一样，此数据仅作为参考值，最终的决定性因素应是必须要满足奥运会篮球比赛的要求。

Q1: The commercial building area in Wukesong Basketball Hall can be increased by 20,000m² to 30,000m², is it included in the 35,000m² of the total building area?

A1: No, 20,000-30,000m² of the commercial facilities is not included in the 35,000m² of the building area of the Basketball Hall. The 35,000m² of the building area is calculated during the Olympic Bidding period, this figure is to meet the basic requirements of the basketball match. Each country may have different design standard, the figure of 35,000m² may only be seen as a reference. The decisive factor is to meet the requirements of the basketball match.

问2：用地的远期发展是否可以安排住宅项目？

答2：该用地是城市总体规划预留了50年的体育用地，规划要求作为城市公共活动中心，安排体育设施、文化设施、商务设施等公共活动设施，并要求有较大面积的集中公共绿地和开放空间，赛后利用不考虑住宅。因为用地周边已经有较密集的住宅区，规划也已安排有大量住宅项目。

Q2: Concerning the long-term development of the land use, can we be allowed to arrange some residential projects?

A2: This land had been reserved for sports facilities since 1950's. According to the City Master Plan, the land will be used for the public center which includes sports, cultural and commercial facilities. Meanwhile, we will arrange a large amount of green area. The residential construction is not our concern because we have got a lot of residential area in the outskirts of this land.

问3：在未来的八到十年内，周边的环境有没有继续发展的计划？

答3：用地东侧、南侧基本上是建成区，北侧是正在规划中的住宅区，西边四环路以外还有一些要发展的地区。

Q3: In the next 8 to 10 years, have you got further development plan on the surrounding environment?

A3: The construction of the land to the east and south of the site has been completed. The north part is the residential area, which is under planning. There are some lands to be developed outside the 4th Ring Road.

问4：近期和远期项目是否可以考虑结合？

答4：欢迎提出近远期项目结合发展的建议。

Q4: Should it be possible to integrate the short-term and the long-term construction?

A4: Yes, you can integrate the short-term construction with the long-term construction.

问5：在用地的西侧和南侧已经分别有100m和50m公共绿化带，用地北侧和东侧没有具体表示出绿化带是多少，是否可以认为，只要根据6.4.2条，建筑退用地红线的要求，各退后10m就可以呢？

答5：根据第6.4.2条建设用地退用地红线的要求，沿用地的东、南、西、北要求分别退红线10m、50m、100m、10m。红线范围以内包括建设用地、规划的集中公共绿地和开放空间等。

Q5: There are existing 100m and a 50m green buffers to the west and the south of the site. But there is no stipulation on the width of the green buffers to the north and east of the site. Can we regard the setback requirement as 10m for the side boundary according to Item 6.4.2?

A5: The setback requirements are 10m, 50m, 100m and 10m for the side boundary respectively on the east, south, west and north side. The lands surround by the redline include the lands of construction, open spaces and green area.

问6：集中公共绿地是否计入"建设用地内绿化率不小于30%"的指标内？体育场地当中的绿地，如足球场中的绿地和屋顶绿化是否计入绿化率？

答6：关于建设用地的绿化指标是按照现行的北京市设计规范提出来的，单体的建设用地内要有不少于30%的绿化率。这个绿地率是在集中公共绿地外的。现阶段给出这个概念，是提醒设计者在公共绿地以外，还应结合建设用地更多布置绿化用地。足球场中的绿地和屋顶绿化不计入绿化率。

Q6: The green area ratio should be no less than 30%, can we calculate the centralized green area into the green area? Is it possible to reckon the grass on the playground or on the roof in the green ration?

A6: The green area ratio was given by the Design Regulation of Beijing. As for the conceptual design, you should only take the green area into consideration. But the green area on the roof can not be reckoned in the green area ratio.

问7：西山风景区与五棵松文化体育中心的关系如何？

答7：西山风景区是北京市总体规划确定的一处位于城市西北部的重要风景区，北京市城市规划中十分注意保护从城市看西山的景观通廊，天气晴朗的时候从五棵松文化体育中心用地内也能看到西山。可以在规划中考虑城市整体景观环境背景这一因素。

Q7: I want to clarify the relationship between the Xishan Scenery Area and the Wukesong Cultural and Sports Center.

A7: Yes, Xishan Scenery Area is the landmark of Beijing, the city plan emphasizes the view of the Xishan Scenery Area from the city, in fine weather, we can have a clear view of the Xishan mountain from the Wukesong cultural and sports center. You could treat Xishan Mountain as a background of the scenery design.

问8：能否统一提供奥运场馆的标准尺寸？

答8：请各设计单位自行解决，但应满足奥运会的需求。

Q8: Could you provide the standard size of all Olympics sports venues?

A8: The size shall be determined by the design firms, but the size should meet the requirements of the Olympic Games.

问9：10万m²商业服务设施需要从一期中考虑出来，还是放在二期，这一点与英文标书中互不相符，应以哪个为准？

答9：可以分期建设，不要求在赛前全部建成，考虑到投资运营的需要，可安排部分设施在近期建设。

Q9: Commercial Services Facilities (100,000 m²) will be picked out from Phase 1 or be considered in Phase 2. This point conflicts with that in English version, which should be followed?

A9: Commercial Services Facilities could construct by phases, some part of them could be constructed in the near future for the commercial concern.

问10：能否得到关于篮球馆、棒球场和垒球场的尺寸和要求？

答10：根据申办报告，篮球馆比赛场地的面积是42m×24m，竞赛区域是28m×15m。棒球场的面积是98m×112m。垒球场包括一块比赛场和两块训练场。

Q10: Would it be possible to receive the official dimensions and requisite concerning an Olympic Basketball, Baseball and Softball stadium/grounds?

A10: The area of the Basketball Hall is 42m×24m, the competition areas are 28m along the left and right sidelines and 15m along the central line. Baseball Competition areas are 98m along the left and right sidelines and 112m along the central line. Each will have two adjacent warm-up practice fields. The softball venue comprises one competition field and two training fields.

问11：1300车位按地下车库设计，其面积是否计入规划总建筑面积？

答11：根据规划要求，地下面积除停车场外，均计入总建筑面积。

Q11: Is the parking lot with 1,300 parking spaces included in the total construction area?

A11: According to the requirement of planning, the underground area should be included in the total floor area except the parking lot.

问12：规划用地西侧及南侧绿化带上可否开车辆出入口？

答12：用地西侧沿四环路辅路可开机动车出入口。南侧临长安街主干道，距长安街和四环路交叉口较近，交通情况复杂，不宜设主要机动车出口。可以结合规划、交通分析，提出建议方案。

Q12: Is it allowed for us to design the entrance and exist for the vehicles in the west greenbelt and south belt of the planning land use?

A12: You can design the entrance and exist for the vehicles in the westside of the planning land along the 4th Ring Road, On the south part of it, which is close to the Chang'an Avenue and cross-section of the 4th Ring Road, concerning the traffic conditions, it is not suitable to design the main entrances or exists for vehicles. In favor of the master plan and traffic analysis, you can offer some reasonable suggestions.

问13：公共文化体育设施是否将在2008年举办奥运后建设？

答13：不要求在赛前建成。

Q13: Please confirm whether the public cultural and sports facilities will be constructed after the Olympic Games.

A13: Yes, they are not required to be constructed before the Olympic Games.

问14：请问100m及50m的绿化带是否计入30%的绿地中？

答14：不计入。

Q14: Are the two green belts (100m&50m) included in the Green Area (30%)?

A14: No, they are excluded from the Green Area.

问15：请确认篮球馆附近的商业/娱乐设施是否在目前仅作出规划，并在2008年举办奥运后再建设？

答15：不要求在赛前建成。

Q15: Are the business and entertainment facilities near the Basketball Hall just a current plan and to be constructed after the Games?

A15: We do not require the business and entertainment facilities to be constructed before Olympic Games.

问16：请明确规划设计必须遵守的北京市有关规范。

答16：主要包括建筑高度控制、建筑退红线、市政站点配套要求、停车位要求、绿化率要求、消防要求、环保要求等，可查北京市相关规范。目前概念性规划阶段，不要求提出各项具体建设控制指标，可从用地规划合理性出发进行设计。

Q16: Can you tell me what relevant criteria in Beijing should be followed during our design?

A16: They include the height control of buildings; set back line control parking space; green ratio requirements; fire fighting; environmental protection requirements; etc. We don't require any specific construction controlling index in the conceptual period. Welcome reasonable suggestions.

评审委员会主席 (Chair of the Jury)

刘太格：
城市规划师,建筑师,新加坡雅思博事务所董事,国家艺术委员会主任,北京市政府城市规划顾问,曾担任2000年北京国际展览体育中心规划设计方案征集活动评委会主席,2001年北京商务中心区规划设计方案征集活动评委。

Liu Thai Ker:
Liu Thai Ker: Architect & Urban Planner, Director of RSP Architects Planners & Engineers (Pte) Ltd., the director of National Arts Committee, Singapore, the urban planning consultant of Beijing Municipal Government, Chairman of the jury committee of International Competition for the Conceptual Planning and Design of Beijing International Exhibition & Sports Center in 2000; the jury member of the International Competition for the Conceptual Planning of Beijing CBD (Central Business District) in 2001.

评审委员会委员 (Jury Members)

楼大鹏：
体育专家,第29届奥林匹克运动会组委会体育部部长,曾担任2000年北京国际展览体育中心规划设计方案征集活动评委。

Lou Dapeng:
Sports Expert, Director of Sports Department of Beijing Organizing Committee for the Games of the 29th Olympiad, the jury member of Assessment of the International Competition for the Conceptual Planning and Design of Beijing International Exhibition Sports Center in 2000.

平永泉：
城市规划师,第29届奥林匹克运动会组委会工程部部长,奥运场馆建设协调委员会办公室主任。

Ping Yongquan:
Urban Planner & Architect, Director of Planning & Construction Department of Beijing Organizing Committee for the Games of the 29th Olympiad. Director of Executive Office of the Olympic Venue Development Coordinating Committee.

吴良镛：
清华大学教授,两院院士,北京市政府专家顾问团专家。

Wu Liangyong:
Professor of Tsinghua University, academician of China Academy of Science and China Academy of Engineering, the consultant of Beijing Municipal Government.

齐 康：
东南大学教授,中国科学院院士,法国建筑科学院外籍院士,曾担任2000年北京国际展览体育中心规划设计方案征集活动评委。

Qi Kang:
Professor of the Southeast University, academician of China Academy of Science and foreign academician of France Academy of Architecture. The jury member of Assessment of International Competition for the Conceptual Planning and Design of Beijing International Exhibition & Sports Center in 2000.

Jim Sloman:
工程师,国际奥委会推荐专家,澳大利亚悉尼奥组委副首席执行官及首席运行官。

评委简历 Background of the Jury Members

Civil Engineer, Expert recommended by IOC(International Olympic Committee). Former Deputy Chief Executive Officer and Chief Operating Officer of the Organizing Committee for the Sydney 2000 (SOCOG).

Chris Johnson:

建筑师,熟悉奥运会组织运行的国际专家,澳大利亚新南威尔士州政府建筑师,奥运项目城市设计评估委员会主席。

Architect, expert of the Olympic Games operation. The NSW Government Architect, Chairman of Assessment Committee for Urban Design of the Sydney Olympic Games Works.

Xavier Menu:

建筑师,法国建筑与城市规划设计国际公司总裁,曾担任2001年北京商务中心区规划方案竞赛评委。

Architect, President of S.C.A.U. Urban and Architecture firm, France, the jury member of the International Competition for the Conceptual Planning of Beijing CBD (Central Business District) in 2001.

Kristin Feireiss(女)(Female):

德国著名建筑、规划评论家,AEDES建筑展览馆馆长,前荷兰建筑师协会会长,前德国建筑博物馆馆长。

Architecture & Urban Planning critic, Germany, Curator of AEDES Architecture Exhibition Center, former chairperson of Architects Association of the Netherlands, former curator of Architecture Museum of Germany.

卢伟民:

城市规划师,美国Lowertown公司总裁,北京市政府城市规划顾问,曾担任2001年北京商务中心区规划方案竞赛评委。

Weiming Lu:

Urban Planner, President of Lowertown Redevelopment Corporation, the urban planning consultant of Beijing Municipal Government, Chairman of jury committee of the International Competition for the Conceptual Planning of Beijing CBD(Central Business District) in 2001.

宣祥鎏:

原首都规划建设委员会秘书长,曾担任2000年北京国际展览体育中心规划设计方案征集活动评委。

Xuan Xiangliu:

The former secretary-general of Capital Planning and Construction Commission. The jury member of Assessment of International Competition for the Conceptual Planning and Design of Beijing International Exhibition & Sports Center in 2000.

王富海:

城市规划师,深圳市规划院院长,曾担任2001年北京商务中心区规划方案竞赛评委,1999年中关村西区规划方案竞赛活动评委。

Wang Fuhai:

Urban Planner, President of Shenzhen Municipal Urban Planning Institute; the jury member of the International Competition for the Conceptual Planning of Beijing CBD (Central Business District) in 2001 and the jury member of the Competition of Planning and Design for West area of Zhongguancun in 1999.

备受国内外关注的北京奥林匹克公园和五棵松文化体育中心两个奥运会场馆集中用地的概念性规划设计方案征集活动已经徐徐拉下帷幕。这是北京2001年7月13日赢得第29届奥运会的举办权之后,为筹建奥运场馆设施举办的第一次大型的、国际性的方案征集活动,得到了境内外规划设计单位的广泛参与和热情投入。

奥林匹克公园是奥运会场馆设施最集中的地区,将有14个比赛项目在这里举行。同时,从城市长远发展的角度,这里也是北京市最重要的一个市民喜爱的、多功能的城市公共活动中心,包括运动场馆、会展博览设施、大型文化设施、旅游和商业设施等四类功能。

五棵松文化体育中心在20世纪50年代的城市总体规划中就被确定为体育中心,经历了半个世纪后,规划终于能够得以实施。这里将成为为北京西部社区服务的、集群众文化、体育和休闲为一体的城市公共活动中心。

北京奥林匹克公园和五棵松文化体育中心规划设计方案评审总结

Assessment Report

1.方案征集活动概况

此次征集活动受北京市人民政府和第29届奥林匹克运动会组委会委托,由北京市规划委员会负责组织。方案的征集活动采用公开方式,在北京市公证处的监督下进行。从2002年4月开始报名到评审结束,历时3个多月,前后有一百多家境内外规划设计单位参加,其中境外设计单位来自美国、加拿大、德国、英国、法国、瑞士、意大利、西班牙、挪威、希腊、奥地利、荷兰、澳大利亚、俄罗斯、日本、韩国、新加坡、马来西亚、泰国、香港、台湾等21个国家和地区,充分显示了此次活动的国际影响。因此,这次活动的目标不仅是要征集到实用并有创意的设计构思,同时,也要向世界展示北京对外开放、积极筹办奥运的新形象。方案征集的全过程秉持公开、公正、公平的原则,参照国际惯例细心筹划和准备。

主办单位共收到应征者按时递交的奥林匹克公园规划设计方案57个,五棵松文化体育中心规划设计方案34个。除去4个方案因为不符合征集书中的程序性规定,被确认为无效方案外,共有有效方案87个。这次方案的评审不仅成为北京,可能也是全国范围内最大规模的一次规划设计方案评审活动。

方案评审的过程分为两个阶段,首先由技术小组进行技术初审,检查相关图纸、文件及校核技术要求,然后由评审委员会进行评审。

评审委员会共由13名委员组成,其中国内评委7人,国外评委6人,包括国际奥委会推荐的奥运会组织运营专家、奥组委的代表、政府部门的代表、国内外著名的城市规划与设计专家等。在评审委员会的预备会上,新加坡著名城市规划设计专家刘太格先生被推举为评委会主席。会议讨论通过了评审办法。两个项目分别进行评审,采用多轮投票的方式确定中奖方案。

2.方案评审结果

根据《征集书》中的有关条款,奥林匹克公园规划设计方案的评审特别强调了以下主要原则:

功能布局应充分满足奥运会的要求,并重视城市的有机生长,有利于城市的长远发展;要体现生态、环保的概念;交通系统及交通组织规划应有利于赛事的组织和赛后的使用,应重视公共交通的规划设计;空间形态应体现北京城市特色,强调城市中轴线的特殊意义;应具有较强的分期建设与市场化融资运作的可操作性。

五棵松文化体育中心规划设计方案的评审参照上述内容,同时充分考虑项目所在的位置及周边环境的特殊性。

为保证客观、真实地评价应征方案,所有应征方案均以评审编号的形式出现,且评审期间对外严格保密。经过评审委员会严肃认真的审核、研究和讨论,通过不记名投票,最终产生了中奖方案。

奥林匹克公园规划设计的中奖方案:

一等奖1个:美国Sasaki Associates,Inc与天津华汇工程建筑设计公司合作的方案。

二等奖2个:北京市城市规划设计研究院与澳大利亚DEM AUST有限公司合作的方案;

日本国株式会社佐藤综合计画的方案。

优秀奖5名:法国AREP公司的方案;

总装备部工程设计研究总院和哈尔滨工业大学天作建筑研究所合作的方案;

北京市建筑设计研究院和美国EDSA规划设计公司及交通部规划研究院合作的方案;

德国HWP公司方案;

北京大学城市规划设计中心和北京大学景观规划设计中心合作的方案。

五棵松文化体育中心规划设计的中奖方案:

一等奖:(空缺)。

二等奖2个:美国Sasaki Associates,Inc的方案;

瑞士Burckhardt+Partner AG的方案。

优秀奖3个:总后勤部建筑设计研究院的方案;

北京市城建建筑设计研究院有限公司和加拿大ABCP设计公司合作的方案;

中国建筑设计研究院的方案。

3.中奖方案介绍

(1)奥林匹克公园规划设计方案

1)美国Sasaki Associates,Inc和天津华汇工程建筑设计公司合作方案

方案布局特点:体育场馆、会展设施、文化和商业设施位于中轴线两侧,中轴线的景观大道和轴线东侧的龙形水系从森林公园向南延伸至现状国家奥林匹克体育中心。

该方案整体性较强,功能分区明确,中轴线的处理刚柔并济,富有创意,绿化、水体和建筑的空间关系灵活。较好地体现了生态、环保的概念,山形水系和现状森林公园有机衔接。赛时各种功能相对集中,有利于赛事组织;便于赛后分期实施,并为单体建筑的设计提供了余地。

不足之处是水系的形状略显具象,还需进一步研究挖掘深层的文化内涵;还需考虑增加地标性建筑物或构筑物;应进一步丰富中轴线的景观设计内容,并注意研究人在空间中实际感受到的尺度。

2)北京市城市规划设计研究院和澳大利亚DEM AUST有限公司合作方案

方案布局特点:功能布局由南向北推进,依次为会展、体育、文化综合区;商务区布置在用地东、西两侧,与周边地区现有商务和居住设施等衔接。

各功能分区明确,有较强的整体序列感和分期建设可操作性。中轴线绿化贯穿,北端自由开敞,融入自然。路网结构清晰,机动车道路全地面,有利于实施;用地内部增加了南北向道路,便于交通组织。不足之处是三个功能分区过渡不

够自然，应改善北部的水体、绿化同中部的建筑空间的衔接关系，中轴线的处理缺乏明确的立意。

3）日本国株式会社佐藤综合计画方案

方案布局特点：体育场馆、会展中心及商业文化设施沿中轴线两侧布置，水面隐喻北京旧城的北海、中海、南海，以升起的平台形成中轴线的骨架。

该方案水体、绿化和中轴线的处理富有变化，浪漫自由；其间设置一些具有象征意义的建筑作为轴线的节点。体育场馆和展览建筑单体形式灵活、统一和谐，同中轴漫步道和水面相互穿插，形成完整、亲切的平面构图。但由于建筑单体个性强烈并与总平面布局关系过于密切，使今后单体建筑设计的余地较小。任何添加都容易改变方案原有的形态。

4）法国AREP公司方案

方案布局特点：纵贯奥林匹克公园分为三个功能区——中轴线上为奥林匹克大道，西侧为会展、商业、办公区，东侧结合起伏的地形、绿化及水面等自然景观布置体育场馆。

该方案功能分区明确，结构清晰；用地东部山形水系与建筑结合，空间灵活；中轴线设计为人性化的步行商业空间；人车分流明确。体育场馆周围空间大小不利于赛时的人流疏散组织。人车交通立体分层，略显复杂，可考虑简化处理。

5）总装备部工程设计研究总院和哈尔滨工业大学天作建筑研究所合作方案

方案布局特点：以体量巨大、宽大平缓的建筑覆盖中轴线，其两侧为体育场馆、商业、文化等各类建筑；体育场馆在北部布置，主体育场位于中轴线上。

该方案气魄宏大，整体性强，建筑形象统一和谐，平面造型优雅，富有创意。中轴线以完整的巨型结构作为高潮，形成轴线北端的新形象。但由于建筑集中布局，体量过大，不利于分期建设。赛时交通和疏散问题有待进一步研究论证。

6）北京市建筑设计研究院、美国EDSA规划设计公司和交通部规划研究院合作方案

方案布局特点：用地从南至北依次布局为文化、体育和展览，用地两侧安排商务设施。

该方案功能分区明确，考虑了分期实施和融资问题，实施性较强。道路同城市路网结合紧密，入口处理简洁明确。在用地内部增加了南北向道路，有利于交通组织。但布局灵活性不足，两侧商业建筑布局与中心区体育建筑关系还需进一步推敲，绿化较少。

7）德国HWP公司方案

方案布局特点：强调城市街道空间，体育设施、商务办公及文化建筑等相间布局，绿化空间集中在用地两侧。

该方案将各种功能的设施集中布置在B区中部四个大屋顶之下，体系严谨，构思奇特。利用绿化作为规划公共设施与现状居住区的缓冲空间。但四个巨型屋面的必要性和尺度有待进一步研究。

8）北京大学城市规划设计中心和北京大学景观规划设计中心合作方案

方案布局特点：方案围绕"荷"的立意展开，中心区各组建筑分别象征花、叶、梗、莲蓬，地下空间象征藕。

该方案功能分区较明确，空间形态规整与自由结合；赛前保留临四环路的绿地作为城市快速交通和建设区的缓冲空间，同时也为城市的未来发展留下了余地；便于分期实施。以"荷"为主题进行构思并不是此方案赢得奖项的原因。除了建筑形态模仿荷花以外，该主题同方案没有更深的内在联系。

（2）五棵松文化体育中心规划设计方案

1）美国Sasaki Associates, Inc方案

此方案功能分区明确，景观规划设计尤为突出；场馆建筑沿四环路的一面被设计为坡地绿化，形成适应于快速路的城市景观，场馆建筑另一侧则亲近于用地中间的城市公共绿地；便于分期实施。不足之处是棒球场和练习场距离较远，商业开发占地面积过大。

2）瑞士Burckhardt + Partner AG方案

方案构思独特，篮球馆和商业设施全部集中在一个立方体内，使用地内留出了大片的城市绿地。主体建筑形式和设计理念新颖，在尺度、结构上进行了大胆尝试。多功能性的建筑有利于商业运营。但在技术和造价上有待进一步研究。

3）总后勤部建筑设计研究院方案

用地布局以方形构图，三个主要建筑连成一体，集中布置在用地对角线方向上，整体造型富有韵律感。方案构思独特，规划具有整体性。不足之处是构图追求形式化，不便于分期实施。

4）北京市城建建筑设计研究院有限责任公司与加拿大ABCP设计公司合作方案

方案构思独特，将全部功能集中在一个建筑屋顶下，建筑造型富有雕塑感。用地内留出了较大的城市公共空间。但建筑一侧立面总长度超过600m，尺度还需进一步研究，不便于分期实施。

5）中国建筑设计研究院方案

整个用地功能分区合理、明确，划分为三部分，中间为绿地和停车，沿四环路为篮球馆等体育设施，东侧为商业、酒店等设施。交通组织流畅，场地内人车分流，人行平台下为机动车道。较好地考虑了沿四环路和复兴路的景观，便于分期实施。不足之处是建筑单体造型缺乏创意。

4. 评审委员会对规划方案的建议

针对应征方案中反映出的问题，评审委员会对奥林匹克公园规划提出了一些建议：中轴线上不一定要以本时期的建筑作为终点标志，可为今后的创造留有余地；建议轴线北端以自然山形水系作为背景；场馆设施应适当紧凑安排，节约使用土地；应注重生态和环境保护，水系和绿化应与北部森林公园有机衔接；交通规划还应进一步深入研究论证，从城市整体交通体系出发解决整个地区的交通问题；注意建筑及城市空间的尺度不宜过大等。

对五棵松文化体育中心的规划，评审委员会强调：五棵松规划用地作为城市公共绿地保留了很多年，定位为群众性文化体育中心，商业性开发要尽量少一些；应多安排群众文化活动及体育健身的场所，尽可能对公众开放，为北京市民所用。建筑单体设计应体现科技奥运的含量，篮球馆的设计在满足篮球比赛要求的同时，应充分展示时代建筑的高科技和环保特征，成为吸引市民和游人的场所。

这次方案征集活动虽已结束，但对于奥运项目的整体规划设计来讲，只是一个开始。而城市的发展更是一个长期、复杂的过程，在以后的规划调整、深化中，还需要有针对性地就其中的诸多问题进行更广泛、全面而深入的研究。

Assessment Report

Assessment Report
of International Competition for Conceptual Planning and Design of Beijing Olympic Green and Wukesong Cultural and Sports Center

The planning and design competitions for the Beijing Olympic Green and Wukesong Cultural and Sports Center, two sites for the Olympic Games, have drawn to an end. It is the first largest event involving domestic and foreign designers to produce designs for the Games facilities in Beijing after the successful bid for the 29th Olympic Games on July 13, 2001.

The Olympic Green has concentrated a large number of facilities as 14 kinds of games will be held there. In terms of urban development, the place also constitutes a very important multi-functional public center for its residents. It will assume four functions as the sports facilities, exhibition centers, cultural facilities as well as tourism and business facilities.

Wukesong Cultural and Sports Center was designated in the 1950s as a sports center in the master plan, and now this plan has eventually been realized. The Center will become a public center comprising cultural, sports and leisure activities in the western part of Beijing.

I. Planning and Design Competition

The Planning and Design Competition was authorized by the Beijing Municipal Government and the Beijing Organizing Committee for the Games of the XXIX Olympiad. The competition was open and supervised by the Beijing Public Notary Office. Starting from April 2002, it lasted 3 months with the participation of over 100 Chinese and foreign design corporations. Participants outside Chinese mainland were from 21 countries and regions including the United states, Canada, Germany, England, France, Switzerland, Italy, Spain, Norway, Greece, Austria, the Netherlands, Australia, Russia, Japan, South Korea, Singapore, Malaysia, Thailand, Hong Kong, and Taiwan. This event, which has an international impact, not only seeks to solicit practical and innovative design concepts, but also tries to establish the new image of Beijing in its opening and active preparation for the Olympic Games. The whole process of competition was in line with the principle of being fair, just and open, and the preparatory work was done in accordance with the international norm.

57 proposals were received for the Olympic Green and those for the Wukesong were 34. Except 4 that violated the procedural specifications and became invalid, altogether 87 proposals were valid. Therefore, the appraisal work could be the largest one not only for Beijing, but also in terms of the whole country.

The evaluation consisted of two parts. In the preliminary stage, a technical panel checked the blueprints, documents and technical requirements. In the second stage, evaluation was done by the Jury.

The Jury comprised 13 members with 7 from China and 6 from other counties. The members included Olympic experts recommended by the International Olympic Committee, representatives of the BOCOG, representatives of the government, and renowned city planning and design experts both from China and abroad. At the preparatory meeting of the Jury, Liu Taiker, famous city planning and design expert of Singapore, was elected Chairman of the Jury. Methods of appraisal were agreed upon. The proposals for the two venues were to be appraised separately, using multi-round of votes to choose the prize winners.

II. Results of the Assessment

According to the clauses in the Competition Documents, the following principles for the Olympic Green planning evaluation were highlighted:

Functional layout must meet the requirements of the Olympic Games, and be conducive to the organic development of the city. Ecological and environmental protection concepts must be demonstrated. Transportation system must take into account of the Games and post-game use. Spatial design should give expression to Beijing characteristics, highlighting the special significance of the Central Axis of the city; and designs can be constructed in phases and able to raise money through financing.

The design evaluation for Wukesong was in the light of the above requirements, but also took into consideration of the special features of the site.

To ensure objectivity, all submitting proposals were processed with numbers, and kept confidential to the public during the evaluation period. After careful evaluation and deliberation, winners were elected by vote of secret ballots.

The winners for the Olympic Green are:
First Prize:
Sasaki Associates, Inc. (USA)+Tianjin Huahui Design Inc. (China)
Second Prize:
Beijing Planning and Design Institute+ DEM AUST Pty (Australia);
Satow Sogo Keikaku (Japan)+Ingerosec Corporation Japan
Honorable Mentions:
AREP (France);
Army General Equipment Engineering Design Institute +Tianzuo Architectural Institute of Harbin Industrial University;
Beijing Architectual Design and Research Institute+ EDSA (USA);
HWP (Germany);
Urban Planning Center, Beijing University+Landscape Planning and Design Center, Beijing University.

The winners for the Wukesong Cultural and Sports Center:
First Prize: vacant
Second Prize:
Sasaki Associates, Inc. (USA);
Burckhardt+PartnerAG (Switzerland);
Honorable Mentions:
Architectural Design Institute of the Army Logistics;
Beijing Urban Development and Architectural Design Institute Ltd.+ABCP Design Inc.(Canada);
China Architectural Design and Research Institute.

III Description of the Winning Proposals

1. Planning and Design for the Olympic Green

1) The Proposal collaborated by Sasaki Associates, Inc. (USA) Tianjin Huahui Design Inc. (China)

Features: Sports, exhibition, cultural and business facilities are located along either side of the Central Axis. The sight on the Axis and dragon-shaped water system on the eastern side of the Axis go from the Forest Area all the way to the existing National Olympic Sports Center.

This planning is strong in overall layout with clear distinction of functions. The relations among greenery, water body and architectural space is flexible, better reflecting the concept of ecological and environmental protection. Also, hills, water system and current forest gardens are linked organically. As functions for matches are concentrated, it is easier for organization of the Games; it is also easy to be constructed in phases after the Games, and leaves leeway for individual unit's design.

The drawback is that the image of the water system is a bit too concrete, thus cultural significance needs to be further tapped; landmark buildings also should be added; sights to be Axis needs to be enriched further and attention be paid to the feeling of human in real spatial dimensions.

2) The Proposal Collaborated by Beijing Urban Planning and Design Institute+DEM AUST Pty (Australia)

Features: The layout goes from south to north, covering exhibition, sports and cultural areas; business sector is scattered on the eastern and western sides of the land used, linking the current business and residential facilities in surrounding areas.

Function division is distinct, gives a strong sense of order and can be easily constructed in phases. The greenery covers the whole Axis with its northern part being open and integrated with nature. Road network is distinct and easy for construction; North to South Roads are added within the land used, providing convenience to traffic. The drawback is the weakness in transitions among the three functional sectors, and the linking relation of water and greenery of the north with the architectural space in the middle part should be improved. The handling of the Axis lacks clear significance.

3) The Proposal by Sato (Japan)+Ingerosec Corporation(Japan)

Feature: Sports facilities, exhibition centers, cultural and business facilities are located along the two sides of the axis, with water implying the North, Middle and South Seas in old Beijing. The raised platform constitutes the framework of the Axis.

The treatment of water, greenery and Axis is rich in variety, romantic and liberal, with some symbolic buildings scattered along the Axis. Each sports and exhibitions facility is flexible but unified and harmonious, and, interwoven with promenade and the water on the Axis, forming a complete and friendly plan figure. But due to the strong character of individual buildings and its too close relation with the overall plan layout, there would be too little room left for future design for each unit. and any addition would alter the original plan.

4) The Proposal by AREP (France)

Feature: Three functional sectors are identified, with the Olympic avenue at the top of the Axis, the exhibition, business and cultural facilities on the west and sports facilities on the east amid the meandering landscape, greenery and water system.

This design is clear in function division and layout. The buildings located in the eastern part are being integrated with the shapes of hills and water, thus providing flexible spaces. The Axis is designed into a commercial space for pedestrians. Pedestrians and motor vehicles are separated. But the space sizes around the stadiums are not convenient for evacuation of people during the matches. And the three dimensional layers of the traffic for man and vehicle are a little complicated, so simplification can be considered.

5) The Proposal Collaborated by Army General Equipment Engineering Design Institute +Tianzuo Architectural Institute of Harbin Industrial University

Features: Buildings of gigantic and sprawling volume cover the Axis, with sports stadiums, business and cultural facilities along the two sides. On the north are concentrated sports facilities with the Main Stadium on the Axis.

Grandness, completeness and harmony of image with graceful and innovative plane figures characterize this design. The huge buildings on the Axis reach their climax, creating a new image at the northern tip of the Axis. However, as the buildings are too concentrated and in huge sizes, construction is difficult to be proceeded in phases. And more research needs to be done on the transportation and evacuation problems during the Games.

6) The Proposal Collaborated by Beijing Architectural Design and Research Institute+EDSA (USA)

Features: Cultural, Sports and exhibition facilities are arranged from south to north, with business facilities on the two sides of the land used.

The design has clear division in functions, taking into account the problems of construction in phases and financing, thus easy for construction. Roads are closely integrated with the city traffic network, and treatment of the entrance is accessible and distinct. As roads from south to north are added within the land used, better organization of transpiration is made possible. But the layout lacks flexibility, and the business buildings on the two sides have an unnatural relation with the sports facilities in the center. Greenery is also lacking.

7) The Proposal by HWP (Germany)

Feature: Street space is highlighted; sports, business and cultural buildings are evenly spaced; greenery spaces are located on the two sides of the land used.

Facilities of various functions are concentrated under the 4 great roofs in the middle of Zone B, providing a disciplined and unique concept. Greenery is utilized as the buffer zone between the planned facilities and existing residential areas. But further study needs to be done concerning the necessity and dimension of the 4 huge roofing.

8) The Proposal Collaborated by Urban Planning Center, Beijing University+Landscape Planning and Design Center, Beijing University

Features: The design is centered around the concept of "lotus" with buildings in the central area respectively representing flowers, leaves, stems, seedpod of the lotus and underground space embodying lotus root.

Clear in functional division, the designed space combines order and flexibility; the green belt along the Fourth Ring Road is retained prior to the Games as a buffer zone for rapid transportation and construction development in the city. It also provides space for future development; construction by stages is feasible; the concept of "lotus" is not the reason why this proposal has won a prize. Apart from imitating lotus in its buildings, this theme has no deeper connection with the design.

2. Proposals for the Wukesong Sports and Cultural Center

1) The Proposal by Sasaki Associates, Inc. (USA)

Clear in function division, this design is conspicuous in its landscape planning. Facilities along the Fourth Ring Road are designed into sloping greenery, forming a landscape fit for the urban expressway; the other side of the facilities border the urban green park in the middle part of the Site. It is easy to be constructed in phases. The drawback is the long distance between the baseball field and practice field. And the land for business sector is too large.

2) The Proposal by Burckhardt+Partner AG (Switzerland)

Unique in concept, the design has the basketball hall and business facilities built within a cubic entity, leaving large stretches for urban greenery. The design concept and form of the main building is novel with brave experiment done in dimension and structure. The multi-functional building can give the basketball hall commercial benefits. But further research needs to be done on the technicalities and costs.

3) The Proposal by Architectural Design Institute of the Army Logistics(China)

The design is a square layout with the three main buildings linked and located on a diagonal line. The overall figure has rhythm. Unique in conception, the plan has a holistic sense. The drawback is that it is too formalistic, and not easy to phased-in construction.

4) The Proposal by Beijing Urban Development and Architectural Design Institute Ltd.+ABCP Design Inc., (Canada)

This design concept is unique, with all functions under one roofing, thus giving a sense of sculpture. Large public open space is saved in the land used. But the total length of one elevation exceeds 600 meters, so investigation needs to be done on dimension. Also, the design is not easy to be constructed by stages.

5) The Proposal by China Architectural Design and Research Institute.

Distinction of the three functions is reasonable and clear. The greenery and parking lot are in the middle, the basketball hall and other sports facilities along the Fourth Ring Road, and business and hotels in the east. Transportation arrangement is smooth with pedestrians and vehicles separated within the site. The landscape along the Fourth Ring Road and Fuxing Avenue is taken into consideration. It is easy to be constructed by stages. The weak point is that the individual buildings lack creativity.

IV Comments of the Jury on the Planning and Designs of the Two Sites

Given the problems in the designs, the Jury put forward some comments: the buildings to be constructed on the Axis may not be regarded as the final landmarks, rather, space should be given for future development. The northern part of the Axis should provide the backdrop of natural hills and water environment, and facilities should be arranged in a compact way so as to save land. Attention should be paid to ecology and environment protection, and water system and greenery should be integrated with the forest gardens in the north. Transportation rules need further research, which should be solved in accordance with the overall transportation system of the city. The scale of architecture and urban space may not be too large.

Concerning the designs for Wukesong, the Jury stressed that as the area had long been designated as an urban greenery, cultural and sports center for the public, commercial construction should be limited. More cultural and sports facilities should be built and open for the public. The design for each unit should be put in hi-technology. So the design for the basketball hall should not only meets the requirement of the Games, but also reflects the contemporary architectural technology achievements and environmental protection features, so as to attract tourists and local people.

The collection of proposals, which has come to an end, only marks the beginning of the overall planning and design for the Olympic Games'programs. And the urban development is even more of a long-term and complex process, which calls for wider and deeper research to tackle various problems in the future planning adjustments.

奥林匹克公园 OLYMPIC GREEN
Contents 目录

■ 一等奖(1名) First Prize (One)

□ SASAKI 设计有限责任公司(美国) — 36
　SASAKI ASSOCIATES, INC (USA)
　天津华汇工程建筑设计有限公司(中国)
　TIANJIN HUAHUIDESIGN INC. (CHINA)

■ 二等奖(2名) Second Prize (Two)

□ 北京城市规划设计研究院(中国) — 48
　BEIJING URBAN PLANNING AND DESIGN INSTITUTE (CHINA)
　DEM 设计有限公司(澳大利亚)
　DEM PTY (AUSTRALIA)

□ 株式会社佐藤综合计画(日本) — 56
　SATOW SOGO KEIKAKU (JAPAN)
　INGEROSEC 技术咨询公司(日本)
　INGEROSEC CORPORATION (JAPAN)

■ 优秀奖(5名) Honorary Mention (Five)

□ AREP 规划设计交通中转枢纽公司(法国) — 64
　AREP (FRANCE)

□ 总装备部工程设计研究总院(中国) — 70
　ARMY GENERAL EQUIPMENT ENGINEERING DESIGN INSTITUTE (CHINA)
　哈尔滨工业大学天作建筑研究所(中国)
　TIANZUO ARCHITECTURAL INSTITUTE OF HARBIN INDUSTRIAL UNIVERSITY (CHINA)

□ 北京市建筑设计研究院(中国) — 76
　BEIJING ARCHITECTUAL DESIGN AND RESEARCH INSTITUTE (CHINA)
　交通部规划研究院(中国)
　PLANNING INSTITRTE OF MINISTRY OF COMMUNICATIONS (CHINA)
　EDSA 规划设计公司(美国)
　EDWARD. STONE, JR. & ASSOCIATES, INC (USA)

□ HWP 设计有限公司(德国) — 82
　HWP (GERMANY)

□ 北京大学城市规划设计中心(中国) — 88
　THE CENTER FOR URBAN PLANNING AND DESIGN (CUPD), PEKING UNIVERSITY (CHINA)
　北京大学景观规划设计中心(中国)
　THE CENTER FOR LANDSCAPE ARCHITECTURE AND PLANNING (CLAP), PEKING UNIVER SITY (CHINA)

■ 参赛作品(48名) Works (Forty Eight)

□ 北京土人景观规划设计研究所(中国) — 94
　TUREN DESIGN INSTITUTE (CHINA)

□ 欧博迈亚工程设计咨询公司(德国) — 98
　OBERMEYER PLANEN (GERMANY)
　园林建筑师施密特教授(德国)
　BERATEN RAINER SCHMIDT LANDSCHAFTSAR CHITEKTEN (GERMANY)

□ 北京清华城市规划设计研究院(中国) — 102
　URBAN PLANNING AND DESIGN INSTITUTE TSINGHUA UNIVERSITY (CHINA)

□ 北京市市政工程设计研究总院(中国) — 105
　BEIJING GENERAL MUNICIPAL ENGINEERING & RESEARCH INSTITUTE (CHINA)
　UKY 公司(澳大利亚)
　URBIS KEYS YOUNG PTY LTD (AUSTRALIA)

□ 奥托·斯泰德勒(德国) — 108
　STEIDLE + PARTNER ARCHITEKTEN (GERMANY)
　方略建筑设计有限公司(中国)
　STRATEGY ARCHITECTURAL DESIGN CO., LTD. (CHINA)

□ 清华大学建筑设计研究院(中国) — 111
　ARCHITECTURAL DESIGN AND RESEARCH INSTITUTE OF TSINGHUA UNIVERSITY (CHINA)

□ ARUP 合伙人有限公司(英国) — 114
　ARUP ASSOCIATES LIMITED (ENGLAND)
　深圳市陈世民建筑师事务所(中国)
　SHENZHEN CHEN SHIMIN ARCHITECTS ASSOCIATES (CHINA)

□ 九源建筑设计有限公司(中国) — 117
　JIUYUAN ARCHITECTLRAL DESIGN CO., LTD (CHINA)
　北京达沃斯景观规划设计中心(中国)
　BEIJING DAVOS LANDSCAPE PLANNING AND DESIGN INSTITUTE (CHINA)

□ PCA INT PICA 恰马拉 R.L 设计事务所(意大利) — 120
　PCA INT PICA CIAMARRA ASSOCIATES.R.L (ITALY)

□ 中国科学院北京建筑设计研究院有限责任公司(中国) — 123
　INSTITUTE OF ARCHITECTURE DESIGN & RESEARCH, CHINESE SCIENCE ACADEMY (CHINA)

□ 沈祖海建筑师事务所(中国台湾) — 126
　HAIGO SHEN & PARTNERS ARCHITECTS AND ENGINEERS (TAIWAN CHINA)
　大地建筑事务所(国际)(中国)
　GREAT EARTH ARCHITECTS & ENGINEERS INTERNATIONAL (CHINA)
　美商迪斯唐工程顾问股份有限公司台湾分公司(中国台湾)
　DESHAZO, STAREK & TANG, INC (TAIWAN CHINA)
　卡瑞尔合伙人工程顾问有限公司(美国)
　CAROL R. JOHNSON ASSOCIATES INC (USA)
　异空建筑师事务所(韩国)
　BEYOND SPACE ARCHITECTS (KOREA)

□ 中元国际工程设计研究院(中国) — 129
　IPPR ENGINEERING INTERNATIONAL (CHINA)
　BCL 国际公司(香港)
　BCL INTERNATIONAL (BCLI) (HANGKONG)

□ 株式会社日建设计(日本) — 132
　NIKKEN SEKKEI (JAPAN)

□ 上海同济城市规划设计研究院(中国) — 135
　SHANGHAI TONGJI URBAN PLANNING & DESIGN INSTITUTE (CHINA)

□ GMP 建筑师事务所(德国) — 138
　ARCHITECTS VON GERKAN, MARG UND PARTNER (GERMANY)

□ 环境设计研究所(日本) — 141
　ENVIRONMENT DESIGN INSTITUTE (JAPAN)
　上海市地下建筑设计研究院(中国)
　SHANGHAI UNDER (CHINA)
　佐藤尚已建筑研究所(日本)
　NAOMI SATO ARCHITECTS, JAPAN (JAPAN)

□ 中国建筑设计研究院(中国) — 144
　CHINA AREHITECTURAL DESIGN INSTITUTE (CHINA)
　kPF 设计公司(美国)
　KPF DESIGN INC. (USA)
　泛亚易道公司(香港)
　FAN YA YI DAO INC. (HONGKONG)

□ (株)川口卫建筑构造设计事务所(日本) — 147
　KAWAGUCHI & ENGINEERS (JAPAN)

□ DECATHLON S.A.规划及工程公司(希腊) — 150
　DECATHLON S. A. PROJECT PLANNING & ENGINEERING CONSULTANTS (GREECE)

- IFB Dr.布拉斯克尔 AG 公司（德国）---------- 152
  IFB DR.BRASCHEL AG CORPORATION(GERMANY)

- 中国建筑科学研究院建筑设计院（中国）---------- 154
  INSTITUTE OF BUILDING DESIGN OF CHINA ACADEMY OF BUILDING RESEARCH (CHINA)
  巴马丹拿集团（香港）
  PALMER & TURNER INTERNATIONAL INC. (HONG KONG)
  百和纽特公司（澳大利亚）
  BLIGH VOLLER NIELD PTY LTD. (AUSTRALIA)

- 北京建筑工程学院建筑设计研究院（中国）---------- 156
  ARCHITECTURAL DESIGN AND RESEARCH INSTITUTE OF BEIJING INSTITUTE
  OF ARCHITECTURAL AND ENGINEERING (CHINA)

- 汉沙杨建筑工程设计有限公司（马来西亚）---------- 158
  T. R. HAMZAH & YEANG SDN. BHD (MALAYSIA)

- 北京市住宅建筑设计研究院（中国）---------- 160
  BEIJING RESIDENTEAL BUILDING DESIGN NETWORK (CHINA)
  北京码维建筑设计咨询有限公司（中国）
  BEIJING MAWEI ARCHITECTURAL DESGIGN CONSULTATION CO., LTD (CHINA)

- 亚太专业同盟有限公司（新加坡）---------- 162
  PROFESSIONAL ASIA-PAC ALLIANCE PTE LTD (SINGAPORE)
  三连庄建筑设计有限公司（中国）
  3HP ARCHITE CTS PTE LTD (CHINA)
  中国对外建设总公司设计研究院（中国）
  INSTITUTE OF DESIGN & RESEARCH, CHINA CONSTRUCTION INTERNATIONAL CORPORATION (CHINA)

- 北方交通大学建筑系 BA 研究室（中国）---------- 164
  ARCHITECTURAL BA STUDIO, NORTHIRN JIAOTONG UNIVERSITY (CHINA)
  东南大学建筑系（中国）
  ARCHITECTURE DEPARTMINT, SOUTHEAST UNIVERSITY (CHINA)
  U & A 设计集团（澳大利亚）
  U & A DESIGN GROUP (AUSTRALIA)
  中厦建筑国际有限公司（中国）
  ZHONGXIA ARCHITECTURE INTERNATIONAL CO., LTD (CHINA)

- 株式会社综合计画研究所（日本）---------- 166
  SYSTEM PLANNING (JAPAN)

- 柏涛建筑设计有限公司（澳大利亚）---------- 168
  PTW ARCHITECTS (AUSTRALIA)

- WORLD QUEST 工程公司（美国）---------- 170
  WORLD QUEST ENGINEERING (USA)

- 利安建筑设计及工程开发顾问（中国）有限公司（中国）---------- 172
  LEIGH & ORANGE DESIGN & PROJECT DEVELOPMENT CONSULTANTS (CHINA) LTD. (CHINA)

- TAM 地域环境研究院（日本）---------- 174
  TAM AREA ENVIRONMENTAL RESEARCH INSTITUTE (JAPAN)

- 中国装饰（集团）公司（中国）---------- 176
  CHINA NATIONAL DECORATION (GROUP) CORPORATION (CHINA)

- 中国武汉市建筑设计院（中国）---------- 178
  WUHAN ARCHITECTURAL DESIGN INSTITUTS CHINA (CHINA)

- 北方工业大学建筑学院（中国）---------- 180
  COLLEGE OF ARCHITECTURE OF NORTH CHINA UNIVERSITY OF TECHNOLOGY (CHINA)
  北京中色北方建筑设计研究院（中国）
  BEIJING ZHONGSE NORTH ARCHITECTURAL DESIGN INSTITUTE (CHINA)

- 北京工业大学建筑勘察设计院（中国）---------- 182
  BEIJING POLYTECHNIC UNIVERSITY INSTITUTE OF ARCHITECTURAL EXPLORATION & DESIGN (CHINA)

- 北京天下原色艺术设计有限责任公司（中国）---------- 184
  TIANXIA CORP. LTD. (CHINA)

- HOK 体育建筑设计公司（澳大利亚）---------- 186
  HOK SPORT + VENUE + EVENT(AUSTRALIA)

- 中国寰球化学工程公司（中国）---------- 188
  CHINA HUANQIU CHEMICAL ENGINEERING CORP. (CHINA)

- 深圳市建筑设计研究总院（中国）---------- 190
  SHENZHEN GENERAL INSTITUTE OF ARCHITETCURAL DESIGN AND RESEARCH (CHINA)
  深圳市景观设计装饰工程有限公司（中国）
  SHENZHEN JINGGUAN LANDSCAPE DECORATION AND DESIGN LTD., (CHINA)

- DP 建筑设计有限责任公司（新加坡）---------- 192
  DP ARCHITECTS PTE LTD (SINGAPORE)

- 中兴工程顾问股份有限公司（中国台湾）---------- 194
  SINOTECH ENGINEERING CONSULTANTS LTD (TAIWAN CHINA)

- 哈尔滨工业大学建筑设计研究院（中国）---------- 196
  ARCHITECTURAL DESIGN AND RESERARCH INSTITUTE OF HIT (CHINA)
  莫斯科第四国立设计院（俄罗斯）
  GUE MNIIP"MOSPROECT-4" (RUSSIAN)

- 同济大学建筑设计研究院（中国）---------- 198
  ARCHITEETURAL DESIGN INSTITUTE OF TONGJI UNIVERSITY (CHINA)
  迪克森·罗斯希尔 PTY 有限公司（澳大利亚）
  DICKSON ROTHSCHILD PTY LTD (AUSTRALIA)

- 景点规划设计公司（澳大利亚）---------- 200
  PLACE PLANNING DESIGN (AUSTRALIA)
  植被生态建筑行（澳大利亚）
  PLANTE & ASSOCIATES ARCHITECTS (AUSTRALIA)
  国际建设有限公司（澳大利亚）
  JIANSHING INTERNATIONAL PTY LTD. (AUSTRALIA)

- 北京中国风景园林规划设计研究中心（中国）---------- 202
  CHINA GARDENING DESIGN & RESEARCH CENTER (CHINA)
  中央美术学院（中国）
  CENTRAL ACODEMY OF FINE ARTS (CHINA)

- 核工业部第四研究设计院深圳设计院（中国）---------- 204
  THE FOURTH DESIGN INSTITUTE OF THE MINISTRY OF NUCLEAR INDUSTRY (CHINA)
  伍凸设计师事务所（深圳有限公司）（中国）
  WUTU ARCHITECTS ASSOCIATES (SHENZHEN LIMITED) (CHINA)

加盟单位:
SPONSOR:
深圳宗灏建筑师事务所（中国）
SHENZHEN ZONGHAO DESIGN ARCHITECT'S OFFICE (CHINA)
北方设计研究院（中国）
NORINDAR INTERNATIONAL (CHINA)
深圳市裕华建筑设计有限公司（中国）
SHENZHEN YUHUA ARCHITECTURAL DESIGN CO., LTD (CHINA)
蒂奥特型建筑科技（深圳）有限公司（中国）
DIAO SPECIAL VUILDING TECHNOLOGY (SHENZHEN) CO.,LTD (CHINA)

- 何弢设计国际有限公司（中国）---------- 206
  TAOHO DESIGN INTERNATIONAL LTD. (CHINA)

- 考克斯集团（澳大利亚）---------- 207
  THE COX GROUP (AUSTRALIA)

北京奥林匹克公园设计方案 Beijing Olympic Green

一等奖　First Prize

SASAKI设计有限责任公司（美国）
SASAKI ASSOCIATES, INC. (USA)
天津华汇工程建筑设计有限公司（中国）
TIANJIN HUAHUI DESIGN INC. (CHINA)

Beijing Olympic Green 2008

1953年，Sasaki公司始建于马萨诸塞州首府波士顿。其创始人为Hideo Sasaki先生。最初的专业人员仅限于同Sasaki先生一起在哈佛大学设计研究院学习的景观建筑师和规划师。之后，随着业务的不断扩展，吸引了一大批建筑师，土木工程师，城市设计师以及环境科学家，为Sasaki公司带来了新的生机。

Hideo Sasaki established the firm in 1953 in Watertown, Massachusetts. The original professional staff included landscape architects and planners who had studied with Mr. Sasaki at the Harvard Graduate School of Design. As the practice flourished, architects, civil engineers, urban designers and environmental scientists brought additional breadth to the firm's capabilities.

天津华汇工程建筑设计有限公司是在天津注册的甲级事务所，由德国留学回来的周恺带领一批毕业于著名大学的建筑师和富有工作经验的国家注册建筑师、结构师和技术人员组成，其中有三位取得了美国的注册资格。过去一年，获得两个国家二等奖和两个国家三等奖。

Tianjin Huahui Architectural Design & Engineering Co.Ltd (HAE) is a Class A design office registered in Tianjin City. The HAE staff, led by Mr. Zhou Kai who once studied in Germany, consists of a group of registered architects, structural engineers and other technicians with rich experiences. Most of them graduated from famous universities. And 3 of them have the registered qualification in the United States. In the past year, we obtained two Second National Awards and two Third National Awards.

■ 规划设计构思

1.设计理念

我们对奥林匹克公园的设计目标在于：寻求和谐性与综合性并存，既充满诗意又考虑到实用性。我们寻求东西方文化，古典与现实，发展与自然，周围已存在的建筑物和奥林匹克公园之间的和谐。我们的设计理念包括以下三个基本因素：

森林公园向南部延伸。

文化轴线向北延伸，作为故宫皇家轴线的终点。

奥林匹克轴线，连接亚运村和国家体育场。

森林公园位于奥林匹克公园中心地带的北部。森林公园面积庞大，在许多地方仿照阿姆斯特丹的Bos公园和纽约Prospect Park公园。森林公园将是一个体现中国几千年传统文化的理想殿堂。

北京城以其南北轴线为基础而建设，城市上百年的发展都是围绕这个轴展开。规划的意图是在轴线的末端建设奥林匹克公园。经过慎重考虑，规划将在轴线的周边地区而不是在轴线上建造新的建筑。这条轴线长达5km的距离，体现中国五千年来的成就与贡献。在文化轴线上，每隔1000m设计一个纪念广场代表一个千年。最后，轴线以其简洁的形式在森林公园的山中消失，代表中国古代文明发源于自然。

构想的奥林匹克轴线起于亚运村体育场，向北穿过国家体育场，到达体育英雄公园，与文化轴线交叉，最后到达森林公园中的奥运精神公园。奥林匹克轴线与文化轴线在周王朝广场交叉，体现了中国对城市建设文化的贡献。规划中，这个交汇点也将是奥运会期间奥林匹克升旗广场的位置。

森林公园、文化轴线和奥林匹克轴线组成了奥运公园的主体构架。这个主体构架非常适应此后每一座奥运会场馆的建设。奥林匹

Introduction

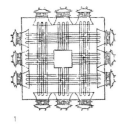   

1　2　3　4

↑ 1~2 商朝王城平面示意图
1-2 THE WANG CITY PLAN OF SHANG DYNASTY
3-4 天安门历史照片
3-4 RECORD PHOTO OF TIANAN MEN

克公园的建设预计于5年后完成。它将最大限度地利用土地，并把北京北部现存的建筑与奥林匹克公园完美结合。

本案总体规划描述了约210万m²的建筑面积发展规划。这部分规划包括：综合型体育设施、会议及展览中心、奥林匹克村、公共建筑和商务建筑以及地下停车场。

2.交通公路网络规划

将重点考虑如何与现存街道格局相联系、奥运期间及奥运后期的车辆进入、交通连接、停车场地、人流的循环等问题。

3.市政概念规划

奥林匹克公园市政概念规划将研究主要市政服务设施，如能源、供水、电力分布、通信、市政管线通道、污水处理等内容。市政概念将紧紧依托生态环境，这六个方面将在现代城市可持续发展中起重要作用。

■ **Planning and Design Concept**

1. Design Concept

Our design proposal for the Olympic Green seeks balance and integration. This balance and integration is both poetic and pragmatic. We seek to balance East with West, the ancient with the contemporary, development with nature, existing surrounding context with the Olympic Green. Our concept has 3 fundamental elements, to include the following:

The Forest Park, and its extension southward

The Cultural Axis, the northward extension and conclusion of the great imperial axis

The Olympic Axis, Linking the Asian Games site with the National Stadium

The Forest Park consists of an area of land to the north of the Central Area of the Olympic Green. The Forest Park is of grand dimension, in many ways evoking in scale and function other great city parks of the world such as the Bos in Amsterdam and Prospect Park in New York. The Forest Park is conceived as an ideal paradise from which Chinese civilization emerged, many millennia ago.

Beijing was founded on the basis of its North-South axis. Development of the city over centuries has been based on this axis. It is the intention of the city to conclude the axis within the Olympic Green. It has been deliberate in choosing to place new buildings at its edge rather than on-axis. In concept extends the axis some 5 kilometers through the Olympic green site. Along the axis, try to seek to commemorate the achievements and contributions of the great Chinese dynasties, beginning 5,000 years ago up to the present day. The Axis of cultural achievement is divided into 1,000-meter segments, symbolizing 1, 000 years whose endpoints lie within commemorative plazas. Finally, the axis concludes with a powerful simplicity, in the hills of the Forest Park, signifying the beginning of Chinese culture and civilization in nature.

The design concept proposes an axis that begins within the existing Asian Games stadium, extending northeast through the proposed National Stadium, continuing onward to a Sports Heroes Garden, intersecting with the Cultural Axis then concluding within the Forest Park at an Olympic Spirit Park. The Olympic Axis intersects with the Cultural Axis at the Zhou Dynasty Plaza, which commemorates the Chinese contributions to city building. In the concept proposes, that this point of intersection also be the site of the Olympic flag-raising square during the Games.

The combination of the Forest Park, Cultural and Olympic Axes provides a framework within which the building development program is arranged. The framework provides extraordinary flexibility in the ultimate sitting and design of each Olympic venue. Our design concept provides for the phased implementation of the Olympic Green during the next 5 years. It maximizes the utilization of land, and provides for the seamless integration of the Olympic Green with the surrounding context of north Beijing.

The master plan depicts a development program of some 2.1 million square meters, as set forth in the Brief. The program includes a mix of uses to include sport facilities, convention and exhibition space, the Olympic Village, public facilities and commercial facilities. Underground parking is also provided, consistent with the requirements of the Brief.

2.Traffic and Road Network Planning

Will emphases take into account how to connect with existing street grid; Olympics and Post-Olympics vehicular access, transit connections, parking, pedestrian circulation.

3.Utility Concept Plan

Conceptual planning for the utility systems to serve the Olympic Green addresses major infrastructure elements of energy, water, wastewater, power supply and distribution, communications, utility corridors and waste disposal. Infrastructure facilities are a fundamental aspect and important basic requirement for the sustainable development of a modern urban economy and society.

英国丘园中的中国塔
CHINA PAGODA IN KEW GARDEN,LONDON

中国画中园林
LANDECAPE GARDEN IN CHINESE PAINTING

北海公园
BEI HAI PARK,BEIJING

ideal paradise

The Forest Park is conceived as an ideal paradise from which Chinese civilization emerged, many millennia ago.

全区总平面图(赛后)
OVERALL AREA MASTER PLAN (POST GAMES)

1	森林公园	FOREST PARK
2	奥运村	OLYMPIC VILLAGE
3	室外展场	OUTDOOR EXHIBITION CENTER
4	会展中心	CONFERENCE AND EXHIBITION CENTER
5	首都青少年宫	CAPITAL TEENAGE PALACE
6	商业服务	BUSINESS SERVICE FACILITIES
7	北京城市规划展示馆	EXHIBITION HALL OF BEIJING CITY PLANNING
8	文化轴线	CULTURAL AXIS
9	奥林匹克轴线	OLYMPIC AXIS
10	国家体育馆	NATIONAL GYMNASIUM
11	国家游泳中心	NATIONAL SWIMMING CENTER
12	观景塔	OBSERVATION TOWER
13	国家体育场	NATIONAL STADIUM
14	奥体中心体育馆	OLYMPIC SPORT CENTER GYMNASIUM
15	英东游泳馆	YING TUNG NATATORIUM
16	奥体中心垒球场	OLYMPIC SPORT CENTER SOFTBALL FIELD
17	奥体中心体育场	OLYMPIC SPORT CENTER STADIUM
18	国家曲棍球场	NATIONAL HOCKEY STADIUM
19	国家网球中心	NATIONAL TENNIS CENTER
20	体育公园	SPORT PARK
21	元大都遗址公园	YUAN DYNASTY CITY WALL RELICS PARK
22	环境教育中心	ENVIRONMENTAL EDUCATION CENTER

The Axes of Human Achievement

文化轴线示意图
CULTURAL AXIS

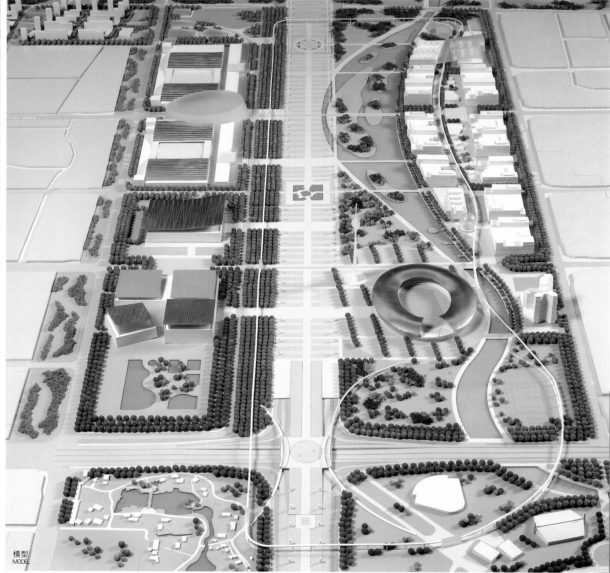

模型
MODEL

露天剧场 AMPHITHEATER

水边步行道 WATERSIDE WALK

自然式水景 NATURAL WATERSCAPE

大草坪 GRAND LAWN

自然式水景 NATURAL WATERSCAPE

自然式景观 NATURAL LANDSCAPE

水边步行道 WATERSIDE WALK

林阴广场 SHADED PLAZA

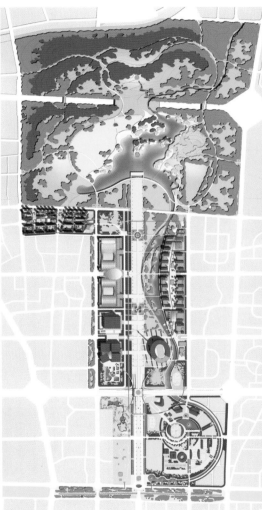

规则式景观 FORMAL LANDSCAPE

入口广场 ENTRANCE PLAZA

总平面图 MASTER PLAN

林阴道 SHADED PROMENADE

商业娱乐 COMMERCIAL AND ENTERTAINMENT FACILITY

广场 PLAZA

庭阴广场 SHADED PLAZA

Beijing Olympic Green 2008

1 全区规划概念图
  OVERALL AREA CONCEPTUAL IMAGES
2 鸟瞰图
  BIRDS-EYE VIEW
3 透视图，从四环路向西北看
  PERSPECTIVE VIEW LOOKING NORTHWEST FORM FOURTH RING ROAD
4 透视图，从国家体育场向西北看
  PERSPECTIVE VIEW LOOKING NORTHWST FORM NATIONAL STADIUM

Design Concept

1	奥运村	OLYMPIC VILLAGE
2	室外展场	OUTDOOR EXHIBITION AREA
3	会展中心	CONFERENCE AND EXHIBITION CENTER
4	首都青少年宫	CAPITAL TEENAGE PALACE
5	商业服务	COMMERCIAL SERVICE FACILITIES
6	北京城市规划展示馆	PLANNING EXHIBITION HALL OF BEIJING CITY PLANNING
7	文化轴线	CULTURAL AXIS
8	奥林匹克轴线	OLYMPIC AXIS
9	国家体育馆	NATIONAL GYMNASIUM
10	国家游泳中心	NATIONAL SWIMMING CENTER
11	观景塔	OBSERVATION TOWER
12	国家体育场	NATIONAL ATADIUM
13	奥体中心体育馆	OLYMPIC SPORT CENTER GYMNASIUM
14	英东游泳馆	YING TUNG NATATORIUM
15	奥体中心垒球场	OLYMPIC SPORT CENTER SOFTBALL FIELD
16	奥体中心体育场	OLYMPIC SPORT CENTER STADIUM
17	环境教育中心	ENVIRONMENTAL EDUCATION CENTER

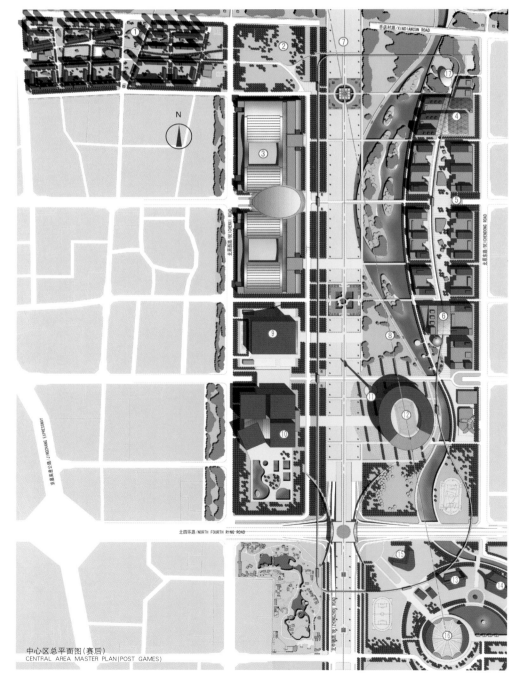

中心区总平面图（赛后）
CENTRAL AREA MASTER PLAN (POST GAMES)

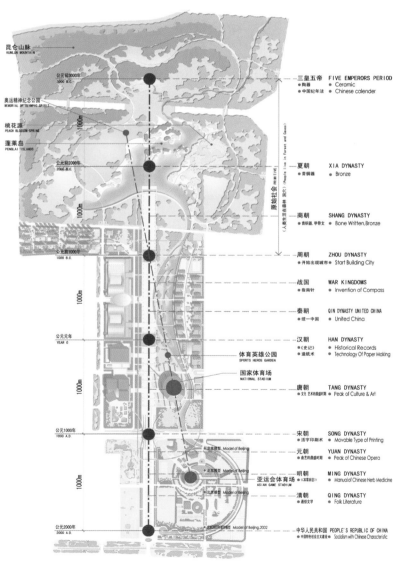

五千年文化成就
5000 YEARS'S CULTURAL ACHIEVEMENT

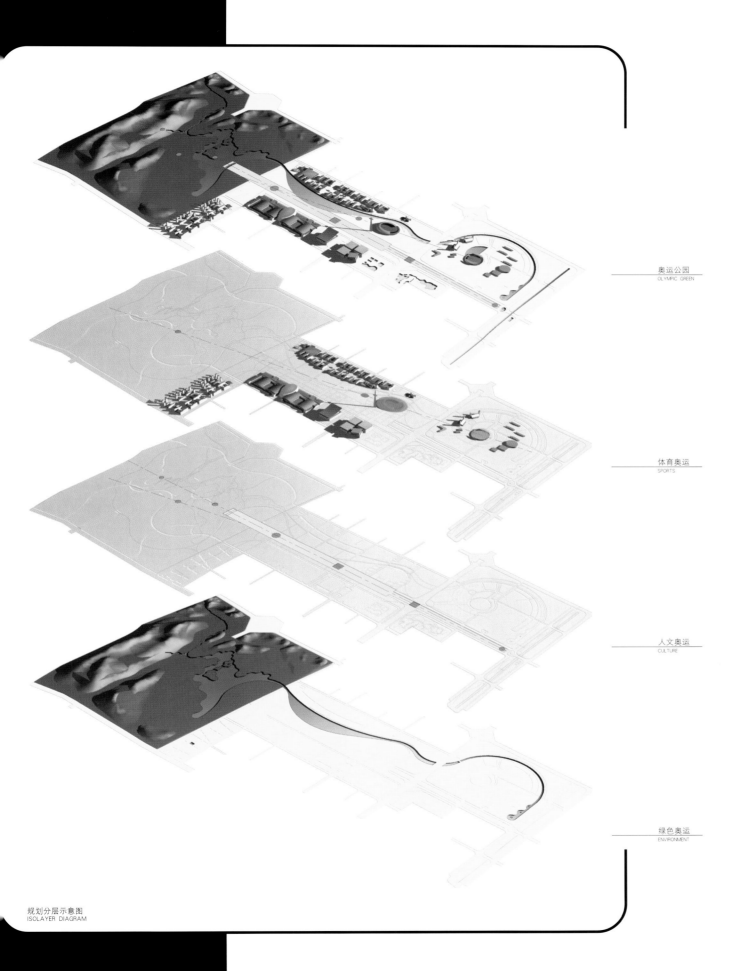

规划分层示意图
ISOLAYER DIAGRAM

奥运公园
OLYMPIC GREEN

体育奥运
SPORTS

人文奥运
CULTURE

绿色奥运
ENVIRONMENT

Circulation

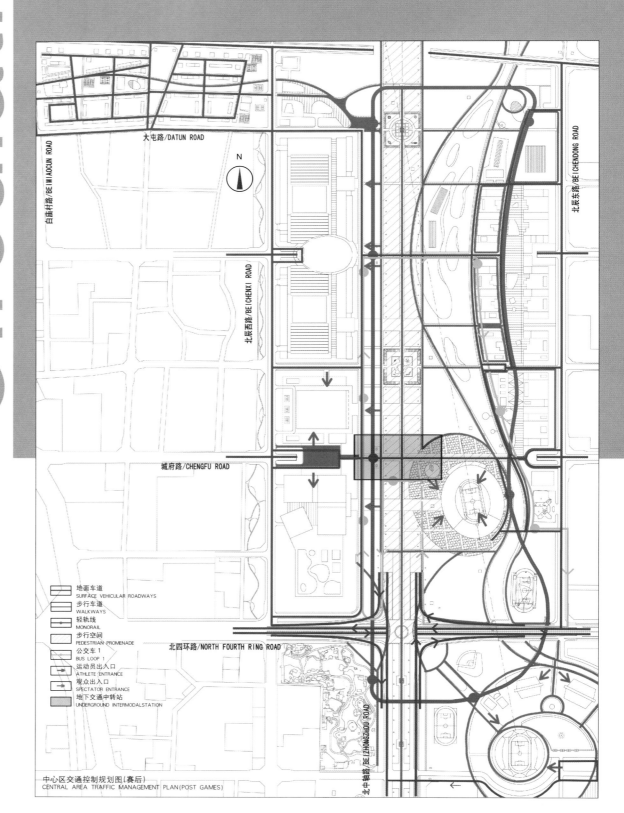

中心区交通控制规划图(赛后)
CENTRAL AREA TRAFFIC MANAGEMENT PLAN (POST GAMES)

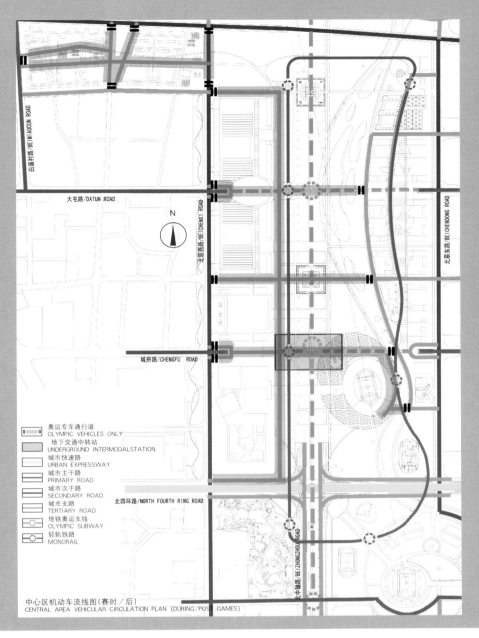
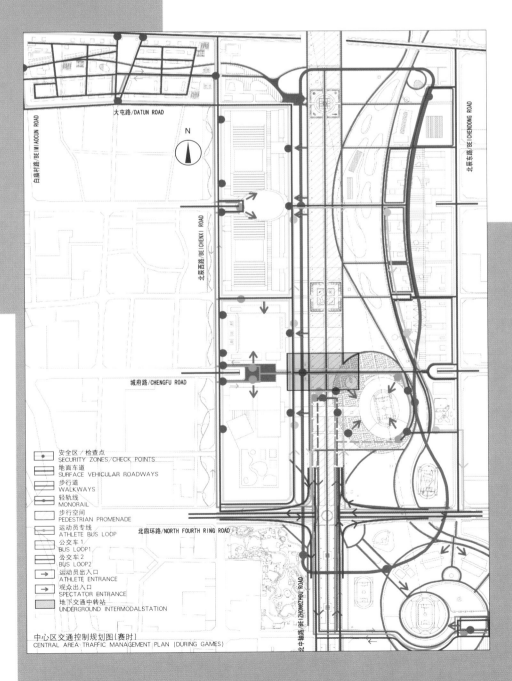

Beijing Olympic Green 2008

北京奥林匹克公园设计方案 Beijing Olympic Green

二等奖 Second Prize

北京城市规划设计研究院（中国）
BEIJING URBAN PLANNING AND DESIGN INSTITUTE(CHINA)
DEM 设计有限公司（澳大利亚）
DEM PTY(AUSTRALIA)

北京城市规划设计研究院是北京市政府授权组织编制城乡各项规划的工作机构，是建设部批准的甲级规划设计研究单位。现有工作人员260多人，科技人员约占82%，其中有高级技术人员58人。
Beijing Municipal Institute of City Planning & Design (BMICPD) is a working organization authorized by the Beijing Municipal Government and engaged in working out various urban construction plans. It is a grade A institute of city planning and design, which was approved of by the Ministry of Construction. Currently it has more than 260 workers, among whom 80% are technological personels and 58 of them are high class technical personels.

澳大利亚DEM公司总部设在悉尼，现有职员75人，提供城市设计、总体规划、规划、建筑、景观建筑、室内设计和各种设施规划的服务。
The Sydney-based firm employs a staff of 75 and leads in providing Urban Design, Masterplanning, Planning, Architecture, Landscape Architecture, Interior Design and Facilities Planning services. Its design teams are committed to interdisciplinary innovation, combined with effective and efficient project coordination.

Beijing Olympic Green 2008

■ 规划设计构思

创造城市庆典——满足赛事功能，空间序列由南向北有节奏展开，逐步达到高潮。策划庆典活动、营造庆典气氛。

强调可持续发展——创造在未来发展中与城市肌理、自然生态的有机融合。

创造生态城市——生态城市的示范区。

提出实施建议——设立组织机构使长期开发与赛事运作良好结合。

1.确定合理框架，使城市有序发展。

A区——纪念性空间的序曲。

B区南部——纪念性空间的发展。赛时用于中小型比赛，赛后为展览综合区。

B区中部——纪念性空间的高潮。奥运三大场馆，大型活动的中心。人群集中，与四环路拉开距离，交通组织合理。

B区北部——纪念性空间的余韵，逐渐融入自然。赛时为临时设施及景观环境用地，赛后为文化综合区。

确定分期建设步骤——建立整体框架，延续完整形象并确保城市有机增长。

2.节约土地资源，给未来预留发展余地。

建设从南向北推进，给未来预留充分发展余地，使后人有机会创造新的辉煌。

6.结合城市肌理，加强内外空间的融合。

赛时中轴线形象完整，各功能区划分明确，观众活动空间充足。

赛后城市设施填充与周围城市肌理结合。

7.建立文化体育城，提供多样的社会服务。

该地区将建成文化体育城，为城市提供多元化的服务，因此需不断填充、完善地区的功能，规划为此留出了可能性。

沿中轴线两侧设立商业娱乐带，连接了展览、体育、文化等设施，形成有活力的城市空间。

8.完善交通系统，利于赛时赛后的使用。

用地内道路全部采用"平交"方式，利于建设与交通组织并节省资金。

地面道路、地铁和观光轻轨形成完善的立体交通网络。

交通设施规模以满足长期需求为准，特殊时期采取特殊管理，以有效利用资源。

中轴线的交通功能向四环路以北延续，缓解市区方向的交通压力；在区内逐渐由城市主干道转为次干道，最终消失于北部森林，避免了大量过境交通穿行该地区。

内部道路的设置与功能布局、景观规划紧密结合，融入内部的肌理中。

Introduction

B区规划用地面积291ha，赛时建设用地为203.4ha，中部预留7.5ha的城市建设用地，北部预留80.1ha，为文化综合用地。

3.适宜的开发强度，与环境容量协调。

该地区现状：周边开发强度较大，交通及市政条件已趋饱和，故不宜进行超强度开发。同时本着节约用地、利于生态保护、空间舒适的原则，确定城市的适当规模。

B区赛时地上总建筑面积205万m²，另有临时设施10万m²。赛后中部预留地可建20万m²，北部预留用地可建56万m²。

4.尊重城市历史，延续并发展人文景观。

继承北京原有轴线的特点，由大型组群建筑构成收放有序的空间序列，强调纪念性；吸取西方城市轴线特点，保持轴线通透，强调开放性。

为保持轴线空间的均衡感，沿中轴两侧的功能对称布局，建筑从体量、形态上协调一致。

自然山水与轴线的关系，暗含北京旧城轴线处理手法。

5.引入自然绿色，进行城市与自然转化。

南北贯穿的中轴绿带：具有装饰性和串联性。严谨对称的城市形态逐渐趋于自由并最终融入自然。

严谨的中轴与活泼的水形是北京传统城市重要特点，水系穿越各种功能用地，形成不同风格的景观线，并与传统对话。

中轴线东立面图
EAST ELEVATION DRAWING OF THE AXIS

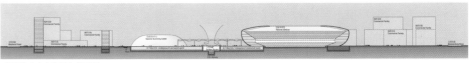

体育综合中心东西向剖面图
SECTION IN EAST-WEST DIRECTIONS OF CENTRAL COMPREHENSIVE SPORTS CENTER

展览综合中心东西向剖面图
SECTION IN EAST-WEST DIRECTIONS OF CENTRAL COMPREHENSIVE EXHIBITION CENTER

中轴线南北剖面图
SECTION IN EAST-WEST DIRECTIONS OF CENTRAL COMPPEHENSIVE EXHIBITION CENTER

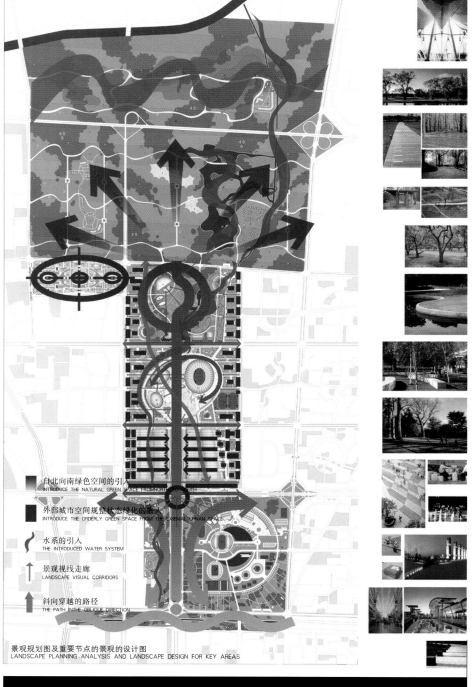

1 山丘 SMALL HILL	21 国家体育场 NATIONAL STADIUM
2 草地 LAWN	22 国家体育馆 NATIONAL GYMNASIUM
3 低矮的灌木 SHRUB	23 国家游泳中心 NATIONAL SWIMMING CENTER
4 树林 WOODS	24 练习场 TRAINING FIELD
5 生物技术学校 BIOTECHNOLOGICAL SCHOOL	25 地铁站 SUBWAY STATION
6 景观性地标灯 SCENERY LAND MARK LANTERN	26 信息中心 INFORMATION CENTER
7 生态农场 ECOLOGICAL FARM	27 会展博览设施 CONVENTION AND EXHIBITION FACILITIES
8 服务设施 SERVICE FACILITY	28 凯迪克大酒店 CATIC HOTEL
9 碧玉园别墅区 BIYU VILLA	29 观光轻轨 LIGHT RAIL FOR SIGHTSEEING
10 仰山大沟 YANGSHANDAGOU	30 公交首末站 BUS TERMINAL
11 居住区 RESIDENTIAL DISTRICT	31 国家奥林匹克体育中心 NATIONAL OLYMPIC SPORTS CENTER STADIUM
12 山坡 HILLSIDE	32 国家曲棍球中心 NATIONAL HOCKEY CENTER
13 首都青少年宫 CAPITAL TEENAGER PALACE	33 国家网球中心 NATIONAL TENNIS CENTER
14 文化中心 CULTURAL CENTER	34 体育公园 SPORTS PARK
15 生态教育中心 ECOLOGICAL EDUCATIONAL CENTER	35 中华民族园 CHINESE ETHNIC CULTURE PARK
16 科技中心 CENTER OF SCIENCE AND TECHNOLOGY	36 元大都遗址公园 YUAN DYNASTY CITY WALL RELICS PARK
17 规划展览馆 EXHIBITION HALL OF BEIJING CITY PLANNING	37 奥林匹克门 OLYMPIC GATEWAY
18 河、湖 LAKE RIVER	38 地景雕塑系列 SERIES OF GROUND LANDSCAPE SCULPTURES
19 多功能设施 MULTI-FUNCTIONAL FACILITY	39 特色铺装广场 THE SQUARE WITH DISTINGUISHING FEATURES
20 商贸设施 COMMERCIAL FACILITY	

景观规划平面图（赛后）
MASTER LAYOUT AND LANDSCAPE PLAN (POST GAMES)

景观规划图及重要节点的景观的设计图
LANDSCAPE PLANNING ANALYSIS AND LANDSCAPE DESIGN FOR KEY AREAS

- 自北向南绿色空间的引入 INTRODUCE THE NATURAL GREEN SPACE FROM NORTH TO SOUTH
- 外部城市空间规整状态绿化的散入 INTRODUCE THE ORDERLY GREEN SPACE FROM THE EXTERNAL URBAN SPACE
- 水系的引入 THE INTRODUCED WATER SYSTEM
- 景观视线走廊 LANDSCAPE VISUAL CORRIDORS
- 斜向穿越的路径 THE PATH IN THE OBLIQUE DIRECTION

9.关注弱势群体，树立相互关爱的典范。

用地内所有设施将符合国际通行的无障碍设计标准，室外避免过多的高差。

赛后所有设施基本不需要改造即可用于残奥会。

10.优化能源结构，建立循环再生的系统。

市政充分利用现有设施，改进挖潜。利用城市水厂，尽量避免使用地下水；利用城市热力网，不在该区产生人为燃烧热；利用城市污水处理厂，集中处理污水。

充分利用天然能源，如地热、太阳能等；加强水的循环再利用，废物的再利用。考虑区域能源使用情况和对能源网的贡献。

11.倡导生态文化，提高市民的生态意识。

建立具有自我修复功能的生态系统，建立生态教育系统。
保持开阔的天穹，减少热岛效应。

用地内规划道路均为平交，不破坏排水系统。尽量采用自重流，减少对自然排水系统的干扰。

不设超高层建筑，地下空间满足基本要求，最多为两层，以解决高层建筑、地下建筑与地下径流的关系。

森林公园建构多层次的综合生态网络系统，规划生态农场、本土动植物栖息地等，将其从简单的绿带上升到生态质量调节系统的高度。

全区总平面图(赛后)
OVERALL AREA MASTER PLAN (POST GAMES)

全区总平面图(赛时)
OVERALL AREA MASTER PLAN (DURING GAMES)

#	中文	English
1	森林公园	FOREST AREA
2	碧玉园别墅区	BIYU VILLA
3	仰山大沟	YANGSHANDAGOU
4	树林	WOODS
5	公寓	APARTMENT
6	地下车库出入口	ACCESS TO AND EXIT FROM UNDERGROUND PARKING LOT
7	配套设施	SUPPORT FACILITY
8	山坡	HILLSIDE
9	公共绿地	PUBLIC GREEN AREA
10	首都青少年宫	CAPITAL TEENAGERS' PALACE
11	文化中心	CULTURE CENTER
12	生态教育中心	ECOLOGICAL EDUCATION CENTER
13	科技中心	CENTER OF SCIENCE AND TECHNOLOGY
14	规划展览馆	EXHIBITION HALL OF BEIJING CITY PLANNING
15	河、湖	LAKE RIVER
16	多功能设施	MULTI-FUNCTIONAL FACILITY
17	商贸设施	COMMERCIAL FACILITIES
18	国家体育场	NATIONAL STADIUM
19	国家体育馆	NATIONAL GYMNASIUM
20	国家游泳中心	NATIONAL SWIMMING CENTER
21	练习场	TRAINING FIELD
22	地铁车站	SUBWAY STATION
23	英东游泳馆	YINTUNG NATATORIUM
24	信息中心	INFORMATION CENTER
25	服务中心	SERVICE FACILITY
26	下沉广场	SUNKEN SQUARE
27	地下空间采光窗	LIGHTING WINDOW OF UNDERGROUND SPACE
28	展厅	EXHIBITION HALL
29	附属设施	ACCESSORY FACILITY
30	凯迪克大酒店	CATIC HOTEL
31	观光轻轨	LIGHT RAIL FOR SIGHT SEEING
32	轻轨车站	LIGHT RAIL STATION
33	观众主入口	MAIN ACCESS FOR AUDIENCE
34	观众次入口	SECONDARY ACCESS FOR AUDIENCE
35	室外展场	OUTDOOR EXHIBITION SPACE
36	公交车站	BUS STATION
37	天桥	FLYOVER
38	奥体中心体育馆	NATIONAL OLYMPIC SPORTS CENTER GYMNASIUM
39	奥体中心体育场	NATIONAL OLYMPIC SPORTS CENTER STADIUM
40	国家曲棍球中心	NATIONAL HOCKEY CENTER
41	国家网球中心	NATIONAL TENNIS CENTER
42	体育公园	SPORTS PARK
43	中华民族园	CHINESE ETHNIC CULTURE PARK
44	停车场	PARKING

#	中文	English
1	森林公园	FOREST AREA
2	碧玉园别墅区	BIYU VILLA
3	仰山大沟	YANGSHANDAGOU
4	树林	WOODS
5	公交车站	BUS STATION
6	安检口	SECURITY GATE
7	天桥	FLYOVER
8	地铁车站	SUBWAY STATION
9	奥林匹克公园安全边界	SAFETY LIMIT OF THE OLYMPIC GREEN
10	观众区域边界	SAFETY LIMIT OF PUBLIC ZONE
11	奥运村边界	BOUNDARY OF OLYMPIC VILLAGE
12	国际区与居住区边界	BOUNDARY BETWEEN INTERNATIONAL ZONE AND RESIDENTIAL ZONE
13	公寓	APARTMENT
14	餐厅	DINNING HALL
15	国家奥委会单元	NOC UNIT
16	娱乐中心	ENTERTAINMENT CENTER
17	多科诊所	POLYCLINIC
18	服务人员用房	STAFF ROOM
19	社区服务中心	COMMUNITY CENTER
20	训练场地	TRAINING FIELDS
21	购物中心	SHOPPING CENTER
22	升旗广场	FLAG-RAISING SQUARE
23	奥运村入口	ENTRANCE TO OLYMPIC VILLAGE
24	制证中心	ACCREDITATION CENTER
25	后勤服务中心	LOGISTIC SERVICE CENTER
26	山坡	HILLSIDE
27	公众活动区域	PUBLIC ACTIVITY ZONE
28	管理区域	MANAGEMENT ZONE
29	奥运村专用停车场	SPECIAL PARKING LOT OUTSIDE THE OLYMPIC VILLAGE
30	奥运村车站	BUS STATION OUTSIDE THE OLYMPIC VILLAGE
31	奥林匹克公园射箭场	ARCHERY RANGE
32	河、湖	LAKE RIVER
33	商贸设施	COMMERCIAL FACILITIES
34	地下车库出入口	ACCESS TO AND EXIT FROM UNDERGROUND PARKING LOT
35	国家体育场	NATIONAL STADIUM
36	国家体育馆	NATIONAL GYMNASIUM
37	国家游泳中心	NATIONAL SWIMMING CENTER
38	信息中心	INFORMATION CENTER
39	下沉广场	SUNKEN SQUARE
40	地下空间采光窗	LIGHTING WINDOW OF UNDERGROUND SPACE
41	展馆A(比赛场馆)	EXHIBITION HALL A (COMPETITION VENUE)
42	展馆B(比赛场馆)	EXHIBITION HALL B (COMPETITION VENUE)
43	展馆C(比赛场馆)	EXHIBITION HALL C (COMPETITION VENUE)
44	展馆D(比赛场馆)	EXHIBITION HALL D (COMPETITION VENUE)
45	展馆E(比赛场馆)	EXHIBITION HALL E (COMPETITION VENUE)
46	主新闻中心	MPC (MAIN PRESS CENTER)
47	国际广播电视中心	IBC (INTERNATIONAL BROADCASTING CENTER)
48	赞助商区域	SPONSORS' ZONE
49	注册车辆停车场	PARKING LOT OF ACCREDITED VEHICLES
50	凯迪克大酒店	CATIC HOTEL
51	观光轻轨	LIGHT RAIL FOR SIGHT SEEING
52	轻轨车站	LIGHT RAIL STATION
53	奥体中心体育馆	NATIONAL OLYMPIC SPORTS CENTER GYMNASIUM
54	英东游泳馆	YINTUNG NATATORIUM
55	奥体中心体育场	NATIONAL OLYMPIC SPORTS CENTER STADIUM
56	国家曲棍球中心	NATIONAL HOCKEY CENTER
57	国家网球中心	NATIONAL TENNIS CENTER
58	绿地	GREEN AREA
59	中华民族园	CHINESE ETHNIC CULTURE PARK

Planning and Design Concept

Creation of City Celebrations - To meet the demands of the Olympic Games; To maintain the spatial sequence from south to north rhythmically and reach the climax gradually; To program celebration activities and foster ceremonial atmosphere.

Emphasis on Sustainable Development - To stress the organic integration of urban tissue and natural ecology in the future development of the site.

Creation of Eco-City - To set up a demonstration district of the eco-city. Suggestions put forward - To set up the organization to integrate long-term development with the operation of the Olympic Games.

Strategy 1 - To propose rational development framework to maintain orderly urban development.

Zone A - To be the prelude to commemorative space.

The southern pan of Zone B - To be the development of the commemorative space. To arrange medium and small events in the Olympic periods and be used as a comprehensive exhibition zone after the Olympic Games.

The middle part of one B - To be the climax of the commemorative space. Three main Olympic venues are located here. It will become the center for large events. A considerable crowd will gather here. The distance to the 4th Ring Road should be kept with rational traffic organization.

The northern part of Zone B - To maintain the lasting charm at the end of the commemorative space and merge into nature gradually. To be used as the land for temporary facilities and landscape environment in the Olympic period and as comprehensive cultural zone after the Olympic Games.

propose staged construction - To set up an integral framework and image, and maintain organic development of the city.

Strategy 2 - To save land resources for future development

The site will be developed from south to north to reserve enough space for future development.

The buildable area of Zone B is 291 hectares, out of which 203.4 hectares are for construction in the Olympic period. 7.5 hectares in the center are reserved for future urban development and 80.1 hectares are reserved for future development of cultural facilities.

Strategy 3 - To maintain rational development intensity in line with environmental capacity.

The current development of the surrounding districts is highly intensified and traffic and utility capacity are almost exhausted. Over-intensified development of the site is therefore unsuitable. In the meantime, the scale of the city should be justified according to the principles of saving land resources, benefiting ecological conservation and creating comfortable spaces.

The gross floor space in Zone B will be 2,050,000 sqm, and, in addition, the gross floor space of temporary facilities 100,000 sqm during the Olympic period. During the post-Olympic period, the reserved land in the center can accommodate a gross floor space of 200,000 sqm and that on the north 560,000 sqm.

Strategy 4 - To respect city history and sustain human and cultural landscapes.

Original characteristics of Beijing's central axis, namely, the spatial sequence created by large-scale building groups and associated opening and closing of space and the emphasis on monumentality, should be succeeded while characteristics of Western city axes that are kept open and accessible as public spaces should be incorporated and emphasized.

In order to stress the sense of balance of the axial space, functional layouts on the two sides of the central axis are symmetrical and the scale and form of the buildings are harmonized.

The relationship between natural hills, waters, and the central axis implies the historical handling of the central axis of Beijing.

Strategy 5 - To introduce natural green into the city to achieve natural transition between the city and the nature.

The north-south green belt throughout the central axis is both decorative and connective. The rigid city form is gradually freed to be eventually melted into the nature.

Rigidity of the central axis and playfulness of water characterize the traditional city of Beijing. The water system passed "through various functional zones to form landscape features in different styles and opens a dialogue with the tradition.

Strategy 6 - To reinforce the integration of internal and external spaces by the incorporation of the existing city tissue.

The central axis will keep its spatial integrity during the Olympic period. The zoning of Olympic functions will be clear out and sufficient space will be provided for spectators' activities.

After the Olympic Games, the site will be enriched by urban facilities and activities, and integrated with the city tissue through landscape design.

Strategy 7 - To create a cultural and sports city and provide diversified services.

The site will be developed as a cultural and sport city to provide diversified services for the city. It is therefore necessary to enrich and improve the functions of the site, which are made possible in the plan.

Linear entertainment and retailing spaces will be established on both sides of the central axis to link up exhibition, commercial and cultural facilities to form dynamic urban space.

Strategy 8 To improve traffic systems for both Olympic and post-Olympic use.

All internal road intersections will be in the form of grade crossing, beneficial to traffic organization and fund saving.

Ground-level roads, subway lines and light rail will consist of a high-quality three-dimensional uamc network.

The scale of transport facilities will be determined according to the demand and special measures will be adopted to meet additional demands during special periods to make most of resources.

Traffic functions of the central axis will be extended northward to relieve traffic pressure from the downtown area. It will be transformed from primary trunk road into secondary trunk road and eventually disappear in the forest area to reduce traffic load across the Olympic Green north-southwardly.

The layout of internal roads will be incorporated with functional arrangement and landscape design integrated with internal grain.

Strategy 9 - To care about powerless groups and make examples for mutual caring.

leave the central space for green areas. The underground space will be two-level high to the maximum and kept as compact as possible to meet the basic functional demands for optimizing the relationships between the high-rises, underground structures and underground circulation.

A multi-layered and integrated ecological network will be established in the Forest Area. A local wildlife reservation area will be

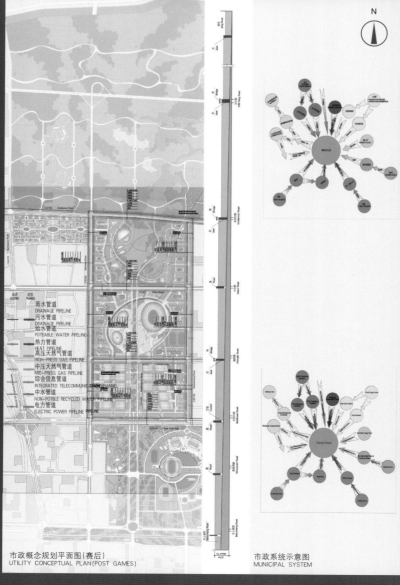

市政概念规划平面图(赛后)
UTILITY CONCEPTUAL PLAN (POST GAMES)

市政系统示意图
MUNICIPAL SYSTEM

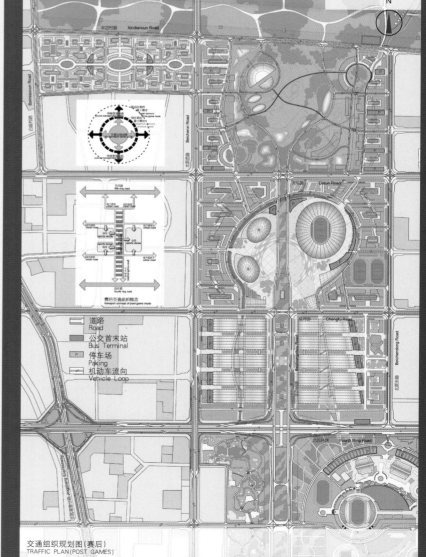

交通组织规划图(赛后)
TRAFFIC PLAN (POST GAMES)

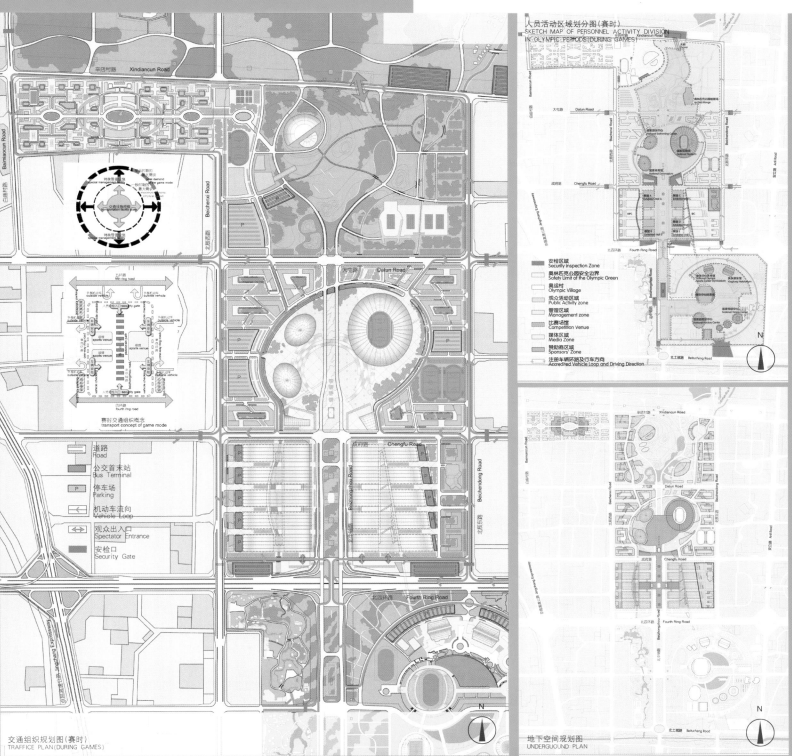

established to change the simple green bell into a self-adaptive and self-recovering system of ecological quality.

All facilities in the Olympic Green should meet international standards for disabled design and outdoor level difference should be avo4ded as much as possible.

All the facilities can be reused for the Paralympic Games without much reconstmction.

Strategy 10 – To optimize energy-consumption structures and establish recycling systems.

Utility development will be taking advantage of the existing infrasffuctures by improvement and upgrading. Water- supply will be from city water supply system to avoid pumping underground water. Heat supply in the area will be kom the city heating system. The wastewater will be massively dealt with by existing wastewater treannent plants.

The utilization of natural energies, such as temestrial and solar heat. will be fully considered. Water and waste recycling will be reinforced. The consideration has been made to fully use local energy and its contribution to the whole eneTgy supply network.

Strategy 11 – To promote ecological culture and improve ecological awareness among citizens.

Self-recovering ecological systems and ecological educational systems will be established.

The sky will be kept open to ground view to increase the visibility of the sky vault and decrease the impact ot heat-island ohenomenon.

High-rising buildings are arranged on the edges to

The road intersections are grade crossing to diminish the interruption to the natural water flowing by weight and natural seepage.

A multi-layered and integrated ecological network will be established in the Forest Area. A locat wildlife reservation- area will be established to change the simple green bell into a self-adaptive and self-recovering system of ecological quality.

鸟瞰图
BIRDS-EYE VIEW

Model pictures

模型 1
Model 1

模型 2
Model 2

二等奖 Second Prize

株式会社佐藤综合计画（日本）
SATOW SOGO KEIKAKU (JAPAN)
INGEROSEC 技术咨询公司（日本）
INGEROSEC CORPORATION (JAPAN)

Beijing Olympic Green 2008

佐藤综合计画，其标志为"AXS"，"A"是指"艺术（Art）"与"建筑（Architecture）"，"S"是指"佐藤（Satow）"和"科学（Science）"，象征着科学和艺术的融合。事务所是日本组织型建筑设计事务所的代表之一。事务所在日本设计了大量政府机构设施、文化设施、医疗设施、住宅、运动设施等公共建筑。

The symbol of Satow Sogo Keikaku is AXS, 'A' indicates 'Art' and 'Architecture', 'S' indicates 'Satow' and 'Science'. AXS means the Combination of science and art. AXS is one of the representatives of Japanese organized architectural design groups, which was founded in 1945. In 1988, the name of ' Takeo Sato Design Atelier' was changed to 'AXS Satow Inc. (AXS)'. AXS Satow Inc. is one of the leading Japanese design companies, which has skills in urban redevelopment planning, architectural design, electrical and mechanical cost estimation, engineering cost estimation and supervision.

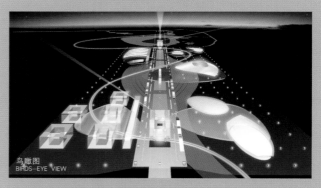
鸟瞰图 BIRDS-EYE VIEW

Ingerosec 是由法国技术咨询公司 Ingerop 和日本 SE 公司的技术软件部门历经 30 年以上、通过许多国际和国内工程以及工程技术方面的硕果累累的合作之后，在近期共同创办的日本技术咨询公司。Ingerosec 里汇聚了经验极为丰富的技术人员，他们都是曾长期活跃于各自的公司，并经选拔后来到这里的优秀专家，他们能够胜任任何技术工作。

Ingerosec is a Japanese consultant recently set up by the French consultant Ingerosec and Soft Engineering Department of Japanese SE Corporation after more than three decades of fruitful collaboration on many international and domestic projects and in many fields of the construction engineering. Ingerosec assembles engineers carrying a rich experience collected during their long activity in both companies and is able to assign the right specialists to any new engineering task.

■规划设计构思

1."未来型生态城市"的提案（高科技与自然的融合）

可持续发展至未来的城市环境需要建立一套生态型的城市体系。本规划采用"生态城市"的规划理念，创造出能降低地球环境负荷的最优良的城市环境，同时又使"绿色、科技、人文"三大奥运主题在北京奥林匹克公园中得以具体体现。

绿色和水是生命之源。本规划中的"生态城市"旨在奥林匹克公园充满绿色和水的优美环境中，布置呈现有机形态的建筑群，形成与自然相协调的新型城市空间。这样的空间环境在具备城市的高密度、高效率、高功能和高速度的同时，能使人们虽然生活在大都市中，却可以时常接触到自然，充分体会到"人工与自然相平衡的环境"的高度惬意和舒适。

未来型生态城市的特征：

（1）与时间共成长

并非在短时期内完成。由于具有柔软的形态并充满了自然（绿化、水）的要素，空间对变化的适应度较宽，可以充分适应时代的需求，构成以自由形态成长的城市。

（2）孕育自然价值

随着绿色植被的成长，环境将变得丰富多彩，同时由于拥有清新的空气和生机勃勃的动植物，土地的价值也将提高。当高度的人工环境与稳步成长的自然环境保持平衡时，就会形成高价值的城市。

（3）具有新时代的高速

因情报功能的集中化和高速化，以及建筑环境的柔软性，可以使时代的需求在城市空间中得到及时地反映和更新，并使这一地区保持最尖端的城市地位。

2.创造象征地球的城市

"生态城市"拥有人与自然共生的新型城市空间。它象征地球，并将"自然、生态系统"具体化。因此，在奥林匹克公园中不但规划有充满绿色、水和光的自然环境，而且通过导入高科技的生态系统，形成由人控制的舒适的城市空间，达到与自然相融合的境界。

（1）全新的自然环境的营造

绿色——森林和大草原

奥林匹克公园用地的大部分以繁茂的树木和同辽阔草原的绿色环境为基础，创造出以绿色为背景的城市景观。

在这一公园般的环境中，建筑群和其他室外的体育、文化设施呈点状分布，构成了充满魅力的景观。

水——大海

奥林匹克公园用地内规划有被称为"新北海、新中海、新南海"的新三"海"。这三座人工湖不但象征着浩瀚的大海，将地球的身姿投影过来，而且它们的形态和北京市中心的北海公园和中南海相类似，可将北京独特的历史景观引入到规划中。

（2）生态系统的利用

太阳能发电——光带

沿位于奥林匹克公园中央的生态漫步道布置繁华热闹的商业设施，并在它们的屋顶上架设太阳能发电板。这些太阳能板形成缓缓弯曲起伏的光带，为城市空间照明、建筑效果照明等提供公共电力，实现城市对自然能源的有效利用。同时，光带的形状又使人联想起中国的龙，充分体现出建筑与环境的地域性。

水的益处——新三海

奥林匹克公园中部由大面积的水空间构成新三"海"。可以利用雨水，使之经常循环，既具有降低城市热岛现象的功能，又可在城市防火中起到一定作用。

3.以北京中轴线为轴构成的城市体系和景观

历史名城北京的独特的城市构造是以紫禁城为中心形成南北延伸的中轴线。本规划将在此中轴线上布置作为奥林匹克公园主干的高密度城市空间——"生态漫步道"。

作为城市骨骼的一部分，生态漫步道如同树干连接着各设施，而一套有机并且高效的城市系统就像树枝一样从这里伸展开来，由此形成一个人工与自然互相融合的生态城市，它能够根据不同情况灵活调整和发展。

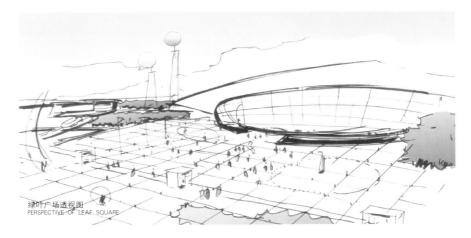

绿叶广场透视图
PERSPECTIVE OF LEAF SQUARE

（1）主干的形成——中轴线上的生态漫步道

生态漫步道：

生命之水（水轴）、通向未来之桥（人轴）以及光带（商业设施群）在中轴线上构成了步行者专用的生态漫步道，形成了城市功能完备且高密度的繁华空间。

和平之果：

在中轴线上布置象征奥林匹克公园历史和特征的设施。

"和平之塔"：

奥林匹克的记忆将结束于北侧的森林公园，因此在这里设置了标志性的塔以象征和平。

"五环信息中心"：

在大体位于B-1区中央的地段上建立五环信息中心，它同时处于国家体育场的广场内。

"城市之门"

本次规划中要求的北京城市规划展览馆以"城市之门"为立意，布置在生态漫步道上。

"绿色之门"：

作为生态城市的标志，"绿色之门"具有生态博物馆的功能。

（2）绿叶设施群的构成

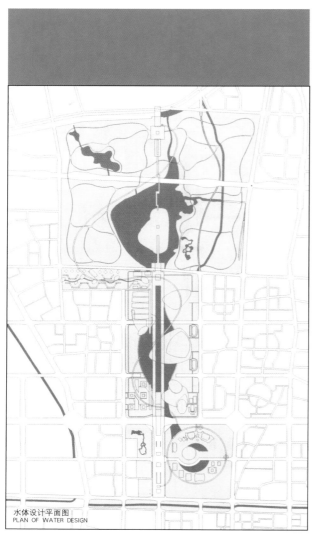

水体设计平面图
PLAN OF WATER DESIGN

cal City can develop freely in the future and always keeps up with the demands of the times.

(2) Cultivating the Natural Value

The verdure and richness of the Ecological City will be enlarged when green grows up, and the land value will increase when the environment fills up with fresh air, flourishing plants and living animals. The Ecological City of high value is composed under the perfect balance between the man-made environment and the peacefully growing natural environment.

(3) Holding A Real-time Speed

With the highly intensive information function and the flexible architectural environment, the Ecological City is able to respond immediately to the demands of the times, and becomes one of the most advanced cities in the world.

2. Creating the City which Symbolizes the Earth

"Ecological City" symbolizes the earth, makes "nature, ecological system" realistic, and contains the new type of urban space where human and nature exist together. In this case, a natural environment with plenty of green, water and light is planned. Meanwhile, with the introduction of the high-tech ecological system, a favorable, man-controlled urban space will be created and combined with nature.

(1) Forming A New Natural Environment

Green - Forest and Grassy Plain

As large areas of the site are planned with flourishing trees and verdurous lands imaging a vast grassy plain, an urban landscape with green for a background will be created. Meanwhile, buildings and external sports and cultural facilities are placed scatteringly on the site, composing an attractive garden environment.

Water - Three "New Seas":

Three "New Seas" which are called as the "New North Sea", "New Middle Sea" and "New South Sea" are planned in Beijing Olympic Green. These three man-made water areas symbolize the sea and reflect the appearance of the earth. The shapes of them are similar to that of the lake Beihai (North Sea) and the lakes Zhongnanhai (Middle & South Seas) which locate in the center of Beijing. This will bring the indigenous characters of the city into the environment of Olympic Green.

(2) Introducing the Ecological System

Solar Power Generation Line of Light

Prosperous commercial facilities are planned along the promenade of the central Olympic Green. On the roof of the facilities there install solar panels to produce electric power. These panels, which are designed in gentle waves, provide public electricity for lighting up the urban space. They not only make the natural energy be utilized effectively inside the Olympic Green, but also create an image of Chinese dragon by which the regional identity can be well expressed.

Bless of Water! Three "New Seas"

The three "New Seas", which are composed by large areas of water in the central Olympic Green, can use rainwater and make it circulating frequently. This can not only restrain the urban heat-island phenomena, but also be useful in preventing disasters of the city.

3. Building the Tree-like Structure and Urban Landscape on the City Central Axis

The urban structure of the historical city Beijing takes the Forbidden City as its center and has a central axis extending to the north and south. This master plan locates the high-dense urban space, Ecological Promenade, on the city central axis, making it the main structure of the Olympic Green. The form and function of the Ecological Promenade is similar to a trunk of a tree from which the organic and efficient urban framework grow up like branches. This kind of system forms an ecological city which combines with nature, and develops flexibly corresponding to different requirements.

(1) Forming the "Trunk" - Ecological Promenade on the Central Axis

Ecological Promenade

Along the central axis, Water of Life (the axis of water), Bridge of Future (the axis of people), and Line of Light (the group of facilities) compose the Ecological Promenade, and create a prosperous, functional space equipped with high-dense urban facilities.

The Fruits of Peace

The facilities which can symbolize the history and characteristics of the Olympic Green are placed on the central axis.

Tower of Peace

A tower symbolizing peace is programmed in the north Forest Area as the ending of the Olympic memory.

Five Rings Media Center

The Five Rings Media Center is proposed in the central area of Site B-1. It is a general information center located on the square of the National Olympic Stadium.

Gate of City

The expected Exhibition Hall of Beijing City Planning is designed as the "Gate of City" on the Ecological Promenade.

Gate of Green

Being a symbol of the Ecological City, an environment museum named "Gate of Green" is proposed as an information center for the worldwide environmental problems.

(2) Composing the Leaves (the Facility Groups)

Leaf Square

Three "Leaf" squares are planned on the Ecological Promenade, which spatial forms are like lotus leaves floating on water. The major facilities of the Olympic Green are located around the squares creating prosperous public spaces for people to communicate.

Leaf Roof

The leaves-like facilities, which can be approached from the Ecological Promenade, are designed in independent and free forms to compose the dynamic landscape of the Olympic Green. The major facilities share a common design code, which is the soft, leaf-like roofs coving the buildings so that an image of leaves floating within the green environment is generated.

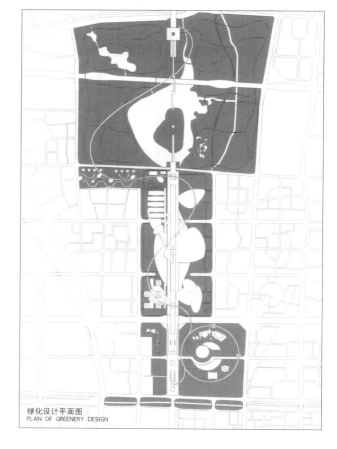

绿化设计平面图
PLAN OF GREENERY DESIGN

绿叶广场：

沿生态漫步道布置三个绿叶广场，广场的空间形状如同漂浮在水面的荷叶。奥林匹克公园的主要建筑以广场为中心布置，创造出一个繁华的、供人们交流的都市广场。

绿叶屋顶：

可由生态漫步道直接达到的设施群宛如绿叶，它们自主、自由的形态构成了颇具动感的景观。其中主要建筑以柔软轻盈的屋顶为主题，在绿色环境中塑造出如同绿叶漂浮般的美丽景象。

■ Planning and Design Concept

1. Proposition of "An Ecological City of Future" (The Harmony Between High Technology and Nature)

An urban environment sustainable to the future needs to set up an ecological urban structure. This planning of Beijing Olympic Green proposes a concept of "Ecological City" in order to create the most advanced urban environment which can not only low down the load of the earth, but also give the vision of Beijing Olympic Games, which is "Green, High-Tech and People", a concrete form in this area.

The urban space of the "Ecological City" will be created within the beautiful environment which has plenty of green and water that life lives on. Inside of the space buildings are designed in organic forms and harmonize with nature. This new type of urban space is not only high dense, built with effective functions and fast speed as a city, but also provides people with places where they can always touch the nature, and feel the amazing comfortableness of "the environment which is well balanced between nature and human works".

The Characteristics of the "Ecological City of Future":

(1) Growing with Time

The Ecological City will not be completed at one time. Instead, since plenty of spaces will be built in flexible form accompanying with natural landscape (green and water) everywhere, the Ecologi-

(The Harmony Between High Technology and Nature)

鸟瞰图(赛后)
BIRDS-EYE VIEW (POST GAMES)

景观设计鸟瞰图
BIRDS-EYE VIEW OF LANDSCAPE DESIGN

1 国家体育场 NATIONAL STADIUM
2 国家游泳中心 NATIONAL SWIMMING CENTER
3 国家体育馆 NATIONAL GYMNASIUM
4 国家奥林匹克体育中心 NATIONAL OLYMPIC SPORTS CENTER
5 首都青少年宫 CAPITAL TEENAGE PALACE
6 绿色之门 GREEN GATE
7 运动场 TRAINING FIELD
8 篮球场 BASKETBALL COURT
9 网球场 TENNIS COURT
10 奥运村 OLYMPIC VILLAGE
11 会展博览设施 EXHIBITION HALL
12 会议厅 CONVENTION HALL
13 五环信息中心 FIVE RINGS MEDIA CENTER
14 商业服务设施 COMMERCIAL SERVICES FACILITIES
15 城市之门 GATE OF CITY
16 中华民族园 CHINESE ETHNIC CULTURE PARK

B区屋顶平面图(赛时)
ROOF PLAN OF SITE B (DURING GAMES)

B区屋顶平面图(赛后)
ROOF PLAN OF SITE B (POST GAMES)

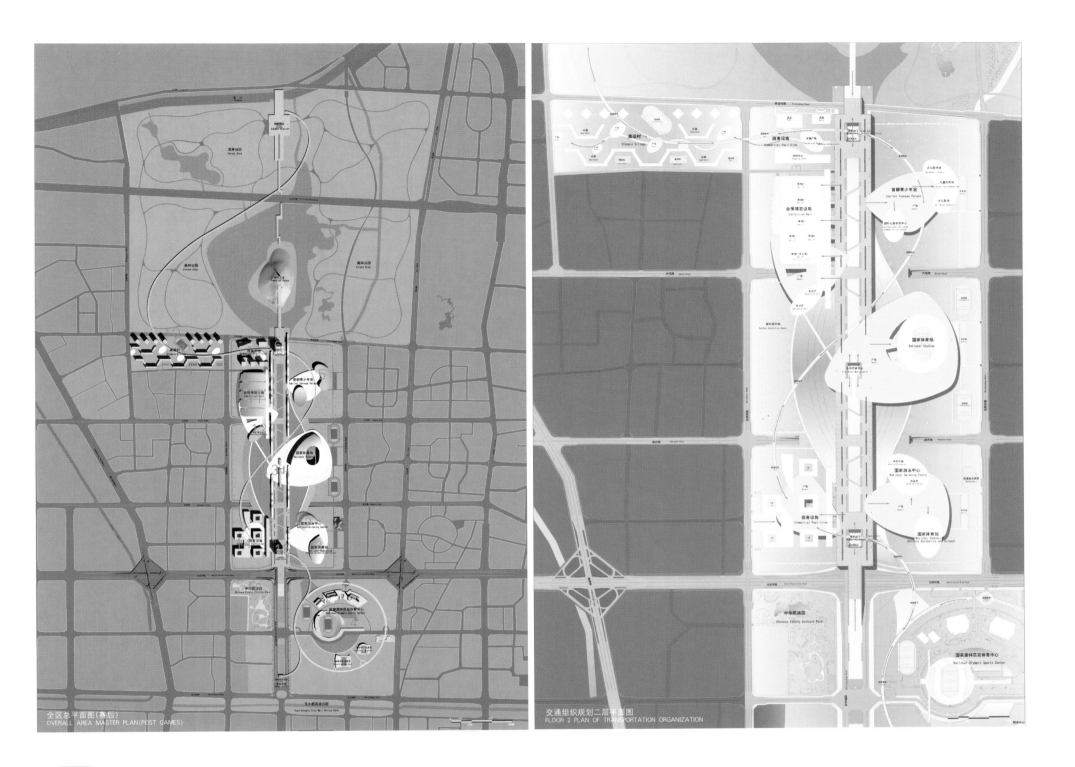

Proposition of "An Ecological City of Future"

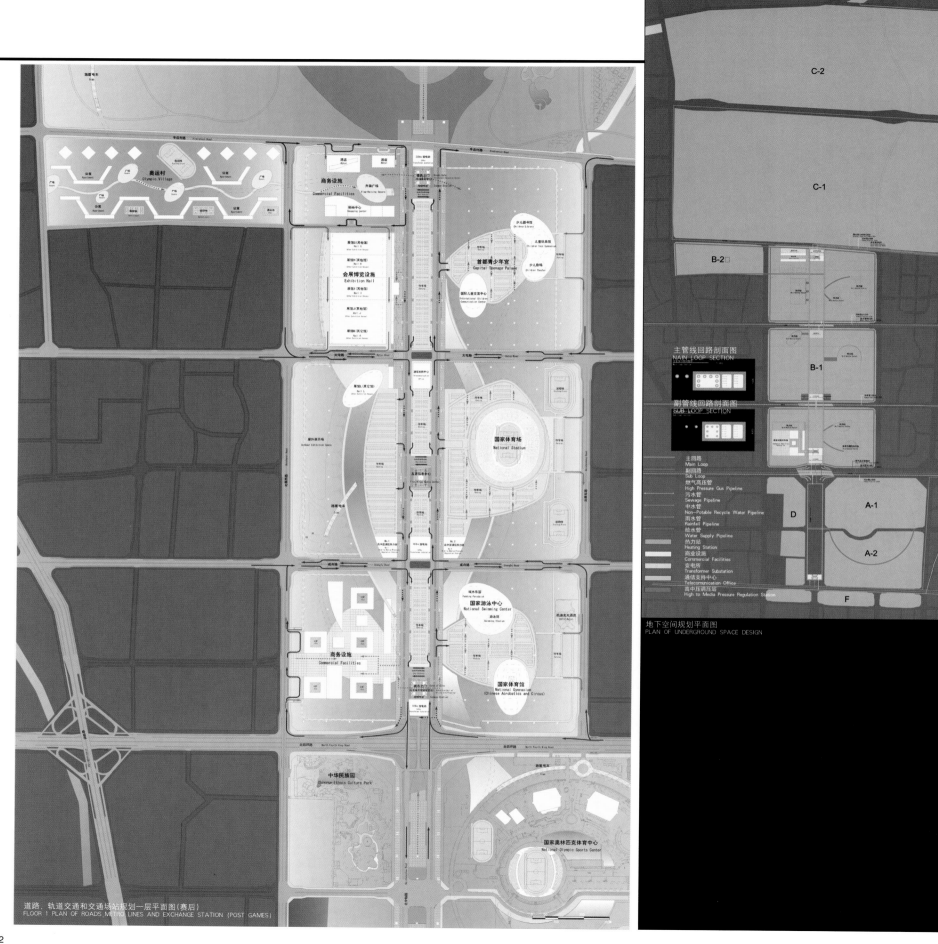

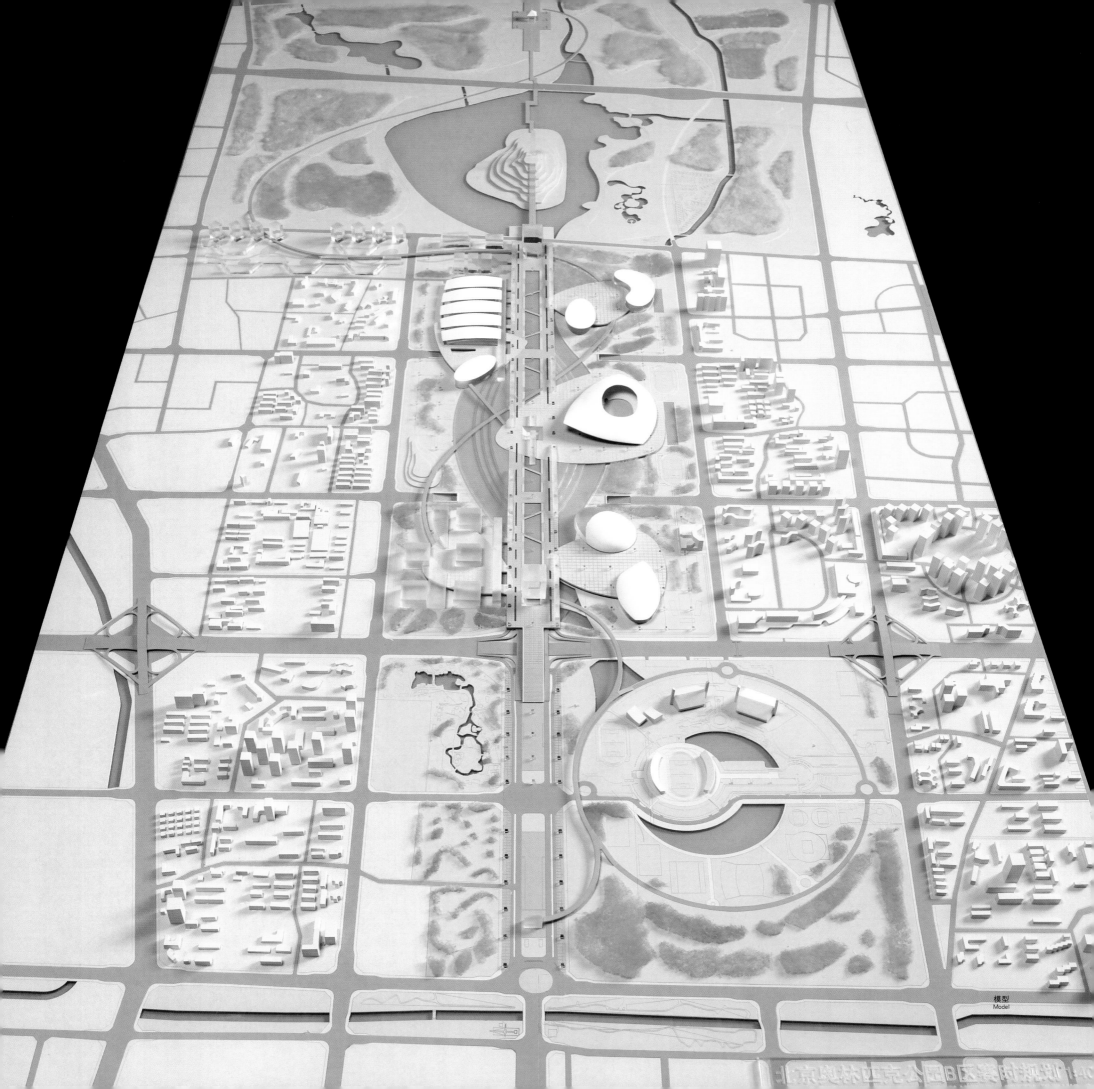

模型
Model

Honorary Mention

AREP 规划设计交通中转枢纽公司（法国）
AREP (FRANCE)

法国AREP公司（AREP规划·设计·交通中转枢纽），为法国铁路总公司SNCF集团的分公司。从多年积累的交通枢纽中心的设计及其周边区域的规划设计经验中，使AREP的设计范围扩展到城市的重新组合，公共设施设计，大型公建，商务中心及住宅设计。

近年来，AREP在与中国业主及建筑师共同探讨城市的未来时，开始了解到中国文化。一年来，实施的合作设计项目有：天津泰达中心金融服务区规划及建筑设计，北京首都博物馆新馆建筑设计，北京西直门站区规划及西环广场建筑设计，上海南站建筑设计等。

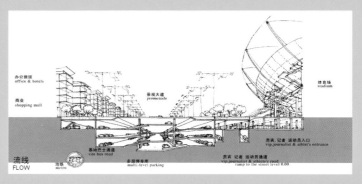

AREP is a branch of SNCF Group (French General Railway Corporation). AREP has rich experience on the design of traffic centers along with the planning of its surroundings. So it's business extended to the re-combination of cities, design of public facilities, large-scale infrastructure, business centers and houses.

In recent years, AREP began to understand Chinese cultures through the dialogue on the future of the city with Chinese architects. Last year, it carried out several projects, including the design of Tian Jin Taida Central Financial Service Area, Beijing Capital Museum's new buildings, Beijing Xizhimen Station Area, Xihuan Square and Shanghai South Railway Station, etc.

Beijing Olympic Green 2008

■规划设计构思

奥林匹克公园的规划项目为北京市带来一个规划1100ha广阔土地的机会。该规划从中轴线展向未来,基于历史,面向未来,成为新城市重要区域的标志。

1. 城市与自然的新关系

在此,不是建设一块通常的城区,而是提出一个和谐组织城市、与地区相协调的、反映北京高速发展的策略。它将成为北京城市发展中的一个重要标志。

建立与地区、与山的新关系

北京由广阔的山脉所护,西依山岳,北靠燕山。理论上,如将历史中轴线延伸至北20km处,将达到燕山山脉第一个海拔1500m高的小山峰。从城北可以很明显地看到,这组连绵起伏的群山轮廓,在与城市环境和谐相处的同时,也界定出了该城市的存在。为了让此特征更明显,并使其与历史中轴线生动结合,我们建议在燕山顶峰设置一个文化建筑,并以一个航标灯突出其要位。

2. 与北京城历史的新关系

很明显,只有将北京与其尊贵的历史相连,人们才能想像出它更加现代的未来。

人民的中轴线

本规划方围绕一个很强、很新的形象:历代皇朝的中轴线而展开。在发扬风水理论精彩之处的同时,又赋予中轴线以新的意义:人民的中轴线。它将服务于人民的奥林匹克以及今后的北京城市生活。

湖

这些湖泊描述了城市规划的历史。如上所述,湖泊与现代建筑之间能够建立起一种新的关系,而湖泊本身的设计,则从北京历史上形成的秀丽湖泊中取得灵感,以建立起城市过去与现代的衔接。

城墙

沿着今天的北辰路,还能找到元城墙的遗址,一条可爱的运河流经元城墙遗址公园。在与中轴线相接处,向南可以看到钟鼓楼的轮廓,向北可以看到整个奥林匹克区域,及其逐渐上升的步行道。

湖岸 LAKE BANK

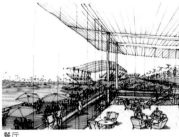

餐厅 RESTAURANT

赛后商业街 MAZE OF STREET

Introduction

国家游泳中心 NATIONAL SWIMMING CENTER

赛后湖畔商住区 LAKE SIDE HOUSING AFTER GAMES

国家体育场 NATIONAL STADIUM

建立与环境的新关系

北京城历来就懂得如何利用这一自然资源,以理水的方式表达城市的特征。然而一直到现在,作为映射天空明镜的水,在中轴线上的出现仅限于紫禁城城墙之内的金水河。这次规划提供了一个独一无二的机会来将北京的两个历史要素:轴线与水结合起来为市民提供一个美妙的漫步空间的机会。从C区已有的小河和湖泊开始,我们建议以京景已有的湖泊为尺度设计一组湖泊。

建立与自然的新关系

为了让本项目的自然和生态特色更臻完美,我们的设计遵循市政府"建立北京周边城市绿化带"的政策。显然,C区的大型公园以广阔的湖面为中心,遍布植被并点缀以形状自由的山丘,构成了规划中最大的一部分。

无论如何,元城墙与中轴线的交点是北京历史的象征。这里还是规划中的一处地铁站,同时也是奥林匹克公园的入口。我们不想复制一座假的历史城门,而想用在中轴线东西两侧对称布置的两栋竖向建筑来点明历史的遗产及通向未来的大门。在这两栋建筑里,可设置"城墙博物馆",讲述北京不同时期城墙的历史。于是,本规划中4000m长的中轴线从南边开始于对历史的追溯,而在北边则以对这个城市未来所表达的敬意而结束。

3. 人文奥运、绿色奥运、科技奥运

北京市为该规划所确定的目标:人文奥运、绿色奥运、科技奥运,已经强烈地表达在对中部的平面组织中。事实上,本项目的建筑形体和空间形态就是按照这三个目标来选定的。

人文奥运

南北贯穿的中轴线象征着人文奥运。皇帝的中轴线由连续的对称的封闭庭院组成。而本规划方案的中轴线向金山岭,向群山,向北京的领土开放,宽阔的步行道服务于所有人,并成为本区域的脊梁。

绿色奥运

中轴线东侧,大自然表现为湖泊、山坡、树林。

科技奥运

与东侧相反,在西侧预留了会展空间、办公区域与商业设施。所有这些建筑严格遵照与高品质的环境相当的原则而设计。本区域使用自然能源,整齐方正的几何形状隐喻着科技。

4.未来城市的新公共空间

在城市规划的历史上,北京懂得并善于利用既对比又互补的手法来组织城市空间。规整的路网方便交通,便于识别方向,正与带来诗意、梦幻、引起人们好奇的胡同相反。在满足现代化的前提下,必须要重新创造出丰富的公共空间。每个主题:人文、绿色、科技均对应于一个公共空间,它们在形式和形体上相互对比。

5.各种活动的和谐布置创造出真正的城市集合体

为了成为真正意义上的城市区域,奥林匹克公园很紧密地设置各种活动,使其成为一个充满吸引力的诸多行为、活动的总汇。

6.区域及交通的合理组织

除去三个主要步行区(中央大道,曲径,漫步区),在地段内,还安排有服务于不同交通类型的特别路径。

地铁在本区内设有4站,其中有3站停在南侧,到达地面后恰好位于中心大道的中心区域。

中心大道的西侧部分比步行道低一层,大巴士和轻轨交通将服务于整个中心大道,每隔340m设一处巴士站,它们与公共空间的联系非常便捷。

在中心大道东侧,与西侧相对应,在赛时,为运动员(直接来自奥林匹克村)、媒体记者及VIP赞助商的专用道路。

有两条横向道路服务于利用北辰西路的观众公交车,公交车站场今后将成为展览厅。

小汽车将从北辰东路和北辰西路进入,它们将停放在中心大道下面的停车库里,停车库既可以从两条横向主路进入,也可以从中心大道西侧间隔170m的横向小路上进入。中心停车场按每120m长度划分为若干段,每两段之间是出入口。

■ Planning and Design Concept

The project to develop the site for the Olympic Games gives us the opportunity to reorganize a vast piece of land, measuring 1100 hectares north of the City which will highlight the historical significance of this city and its roots, around its axis and its view towards the future.

1 - A New relationship between territory and nature

We are not talking about building a new urban group like other ones; we are proposing an organizational strategy in harmony with the city and the territory which will make a strong mark on the urban development of Peking.

Upbuild a new relationship with the territory, the mountains

The plain of Peking is dominated by a large chain of mountains, the western hills on one side and the YAN SHAN mountains in the North. Theoretically, if one were to lengthen the Historical Axis, one would meet, 20 kilometers down the line, the YAN SHAN foothills

鸟瞰图(赛后)
BIRDS-EYE VIEW (POST GAMES)

and notably a small hill, 1500 meters high. Very visible in the north of the city, the outline of these foothills seems to be in harmony with the urban structure and marks the presence of the large Chinese territory in relationship to the City. In order to reinforce this idea and to bring it alive in conjunction with the historical axis, we propose to set up, on the summit of the YAN SHAN, a cultural edifice of the City whose presence would be signaled by a searchlight, beaming out over the plain between the mountains of the city and

can imagine its more modern future.

The people's axis

The project is organized mainly around a strong but new image: the Historical Axis of the Emperor. While taking advantage of its superb cosmogonic geometry, it also gives it another significance: the People's Axis, at the service of the "People's Olympics" and, later on, the life of the city of Peking.

The lakes

Lakes play an important role in the urban history of the city. As we saw before, these lakes introduce a new relationship with the modern buildings, but their design was inspired by some magnificent

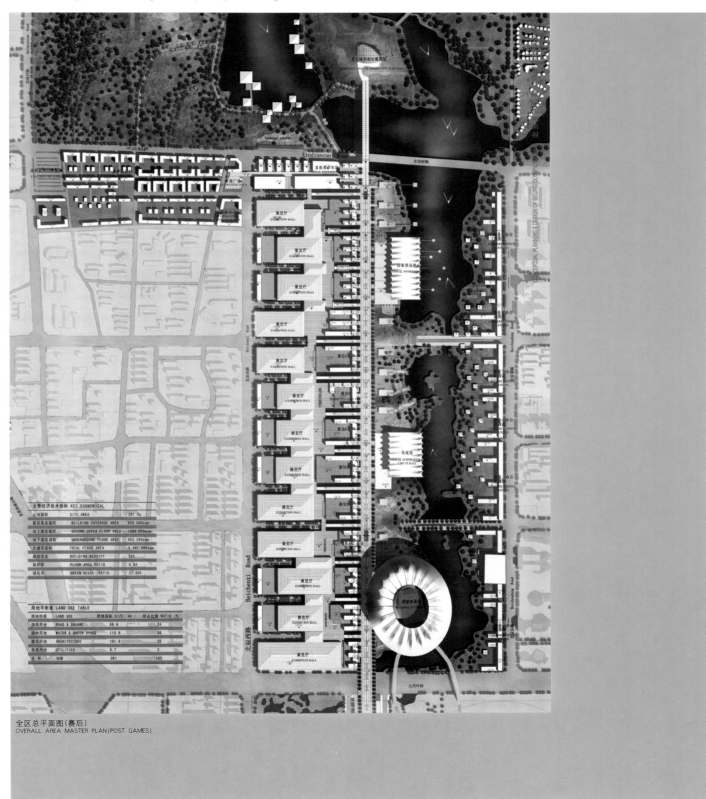

全区总平面图(赛后)
OVERALL AREA MASTER PLAN (POST GAMES)

over the city.

Upbuild a new relationship with the territory: water

Historically, the city knew how to take advantage of these phenomena and the presence of numerous lakes and waterways gives this city an identity.

Nevertheless, up until now, this water, reflecting the sky has only had a limited presence on the axis in the perimeter of the Forbidden City under as the golden river (to be verified). This development offers the unique opportunity to bring together 2 historic elements of Peking, the axis and the water to provide its citizens with a wonderful place to stroll from the existing canal and the lakes in the Part C, we propose to create new lakes proportionate to those already cited.

Upbuild a new relationship with nature: greenspace

To perfect the natural and ecological characteristics of the project, we propose to develop a natural space around the city of Peking. Obviously, the large park in the Part C, built around the large central lake, with wooded and grassy surfaces, marked by freeform hills make up the largest part of the plan.

2 - A New relationship with the City's History

It is clear that if Peking reaches back to its prestigious past, it

historical designs of Pekinese lakes in order to create the dynamic dialogue between the City's past and the present.

The ramparts

The traces of ramparts in the city of YUAN are still present along BeiTucheng Road. The linear park of "Yuan Dynasty City Wall Relics Park" has a lovely canal running beside it. At the intersection of the historic axis one can see both the outlines of the drum and bell towers to the south which make up the center of the city of YUAN and, to the north, the organization of the Olympic Village with its large avenue climbing the hill. Nonetheless, the intersection of the North-South axis and of the YUAN rampart makes up a symbolic convergence reflecting the history of the city of Peking. The site of a future metro station, it makes up the main entrance to the Olympic Village. Without wishing to create a false historical door, we think it would be interesting to signal this souvenir of history and this opening towards the future by building two symmetrical vertical constructions on either side of the axis. In these constructions would be a "ramparts museum" which would tell the story of the different walls of the city. As such, the 4 000 meters of the Historic Axis of this project start out in the south with a call to the past history of the city and end in the north with a nod to the Future of the City.

3- The people's olympics: nature and technology

The goals fixed by the city of Peking (to make these games the People's Olympics of Nature and Technology) have been strongly expressed in the organization of the central plan. In fact, the geometry and the spatial layouts of the project were chosen to mark the land with these three goals.

The people's olympics

This is symbolized by the North-South axis, which runs through the site. Just as the Emperor's axis was made up of a succession of symmetrical closed and reserved courts, the project's axis is open on the hill, the mountains, the territory and its treatment with a large pedestrian street which is the backbone of the project, a place where all the visitors shall pass through.

The nature olympics

To the east of the axis, nature is ever-present in the form of lakes, hills and trees.

The technological olympic

The area uses soft energy. This is set up according to a regular orthogonal geometry which would be a metaphor for TECHNOLOGY.

4-New public spaces for the future city

In the history of its planning, the city of Peking knew how to organize its public spaces in a way both contrasting and complementary. The large grid-pattern of streets facilitates traffic and navigation; in contrast, the small streets offer poetry, fantasy and discovery. While still maintaining its modern requirements, we must recreate this richness in these public spaces. Each theme: PEOPLE, NATURE and TECHNOLOGY provides the public spaces with a contrast in their form and geometry.

5- A Harmonious spread of activities crertes a true urban mix

In order to be an authentic city neighborhood, the Olympic Games neighborhood includes a mixture of activities all brought together in one place to create an attractive ensemble.

6-A Rational organization of zones of vehicles

Other that three main pedestrian zones described above (axis, maze and strolling area) the site offers some particular itineraries for different modes of transportation.

The metro has 4 stops at the site, 3 stops at the south which come out just next to the central part of the axis.

In the western part of the axis at a level under the pedestrian street a shuttle bus or rail vehicles will serve the entire axis. A bus stop will be situated every 350 meters, which also offers excellent access to the public space.

At the other end of the eastern part of the axis the same space is reserved during the Games for the athletes circuit (coming directly from the Olympic village), journalists, VIPS and sponsors.

Buses with visitors use Beichenxi road, 2 lanes reserved for them to get to the stock area situated at the future site of the exhibition halls.

Cars come in by Beichenxi and Beichendong and can then get to the parking areas of the central axis either by two transversal roads or by the west of the roads situated every 170 meters. The central parking lot is sectioned into 120 meters long spaces with access between each section.

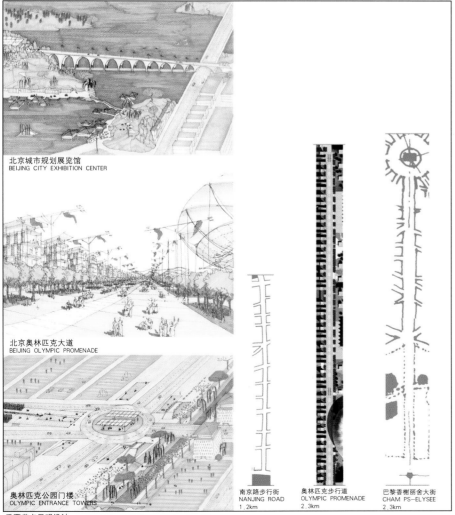

重要节点景观设计
LANDSCAPE DESIGN FOR NODES

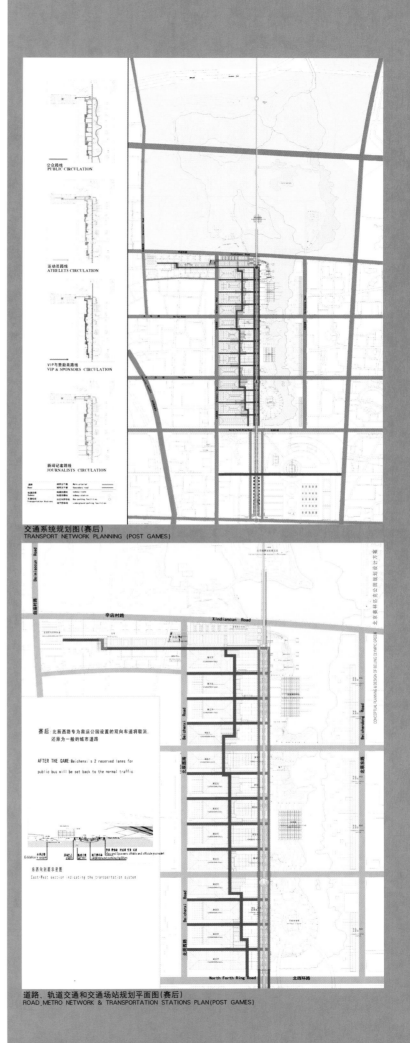

北京奥林匹克公园设计方案
Beijing Olympic Green

优秀奖 Honorary Mention

总装备部工程设计研究总院（中国）
ARMY GENERAL EQUIPMENT ENGINEERING DESIGN INSTITUTE(CHINA)

哈尔滨工业大学天作建筑研究所（中国）
TIANZUO ARCHITECTURAL INSTITUTE OF HARBIN INDUSTRIAL UNIVERSITY(CHINA)

Beijing Olympic Green 2008

总装备部工程设计研究总院是全军最大的综合性甲级设计院，1958年创建于北京。现有员工600余人，其中：专业技术人员500余人（高级工程师170人、工程师302人、助理工程师60人）。设有城市规划、园林、建筑、结构、给排水、采暖通风、环境保护、强电弱电、室内设计、概预算、工程管理、总体工艺、机械、通信、自动控制、供油供气、工程地质、工程测量、道路桥梁、情报信息和计算机科学等20多个专业。

Army General Equipment Engineering Design Institute is the largest comprehensive design institute of class A in the whole army. It was established in Beijing in 1958,and now has a workforce over 600,more than 500 of which are professionals, including about 170 senior engineers, 302 engineers and 60 assistant engineers, cover-

ing twenty specialities such as city planning, gardening, architecture, structure, water supply and sewage, heating and ventilation, environmental protection, power supply and distribution, interior design, technological economy, engineering management, general technology machinery, communication, automatic control, fueling and gas supply, engineering geology, engineering survey, road and bridge engineering, technical information, and computer science.

哈尔滨工业大学天作建筑研究所成立于1997年。在进行理论研究的同时，致力于探索建筑创作规律，并在实践中逐步提高。现已发展成以中青年教师为骨干，以博士和硕士研究生为新生力量的学术性创作群体，现有人员40余人，平均年龄30岁。其负责人和主创人员均为30岁左右的青年建筑师。

Founded in 1997, Tianzuo Architectural Institute of Harbin Industrial University devotes to studying architectural theory and exploring architecture creation regulations, and has made great progress in practice. There are more than 40 research fellows in the research institute, most of whom are postgraduates applying for master or doctor degrees, and the main members are young teachers of School of Architecture. The person in charge and main creation members are all young architects, aged about 30.

■规划设计构思

轴线：从北京旧城传统中轴线的空间序列处理中得到启示，对规划区域内中轴线上的景观序列做出如下构想：序曲（从元大都遗址公园到北中轴路绿化带）——开端（喷泉广场）——铺垫（生态谷）——高潮（圣火广场）——回响（奥运纪念广场）——尾声（森林公园小山）整个景观序列主次分明，结构严谨，变化丰富。

标志：中轴线景观序列中的主要部分几乎都与"奥林匹克山"这个超大尺度的地标紧密地结合在一起，具有鲜明的标志性，突出体现了21世纪北京的新风貌。

绿色：将城市边缘的绿化引入市区内，以形态丰富的植被和水体作为景观设计的主旋律，来控制整个区域。

整体：从区域整体设计的概念出发，使整个规划区域成为城市的有机组成部分，并进一步丰富城市空间。区域的交通规划从城市的整体出发，尊重原有的道路网络，并采用多层次的立体交通模式而使整个区域的交通状况得到改善。

重点区域

构想一：B区主体建筑群被覆盖在巨大的草坪之下，大面积的绿化植被突出了"绿色奥运"的主题。草坡中部劈开的峡谷和表皮上大大小小的圆洞是生态性构思的表达，不仅使得下部空间获得良好的自然采光和通风，同时辅以高科技的光感、温感控制系统，使得整个群体成为一座可以"呼吸的"有生命的巨型结构。

构想二：将"奥林匹克山"这个超大尺度的地标作为城市中轴线的高潮和结束，使得建筑群在与自然环境相融合的大前提下，具有了鲜明的标志性。形态自由的草坡和漂浮其上的流线型的"绿叶"屋面是对自然环境的回应，中轴线上的"生态谷"和作为区域制高点的火炬则是对城市肌理和"绿色奥运"的提示与强化。

构想三：将整个区域作为一个巨型结构来考虑，使其成为城市整体的一部分，为城市提供了"呼吸的器官"，极好地丰富了城市特色。

Introduction

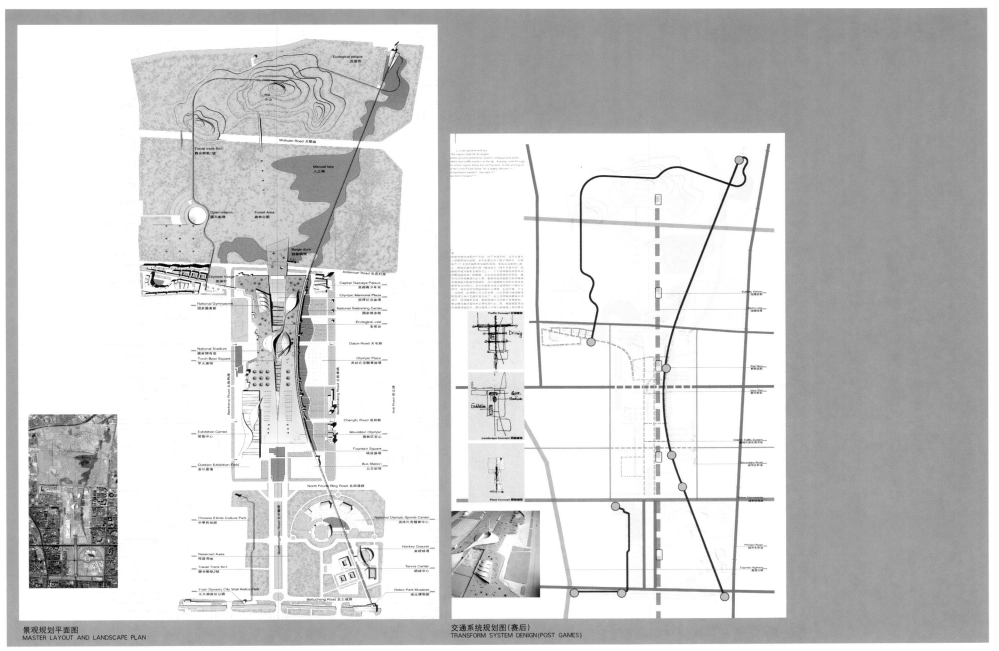

景观规划平面图
MASTER LAYOUT AND LANDSCAPE PLAN

交通系统规划图（赛后）
TRANSFORM SYSTEM DENIGN(POST GAMES)

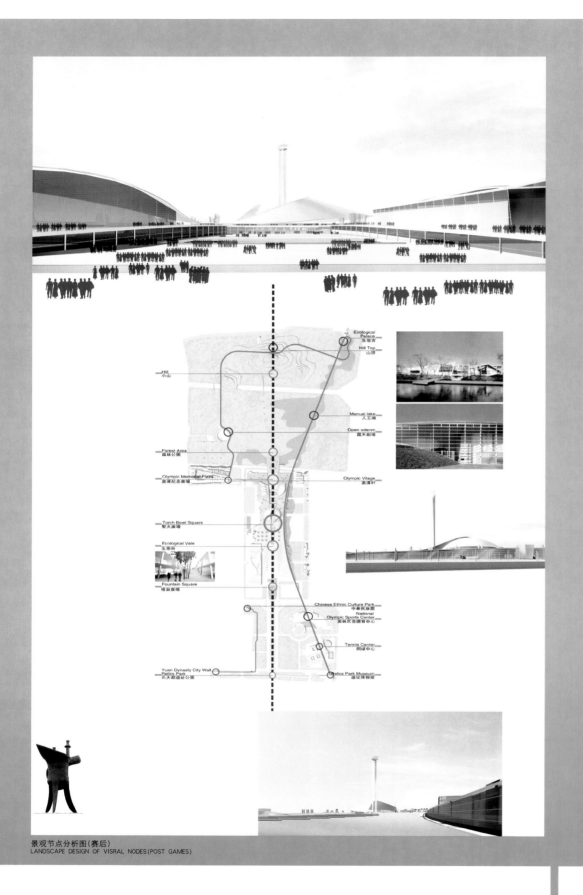

景观节点分析图（赛后）
LANDSCAPE DESIGN OF VISRAL NODES (POST GAMES)

景观规划鸟瞰图
MASTER LAYOUT AND LANDSCAPE BIRDS-EYE VIEW

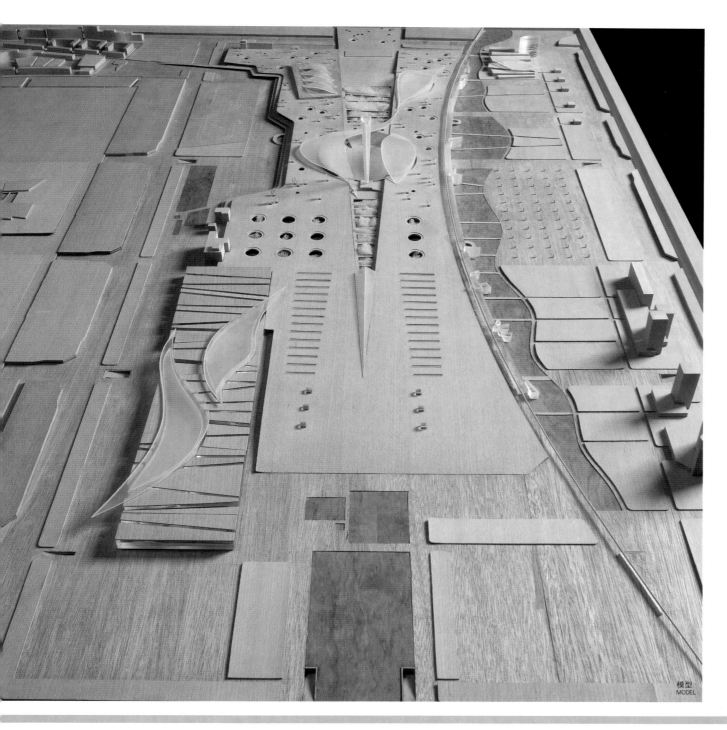

■ Planning and Design Concept

Axis: The inspiration is from the traditional spatial sequence along central axis in the old Beijing City. Conception from this traditional design is elaborated as followed:Sinfonia (From the Yuan Dynasty City Wall Relic Park to the green belt near North Central-axis Avenue)-Outset(The fountain square)-Foreshadowing(Eco-valley)-Climax(The Olympic Flame Square)-Echo(Olympic Memorial Square) -Epilogue(hill in the Forest Park),The whole landscape has clear sequence, precise structure and affluent variety.

Symbol: The major sights along the central axis almost all link with the Olympic Mountain, the super-sizable landmark, hence have outstanding features. They are the embodiment of the new Beijing in the 21st century.

Green Color: Our conception is to introduce the planting at peripheral city into urban area. The theme of the design for the region is the variety of planting and the water to control the whole site.

The Entirety: From this conception, the site becomes a component of the city, enriches the city and boosts the development of the peripheral areas.Traffic planning for the site is based on the previous road net.Three-dimensional transportation is adopted to improve the traffic conditions.

Important flied

Conception one: The major building complex in Site B is covered by vast lawn. This large-area planting reflects the theme of "Green Olympics". The gorges rived in the middle of the lawn slope and the holes on the surface are the embodiment of ecological environment, which provide natural lighting and ventilation to the underneath space, where there is hi-tech control system sensitive to both light and temperature. The whole complex hence becomes an organic giant structure that is able to "breathe".

Conception two: Making Olympic Mountain, the super-sizable landmark the climax and the end of the city central axis bestows an outstanding feature to the complex which perfectly blends in the nature. The natural-shaped lawn slope and the streamline leaf-like roof floating on it echo to the nature. The Eco-valley on the axle and the torch, which is the vertex of the region, strengthens the city texture and the theme of 'Green Olympic'

Conception three: The whole region is perceived as one giant structure and part of the city entirety. It is then the breathing organ of the city and enriches the city style. Meanwhile, the individual functions of Site B are more closely integrated.

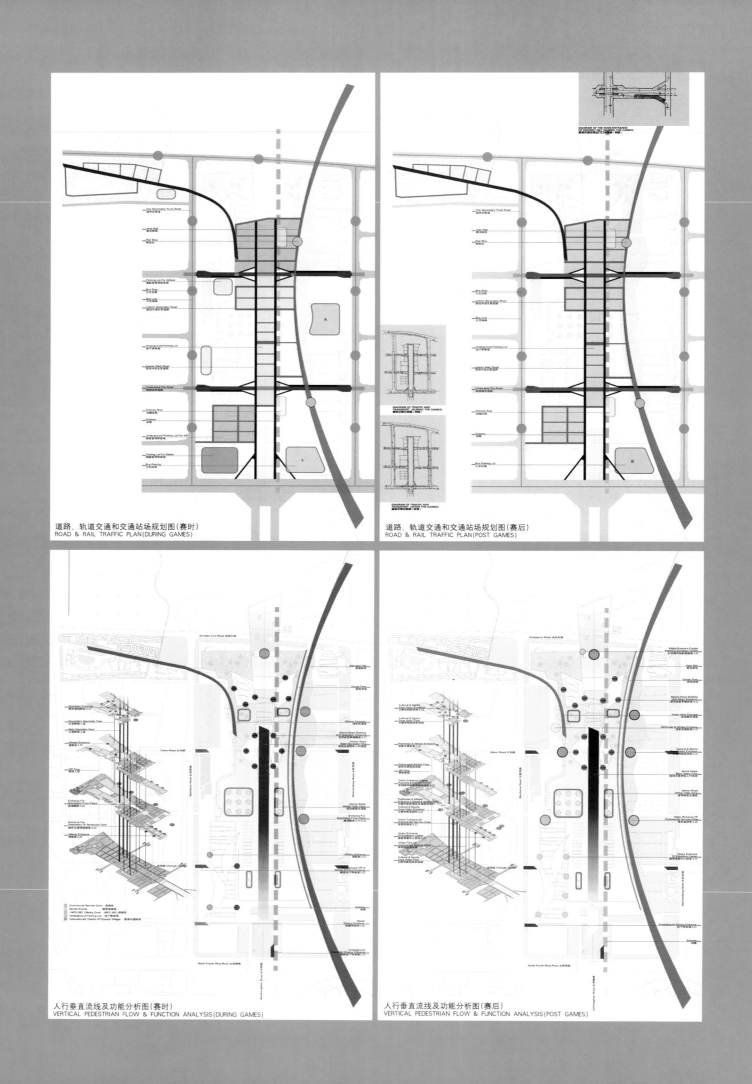

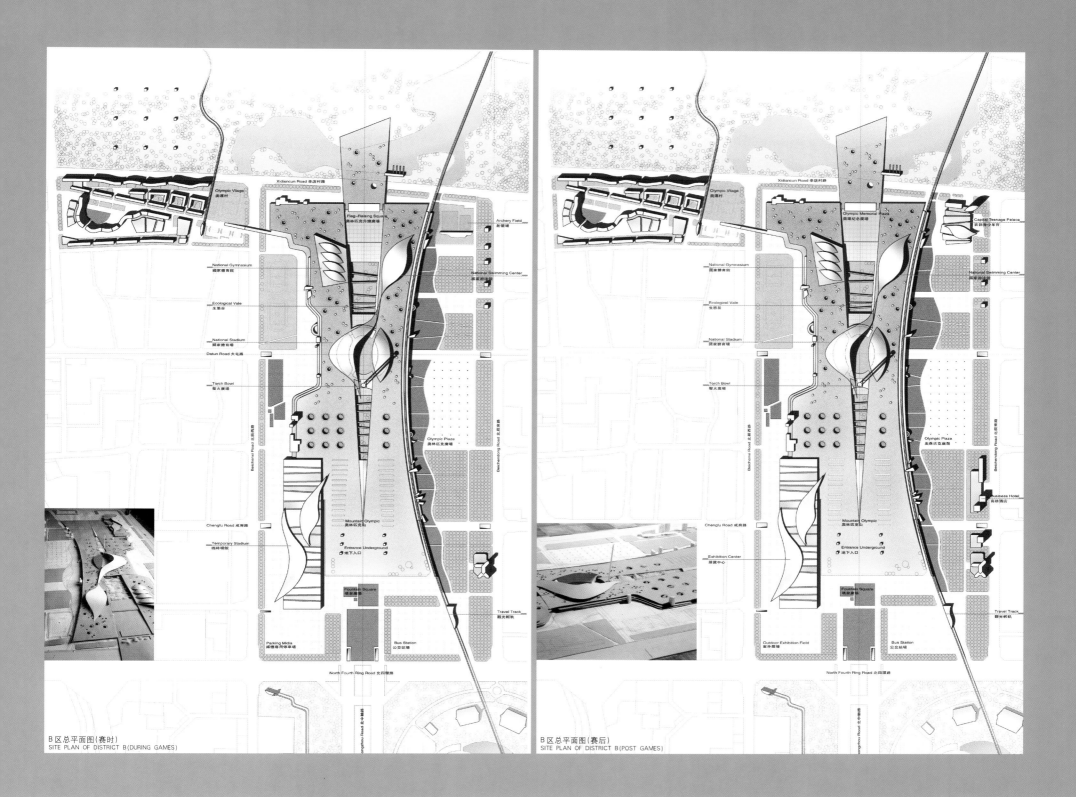

B区总平面图(赛时)
SITE PLAN OF DISTRICT B (DURING GAMES)

B区总平面图(赛后)
SITE PLAN OF DISTRICT B (POST GAMES)

Axis | Symbol Green Color
The Entirety

Honorary Mention

北京市建筑设计研究院(中国)
BEIJING ARCHITECTUAL DESIGN AND RESEARCH INSTITUTE(CHINA)

交通部规划研究院(中国)
PLANNING INSTITUTE OF MINISTRY OF COMMUNICATIONS(CHINA)

EDSA 规划设计公司(美国)
EDWARD. STONE, JR. & ASSOCIATES, INC (USA)

Beijing Olympic Green 2008

北京市建筑设计研究院作为北京市第一家国有建筑设计单位1989年正式定名为北京市建筑设计研究院。1992年在由建设部等机构评选的首届"中国勘察设计单位综合实力百强"中,该院名列民用建筑设计单位第一。

全院现有职工1125人,其中技术干部896人,包括中国工程院院士1名、国家设计大师5名。

Beijing Institute of Architectural Design and Research was the first state owned architectural design unit and named as what it is in 1989. It was listed as the No.1 architectural design unit among the top hundred design units in the evaluation made iointly by the Menistry of Construction and other authorized organizations in 1992.

The Institute has 1125 employees, of whom 896 are technical persons an academician of the Engineertng Academy of China, 5 national design masters.

交通部规划研究院成立于1998年3月,主要承担:公路和水运交通的发展战略研究,政策研究和规划研究工作,编制有关规章制度等技术业务管理工作。院现有职工140人,其中高级专业技术人员60多人。

Found in March 1998, Transport Planning and Research Institute (TPR) is a major advisory body to the Ministry of Communications (MOC) with outstanding track record. Its main responsibilities include policy research and formulation, strategic planning of highway and waterway sector. TPRI has a total of 140 staffs, of which 60 are senior professional staff.

EDSA是一家从事规划和景观建筑设计的公司。公司的工程范围包括饭店设计、度假村和社区规划、城市设计、规划以及生态旅游和资源规划等等多个方面。

目前,EDSA的员工团队由140余位规划师、景观建筑师、图形设计师和支持人员组成。

EDSA has established itself as one of the leading planning and landscape architectural firms in the world. Our projects range from hotels and resorts, community planning, attractions and entertainment to urban design and planning, ecotourism and resource planning.

EDSA staff consists of approximately 140 planners, landscape architects, graphic designers and support personnel.

■ **规划设计构思**

新世纪的发现：将北京传统文化与现代设计理念,以一种高度洗炼的方式加以融合,形成与老北京"实"中轴线相对比"虚"的意念,并展示出一个可以超越国境交流、充分体现奥林匹克"广泛和参与的可能性"。有序、开放的空间,高耸的"世纪之塔",在那里人们可以信步其中,仰首想像,从而实现肉体与灵魂、力量与精神的统一,描绘出另一个崭新的世界。

场所与文脉：设计的思考在于认识时间与空间连续性的概念。从过去到现在,从现在到未来,这一连续的时间轴贯穿了城市发展的空间轴,奥林匹克公园处于这些时空要素交错、冲突、集合的场所。场所在这里不是一个简单的巨大容器,而是召集市民、感动市民、唤起某些情景的空间。对场所文脉深入的观察与解读,思考设计对现有城市空间的影响,塑造新与旧的冲突,钢和玻璃与红墙绿

鸟瞰图
BIRDS-EYE VIEW

瓦,激发城市的活力,场所脉络之间的差异形成了城市的多样性与丰富性。

总体设计构思要素：

功能：充分考虑奥运会时及会后的长期发展,结合建筑使用功能,城市道路交通情况以及周围用地性质,将B区用地分为四个部分,分别为青少年文化区、体育场馆区、会展博览区和商务设施区。每个区都能形成相对独立,城市生态景观良好的区域,为今后的分期建设创造了非常便利的条件。

绿树环抱的青少年活动中心：与其南部中华民族园共同构成文化设施区。

中部平阔宽敞的体育设施区：与现国家奥林匹克体育中心形成相互对话的体育公园。

北部平缓舒展的会展区：有利于物流交通的组织。

东西两侧商业服务设施：有利于商业开发的分期建设。

空间：把体量相对较小的商业服务设施放在两侧,并将体量较大的体育展览设施放在中央,形成与现状城市空间的过渡,实现了大体量建筑在城市空间的"软着陆"。

交通：在用地内设两个平行于北辰东西路的辅路(奥运东路、奥运西路),有效疏导内部交通,缓解城市压力。

标识：靠近四环路的文化设施区采用下沉方式,可以使人们在四环路上清晰地看到位于中央的体育场馆,增强了奥林匹克公园对城市的标识性,并与现状国家奥林匹克体育中心在视觉上、空间上形成呼应。

发展：设计充分考虑奥林匹克公园的滚动发展和分期建设的可能性。由大屯路、成府路及两条辅路分割成适合于商业设施建设的地块,便于分期开发建设。

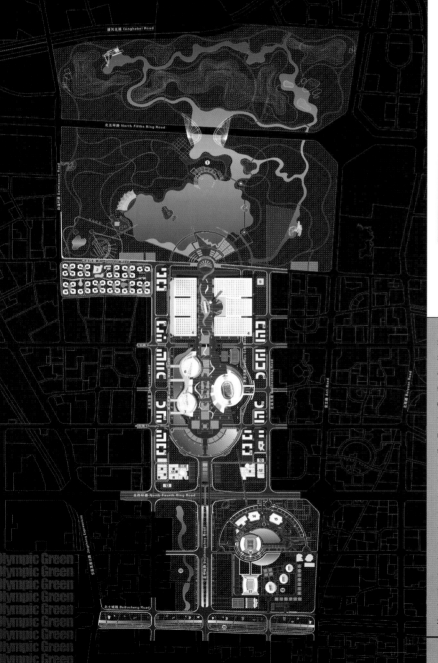

1 洼里森林公园 WALI FOREST AREA
2 住宅区 RESIDENTIAL QUARTER
3 商务服务 COMMERCIAL SERVICES
4 会展博览中心 EXHIBITION CENTER
5 变配电站 ELECTRICITY TRANSFORMER AND DISTRIBUTION STATION
6 国家游泳中心 STATE NATATORIUM
7 国家体育馆 STATE GYMNASIUM
8 国家体育场 STATE STADIUM
9 凯迪克大酒店 CATIC HOTEL
10 环境教育中心 ENVIRONMENTAL EDUCATION CENTER
11 首都青少年宫 CAPITAL TEENAGE PALACE
12 北京城市规划展览馆 BEIJING URBAN PLANNING EXHIBITION CENTER
13 电信局 TELECOMMUNICATION BUREAU
14 国家奥林匹克体育中心体育馆 GYMNASIUM OF NATIONAL OLYMPIC SPORTS CENTER
15 国家奥林匹克体育中心英东游泳馆 YINGTONG NATATORIUM OF NATIONAL OLYMPIC SPORTS CENTER
16 国家奥林匹克体育中心体育场 STADIUM OF NATIONAL OLYMPIC SPORTS CENTER
17 国家曲棍球场 STATE HOCKEY FIELD
18 国家网球中心 STATE TENNIS CENTER
19 中华民族园 CHINESE ETHNIC CULTURE PARK
20 元大都遗址公园 YUAN DYNASTY CITY WALL RELICS PARK

景观规划总平面图(赛后)
LANDSCAPES PLANNING LAYOUT(POST GAMES)

■ **Planning and Design Concept**

Discovery of the new century: Beijng's traditional culture and modern design concept are integrated in a highly succinct way. A vision is formed in correspondence to the city's axis line of Beijing. Here people can have ex-changes beyond national boundaries and mass participation becomes possible. Fair plays, open spaces and century towers provide people freedom for leisure walking and imagination. Here people enter into a new century for unity of physical comforts and soul content, strength and spirit

SITE AND CULTURE HERITAGE: on realization of continuation of times and spaces. The continuation comes a-long with the spacious axis of the whole city of Beijing from past to present, from present to future. The Olympic Green is situated at a site where elements of times and spaces interlock, interact and combine. It is not simply a huge container, but rather a place that convenes the citizens, greets the citizens and arouses the citizens. After deep observation and thorough understanding of the cultural heritage of the site's effects on urban spaces by the planning, contradictions between the new and the old, between steel and glass, between red walls and green tiles have been taken into consideration. The planning design is to activate the vigour of the city, to create variety and colors.

The elements in the overall design:

Functions: The use during the Olylmpic Games and the long-term development after the Games should be given full consideration. In light of the functions of the facilities, traffic conditions and purpose of the surrounding, the facilities to be built along the central axis in the Area B are Youth culture centre, sports facility area, exhibition and business area. Each area is a comparatively independent area and provides a sound visual effect and convenience for future constnlction in stages.

In the spacious central area is the sports arena area. It forms a sports garden with the existing State Olympic Sports Center.

In the north is the flat and open convention and exhibition area, where logistic and traffic now will be organized smoothly.

The commercial service facilities along the two sides could be built in different phases.

Space: The commercial service facilities with relatively smaller volumes will be placed along the two sides while the sports and exhibition facilities with larger volume will be built in the middle. A space transition is designed from the new to current buildings so as to realize a soft landing for the large-volume buildings in the city.

Transport: Two side roads will be built parallel to the Beichen Xilu and Beichen Donglu to effectively dilute the traffic and ease the traffic pressure on the city.

Identification: Cultural facility area near the Fourth Ring-Road is sunk into the basement. From the Fourth Ring Road, one could directly see these facilities. This will increase the indenticalness of the Olympic Green in the city and make the park well echo the existent State Olympic Sports Center form vision ana space.

Development: The design will give full consideration to the progressive development and stage construction. The land plot separated by Datun Lu, Chengfu Lu and the two side roads is suitable for the construction of commercial facilities in stages.

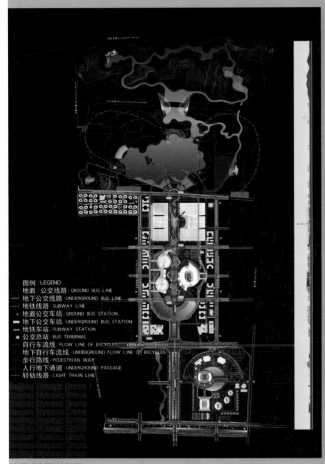

交通系统规划(赛后)
LAYOUT PLAN OF TRASPORTATION SYSTEM (POST GAMES)

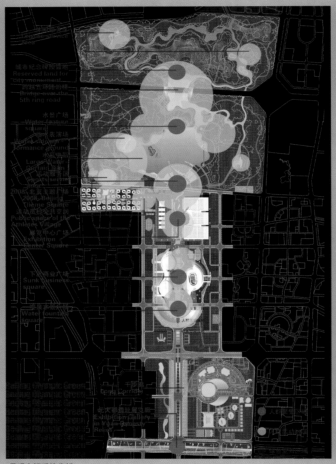

景观空间系统分析
SYSTEMATIC ANALYSIS OF VISUAL SPACES

透视效果图
PERSPECTIVE

Introduction

Brief Introduction of the Planning and Design of Beijing Olympic Green

2008—城市与自然的交融
2008—INTEGRATION OF URBAN AREAS WITH NATURE

B区鸟瞰图(赛时)
BIRDS-EYE VIEW OF AREA B (DURING GAMES)

元大都遗址——新老城区的契合点
YUAN DYNASTY RELICS JOINT BETWEEN OLD AND NEW CITY AREAS

B区鸟瞰图(赛后)
BIRDS-EYE VIEW OF AREA B (POST GAMES)

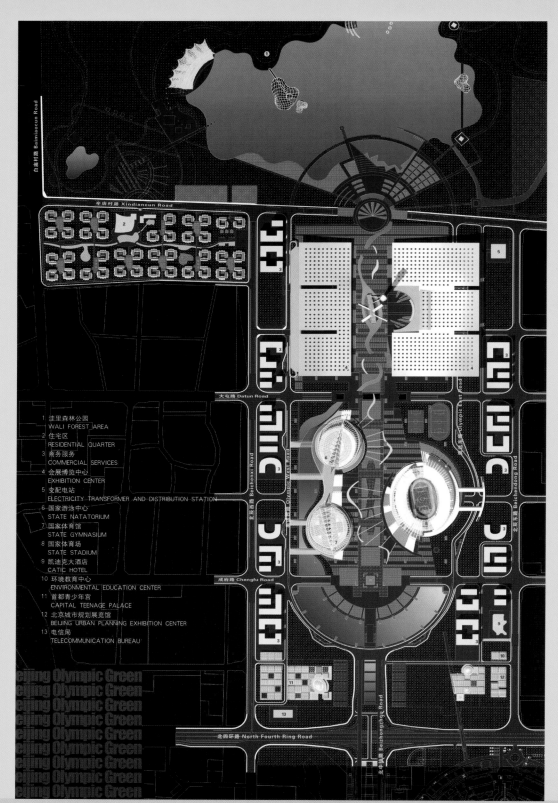

B区总体规划平面（赛后）
GENERAL PLANNING LAYOUT OF AREA B (POST GAMES)

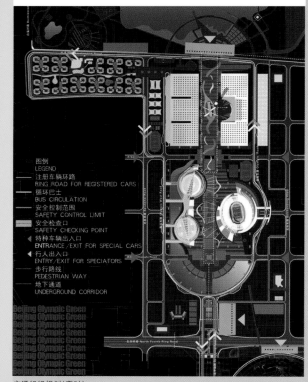

交通组织规划（赛时）
TRANSPORTATION ORGANIZATION (DURING GAMES)

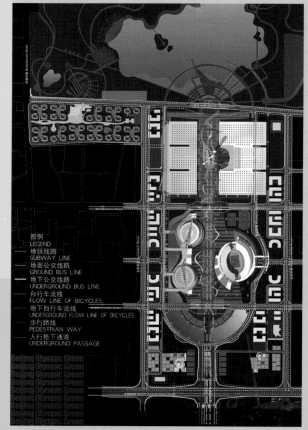

交通组织规划（赛后）
PLAN OF TRANSPORTATION ORGANIZATION (POST GAMES)

透视效果图
PERSPECTIVE

模型
MODEL

Beijing Olympic Green 2008
Brief Introduction of the Planning and Design of Beijing Olympic Green

Beijing Olympic Green 2008

优秀奖 Honorary Mention

HWP 设计有限公司（德国）
HWP (GERMANY)

HWP设计有限公司为各种建设项目提供广泛的业务服务,范围包括：从医疗卫生事业到科学教育研究；从工业企业到社会公共事业,从交通建设到娱乐休闲、住宅建设等。30多年来的历史使HWP在各种有关设计、研究、建造等方面积累了丰富的经验。目前公司由近250多个建筑师、工程师、计算机专家、医学专家、经济学家及社会管理专业人员组成。

The HWP Planungsgesellschaft mbH has been planning and advising companies, providing objective and technical expertise in a multitude of areas for over 30 years. We work as an interdisciplinary workshop, forming our team for your project from a staff of more than 250 architects, engineers, computer specialists, physicians, economists, social and administrative experts.

■规划设计构思

方案以体现"绿色奥运,科技奥运,人文奥运"为宗旨,塑造有中国民族风格,生动活泼,灵活多样的城市新空间,建设一个能为市民长期提供多功能服务的现代化的城市公共活动中心.

1.城市空间总体规划

延续北京城已有的城市中轴线,定义中轴线的新高潮与终点.

在北四环路与北中轴路立体交叉路口明确勾画奥林匹克公园主大门.

奥运中心区(B-1区)主要由三大部分组成:

南部,集中大跨度多功能建筑群,赛时为临时比赛场所和新闻中心,赛后为展览场地.

中部,集中国家体育场、国家体育馆、文化贸易中心及各种主要商业娱乐设施,分布在四块大小相同却各具特色的面积内,由四个大面积轻结构屋顶覆盖,地势高于周围地区,为城市中心步行广场.

北部,由国家游泳中心,首都青少年宫,奥运村国际区和其他商业服务设施组成.

奥运中心区东西两边分别布置绿化带,为未来奥林匹克公园与周边地区相互融合发展(如居住小区的营建等)留下了余地.北部为雨水形成的奥林匹克湖.

奥运村北部空间向森林公园敞开,并与B-1区有良好的联系.

A-2区内新建的体育设施布局延续已建成的A-1区内规划结构,景观上讲究与南面元大都遗址公园相呼应.

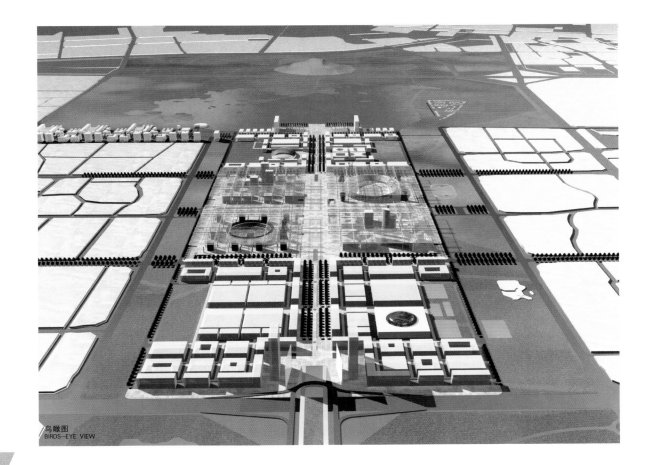
鸟瞰图
BIRDS-EYE VIEW

Introduction

2.景观规划

奥运公园的整个地形总的来说由南向北缓缓降低.

奥运中心广场聚集体育、文化、商业等中心设施,伴随喷泉、雕塑、灯光等艺术装饰,形成奥运区高潮点.

B1区与B2区的建筑空间通过中轴线来组织.

奥运城与奥运村北部直接与森林公园和奥林匹克湖相连.

中轴线向北延伸穿过奥林匹克湖.

森林公园中轴线上的奥林匹克山呈圆锥形,高80m,山上不种树木.

C2区的森林带作为北京环城绿化带的组成部分之一.

对奥林匹克地区进行彻底的改造,在五环路处建造一个面积为540ha的大公园,其中150ha水面,320ha森林用地,对周边地区有良好的气候环境影响.

收集利用雨水,并利用奥林匹克湖对雨水进行循环净化.

3.结构设计

奥林匹克公园中心区由四大轻结构屋顶突出表现,顶下覆盖了各种体育商贸办公设施及城市公共场所.每个轻结构顶为方形350m×350m,分为81个单元,每个单元为38.9m×38.9m.每个屋顶由16根高45m的钢柱支撑.钢柱内浇筑钢筋混凝土,增加钢柱承重强度.

水平方向的加劲主要由桁架大梁上下两根主梁之间的对角斜梁来完成.对角斜梁由被拉紧的金属条组成.

屋顶覆盖物为ETFE气垫式塑料薄膜顶,固定在桁架的主梁之间.气垫两面加细金属网进一步加强薄膜强度.气垫与气垫之间相互水平牵引,保持平衡.水平方向的边缘梁保证边缘处气垫的强度.

4.交通规划

行人与自行车将由南部沿北中轴线引入奥运地区,并在中心区北端的升旗广场处左右分流向北进入森林公园.

沿中轴线地下为地铁8线,分别设置四个地铁站.

公共汽车将从安立路到白庙村路横向连接地铁4线、8线和5线之间的地区.

城市主干道:安立路、北辰东路和白庙村路、北辰西路构成奥运地区的外围道路网.大屯路与成府路表现为横向交通,两路的机动车道采取隧道形式从地下穿过奥运中心区.

建议取消白庙村路与北辰东路之间的一段辛店村路,使奥运中心区与北部的森林绿地有紧密联系,随着城市发展的需要,可以考虑用隧道来连通两边地区.

中心区和B-1区的地下车库通过平行于北辰东路和北辰西路的两条区内道路来连接.区内交通,货物运输等主要由区内道路和地下车库完成.

5.市政概念规划

(1)供水规划

建立一个独立的水净化循环系统,回收利用可再生能源,做到节能节资.采用新技术、新设备、引进物流水流分离法、废物净化再利用等技术,减少废物废水的生产,减轻环境污染.同时进行雨水收集利用,生物垃圾收集处理,建立供暖供电辅助系统.

能源供应系统将主要利用可再生资源.现有的饮水系统需要得到进一步的发展和改善,以适应新系统的要求.另外把雨水变为饮用水的经济效益及采用先进的节水浴厕设施都是值得考虑的方面.

(2)供电与供热规划

1)供电

奥林匹克公园区域内的高压电供应将利用已有的220KV及110KV电网.电力分布情况也将按照地区内已规划的输电线路并结合其他能量如太阳能、风能和热能的利用来分配.通过联网系统能够充分满足能源需要.

供电网由高压电环路、中压电环路(直接满足大客户的要求)和低压电环路组成.一般性的城市用电如街道照明,外部环境照明等将利用以上提及的其他能量来满足需求.供应区将按照规划的区域划分,供电线路为地下电网形式.供电局可为单独的建筑物,也可以设在公共建筑物内。
太阳能吸收系统(Photovoltaik System)将运用于奥运村建筑群和体育场馆等的屋顶上,在森林公园,人工湖附近安排各种风车(同时作为景观标记),以获取足够的太阳能和风能。

2)供暖
已规划的奥运区供暖系统为城市近距离供暖系统。远距离供暖主要利用地区外地热资源或是利用其他的能源生产余热如燃气制造、垃圾处理和太阳能等。建筑物均配备热水供应,多余热量将由地区供热系统储存。

■ Planning and Design Concept

1. Design Concept for Urban Space Planning

Upgrading of the central main axis of Beijing and definition of its conclusion.

Unambiguous definition of the main entrance to the Olympic area.

The B-1 area is divided into three main zones

Southern zone, Halls for small area sport events later used as trade and exhibition halls as well as buildings for business service facilities.

Central zone, National Stadium and National Gymnastic Hall with the central area for cultural and business service facilities in four equally sized but differently designed areas. This zone is to be on a plateau, characterised and emphasised by four primary roof structures.

Northern zone, National Swimming Center and Capital Teenage Palace as well as the International Zone Facilities with connection to the Olympic Village, accompanied with business service facilities.

The Olympic and urban centre is outlined by a plateau and covered by four large transparent roofs.

Large green belts are integrated into the urban structure on the eastern and western sides of the B1 area. Water courses feed the planned lake in the north with rain water.

The Olympic Village is open onto the planned forest area and to the lake and is well connected with the B-1 area.

The new sport facilities in the A-2 area complement the existing ones in A-1. The landscaping underlines the direct adjoining of the Yuan Dynasty City Wall.

2. Landscape concepts

Gently sloping terrain falls from south to north, completely regraded in the Olympic area.

The central plateau with sport, cultural and commercial facilities, fountains, art and light objects as the high point of the Olympic area.

Open space structure of the Olympic city (B1 and B2 areas) organised through the wide central Olympic axis.

Olympic City and the Olympic Village border in the north directly on an extensive (without roads) large park landscape.

The central Olympic axis with its visual extension into the lake
Unwooded spherical shaped Olympic Hill, ca.80m high, as the conclusion of the central axis.

The Forest area (C2 area) as a part of the primary green structure planning of Beijing.

Removal of all urban structures and agricultural areas for a new large park with a total area of ca. 540 ha containing a 150 ha lake and 320 ha forest belt north of the Fifth Ring Road, with a relevant, positive climatic influence reaching into the neighbouring developments.

The use of rain water: Continual recirculation of the lakes water and the utilisation of the rain water as usable water.

3. Structural Concept

The roof over the central zone in B-1 Area consists of four main structures, each 350x350m, housing the main public areas. The primary load-bearing structure is a grid consisting of 81 elements 40 x 40 m, supported by 16 hinged columns approximately 45 m length, filled with concrete.

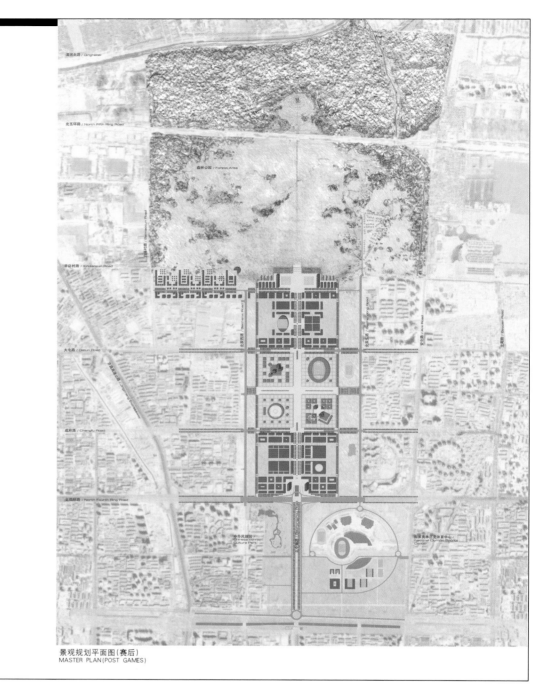

景观规划平面图(赛后)
MASTER PLAN (POST GAMES)

Horizontal stiffening is provided by diagonal bracing between each pair of columns.

The roof is covered by inflated, transparent cushions made of ETFE-foil. Additional stiffness for the inflated membranes is provided by cable-nets on both sides of cushions. Horizontal edge girders ensure sufficient frame stiffness for edge cushions.

4. Transport and traffic planning

The pedestrian and cycle traffic is directed from the traditional Central Axis over the North Central Axis onto the Beizhongzhou Road - into the Olympic centre. In the northern continuation, the Central Axis ends at the Place of Nations, the pedestrian and cycle route axis divides into two main routes which connect the green area between the lake and forest.

The subway line 8 runs under the Central Axis with subway stations.

The buses connect the areas between the subway lines 4, 8 and 5 in the Anli Road and Baimiaocun Road / Beichenxi Road and link the subway stations in the east-west axis to each other.

The Primary Roads, Anli Road and Baimiaocun Road border the Olympic Site and form together with the Beichenxi-Road and the Beichendong Road the framework for the external connections. The Datun Road and the Chengfu Road link this framework perpendicular to the Central Axis, where they both travel under the Olympic site and the Central Axis. To connect the urban area in Site B with the green area in site C1 closely with another, it is proposed, to discontinue the Xindiancun Road between Baichendong Road and Baimiaocun Road. The tunnel connection can be formed later, if required.

The basement car park entrances and exits on site B are organised over the connecting road parallel to the Beichenxi Road and the Beichendong Road. The internal area connection and delivery occurs in the road area i.e. through the basement car parks.

5. Utility conceptual planning
(1) Water supply

The objective of the planning is an extensive independent solution by the use of a self-contained water circuit and the rational utilisation of reusable sources of energy. It is expected that the proposed solutions will considerably reduce the cost of maintenance.

Within the framework of the development project there is the

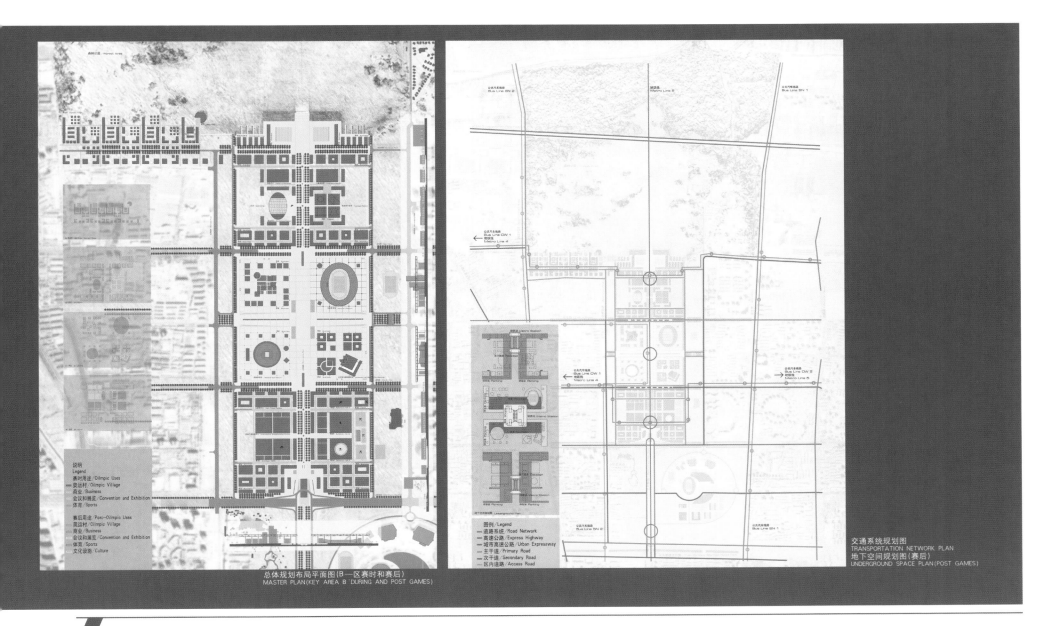

总体规划布局平面图(B—区赛时和赛后)
MASTER PLAN(KEY AREA B DURING AND POST GAMES)

交通系统规划图
地下空间规划图(赛后)
TRANSPORTATION NETWORK PLAN
UNDERGROUND SPACE PLAN(POST GAMES)

Introduction

possibility to use new technologies and concepts to reduce the effluent yield, to separate material and water streams, as well as to purify and reuse the sewage. At the same time, the potential of rainwater harvesting could be optimally utilised. The planning of the decentral energy and heating supply systems could include the collection and waste management of organic waste.

The energy supply system should therefore be reconverted to make use of renewable sources of energy. The objective should be the redevelopment or re-commissioning of a drinking water supply system suitable for this project. The economic advantage of using rainwater as a substitute for drinking water as well as water conserving sanitary engineering should also be taken into consideration.

(2) Power and Heat Supply
1) Principle Power Distribution

For the energy supply of the area, the high voltage supply is planned from the existing 220kV-, i.e. the 110kV grid.

The order of the energy distribution occurs by means of the performance requirements in analogue with the planned routes in the area as well as in combination with following alternative energy supply techniques:

Solar Energy
Wind Energy
Combined Heat and Power System

The basic energy requirement guaranteed through the appropriate integrating of the grid is put surely. The grid structure plans a high voltage ring as well as middle voltage rings with direct supply of high power consumers and low voltage grids.

The peak loads and defined basic loads (street lighting, general external lighting), are covered by the above mentioned sources of energy.

The power supply units are conformed with the external appearance within the area, the supply routes run in underground media channels. Power stations are proposed as independent or integrated buildings on the Olympic Site

Photo voltaic systems are integrated in the Olympic Village and the roof surfaces of the Sport Halls as well as wind turbines as park ancillary structures planted on the lake plateau landscape.

2) Primary heat supply

The thermal energy supply is defined in the infra-structure plan as a local heat concept.

On the one hand the supply to the district heating mains is carried out through the geothermal heat extraction of processes outside of the Olympic site and on the other hand through the use of the waste heat from the following processes:

Combined Heat and Power Plants
Waste Heat
Solar Heat

Warm water will be used both within the building network, or in the case of a surplus energy, injected into the district heating supply system.

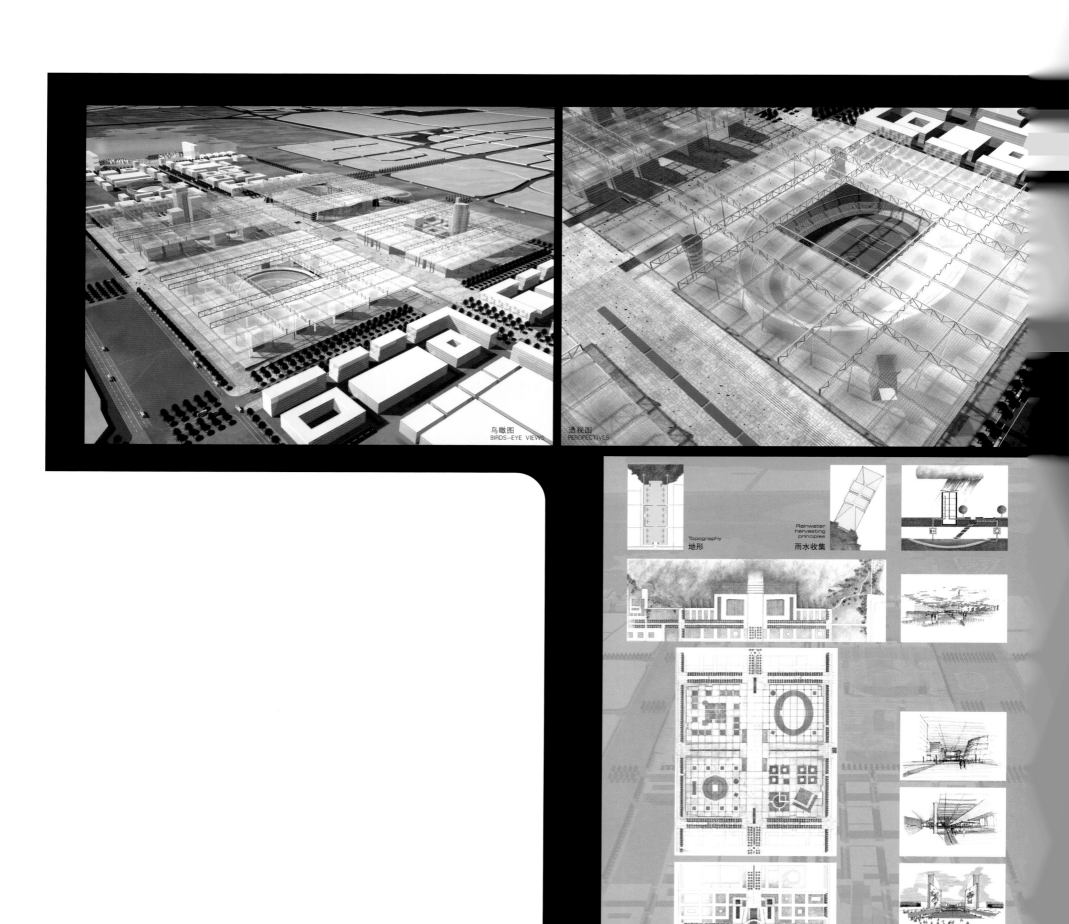

Olympic

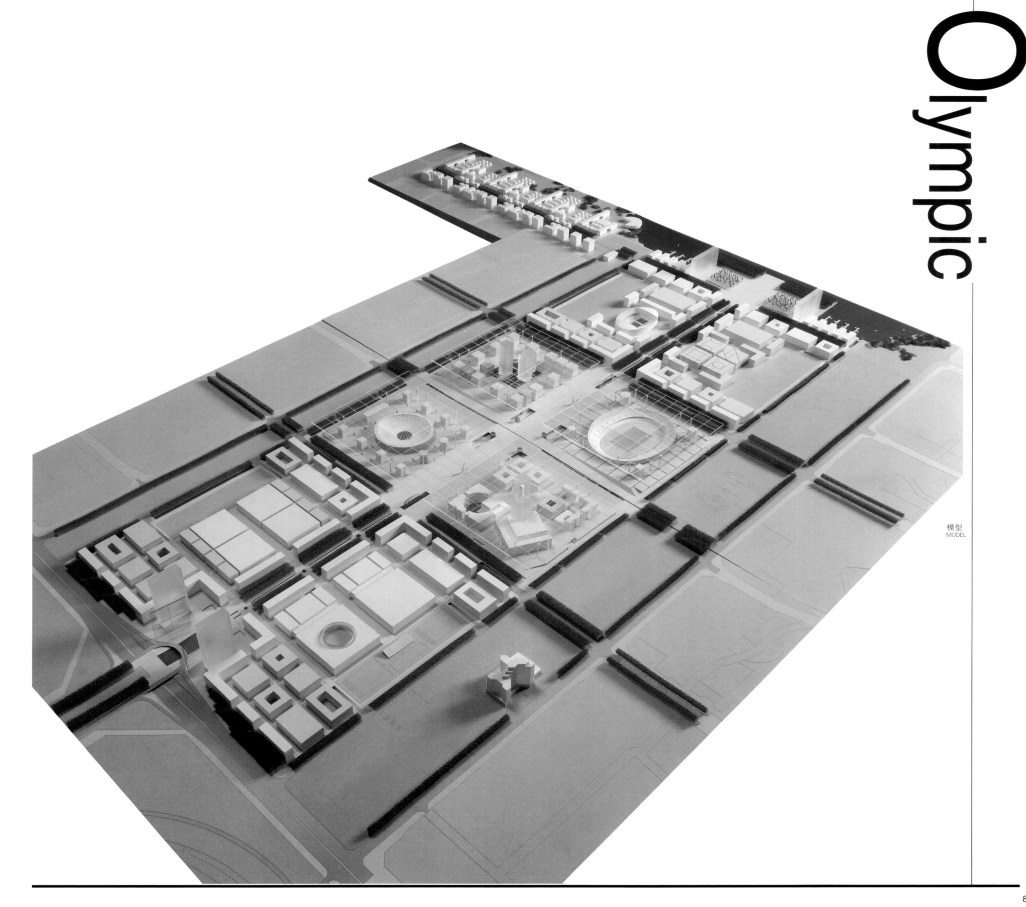

模型
MODEL

Beijing Olympic Green 2008

优秀奖 Honorary Mention

北京大学城市规划设计中心（中国）
THE CENTER FOR URBAN PLANNING AND DESIGN(CUPD), PEKING UNIVERSITY(CHINA)

北京大学景观规划设计中心（中国）
THE CENTER FOR LANDSCAPE ARCHITECTURE AND PLANNING(CLAP), PEKING UNIVERSITY(CHINA)

北京大学城市规划设计中心是集教学、科研、规划、设计为一体的甲级城市规划设计机构。依托北京大学综合性大学的学科优势，在传统的规划理论、方法的基础上，加入了生态学、环境学、经济学、法律学、社会学、房地产学等相关学科的内容，从而为城市和风景区的规划、建设、管理带来了较好的实用性和可操作性。

The Center for Urban Planning and Design (CUPD) of Peking University is a first ranking institution that focuses on teaching, research, planning and design. Depending on the subject advantages of Peking University are wide range of knowledge form the following fields: ecology, environmental science, economic, law, sociology, real estate and other relative subjects added to the traditional theories and methodologies on city planning. So it can make the planning, design and management more practicable and exercisable.

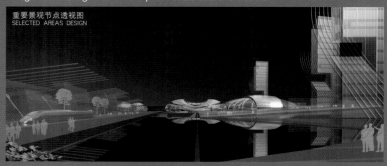
重要景观节点透视图
SELECTED AREAS DESIGN

北京大学景观规划设计中心（简称CLAP）成立于1997年3月，旨在针对中国城市化过程中出现的一系列土地开发与资源利用问题，重点研究景观综合体的空间格局及其景观过程，探讨如何通过对城市及区域环境的设计，协调人类活动与自然过程、现代化与传统文化的关系。

The Center For Landscape Architecture and Planning (CLAP) of Peking University, was founded in March, 1997. Dealing with series of problems associated with development and resource in the process of urbanization in China, the center emphasizes on the processes and patterns of landscape, tries to harmonize man and nature, modernization and tradition through urban and landscape design and regional planning.

■ 规划设计构思

本规划中，中央的步行区域和必要的交通设施是关键。人流从地铁中出来好像穿越荷塘水面，精心设计的南主入口将设计成最大的一片荷塘，并依此形成基本的几何构图，一直延伸到各个主要的体育场馆。

对荷花的抽象被应用到各个层面和阶段的设计中。人们能够从奥林匹克公园的基本布局（城市设计）中发现其基本形式。在南侧，从元大都北护城河开始引进荷花到现有水面。沿着从南向北的垂直轴线进入场地，各个规划项目布置在轴线的两侧。东侧的平台由荷叶抽象而来，从中轴线开始缓缓抬升到场地的东缘，为盛开的荷花提供欢趣的基础。荷叶之上盛开的荷花即三个主体育场馆（国家体育场、国家游泳中心和国家体育馆）和赛后位于场地西北的北京城市规划展览馆。中轴线从故宫向北延伸（进入森林公园的南部），然而主要体育场馆的自东南向西北延伸的弧线形布局使其重新回到城市中。奥运会后，设计外形呈螺旋状的北京城市规划展览馆将使这一回转得到强化，连续的和无限延伸的螺线运动被安置在一个玻璃的容器中，通过这种解决方案，奥林匹克公园的中央区域成为中轴线北端的高潮。另外，从更大的尺度上，由北京西山和北山形成的山体轮廓线，与从南二环延伸到奥林匹克公园转而向西的城市空间轴线呈合抱之势，城市空间轴线和地理轮廓完美协调。

景观设计结合现状景观及其特性：林地在北部，重新设计河溪、林地斑块和树丛、荷塘，并围绕荷花的主题。环境的生态状况被保持和改善。广义奥运区域的土地利用包括林地、草地、湿地（荷塘、主要河溪两岸和清河南侧的低洼地）、耕地，以及奥林匹克中心区域规划的开放空间。

■ Planning and Design Concept

Olympic Green Area is designed like a lotus pond in general and plant in particular. Interaction of spectators, sportsmen and other participants can be compared with the rich life under the lotus leaves,

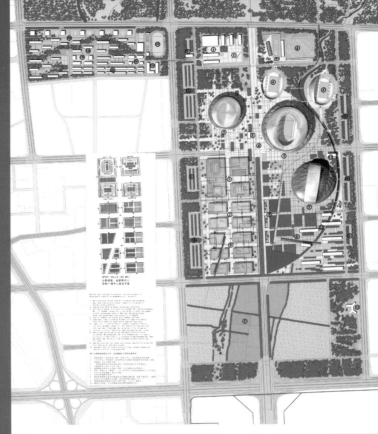

全区总平面图（赛时）
OVERALL AREA MASTER PLAN (DURING GAMES)

1 国家体育场 NATIONAL STADIUM
2 国家游泳中心 NATIONAL SWIMMING CENTER
3 国家体育馆 NATIONAL GYMNASIUM
4 训练场 TRAINNING FIELDS
5 比赛场馆 SPORT HALLS
6 射箭馆 ARCHERY RANGE
7 国际广播中心 IBC
8 主新闻中心 MPC
9 国际区 OLYMPIC VILLAGE
10 居住部分 RESIDENTIAL AREA
11 商业娱乐中心 ENTERTAINMENT COMMERCIAL CENTER
12 国际奥委会社区服务、办公 IOC, SERVICE, OFFICES
13 安检及等候区 SECURITY AND WAITING AREA
14 商业服务区 COMMERCIAL AND BUSINESS BUILDINGS
15 凯迪克酒店及花园 CATIC HOTEL AND GARDEN
16 新酒店 NEW HOTEL
17 荷花塘广场 LOTUS POND PLAZA
18 主体育场广场 NATIONAL STADIUM PLAZA
19 升旗广场 FLAG RAISING SQUARE
20 地铁南站 SOUTH SUBWAY STATION
21 地铁北站 NORTH SUBWAY STATION
22 现状绿地 EXISTING FIELDS AND GRASSLANDS
23 停车场 PARKING SPACE
24 轻轨电车 LIGHT RAIL

Introduction

重要景观节点透视图
SELECTED AREAS DESIGN

in the water and above it. Central pedestrian area with needed traffic infrastructure is crucial. People come out of the subway system through the lotus ponds (in exact same location as existing pond) and through South Gate where the largest lotus pond is designed as an entering geometric motif and spread around to get to the main sport objects.

An abstraction of lotus plant is used in different layers and stages of design. One can find its structure in a general layout of Olympics Green (urban design). In South, the program starts as lotus in existing water of Hu Cheng He river in Yuan Dynasty City Wall Relics Park. It follows then a vertical axis toward North along which required program zones are placed on both sides. Main eastern platform is an abstraction of green lotus leaf. It raises from the central axis toward eastern edge and provides an interesting base for blossoms: three main sport objects (National Stadium, National Swimming Center and National Gymnasium) and The Beijing Exhibition Hall after the games north-west of it (as a lotus fruit). Main objects are placed at the end of the Central Axis that starts in the Forbidden City, continues to the North (into the southern part of Forest Area), but locations of main sport objects turn it back to the city. Turn of

景观规划平面图(赛后)
MASTER LAYOUT AND LANDSCAPE PLAN (POST GAMES)

1 环境教育中心
   ENIRONMENT EDUCATION CENTER
2 休闲公园
   LEISURE PARK
3 北京城市规划展览馆
   BEIJING CITY PLANNING EXHIBITION HALL
4 国家体育场
   NATIONAL STADIUM
5 国家体育馆
   NATIONAL GYMNASIUM
6 国家游泳中心
   NATIONAL SWIMMING CENTER
7 热身训练场
   WAITING AND WARM-UP SPORTS FIELDS
8 主展览区
   MAIN EXHIBITION AREA
9 主体育场广场
   NATIONAL STADIUM PLAZA
10 荷塘广场
   LOTUS POND PLAZA
11 商业办公区
   COMMERCIAL AND BUSINESS
12 首都青少宫
   CAPITAL TEENAGE PALACE
13 休闲公园
   ENTERTAINMENT PARK
14 曲棍球场
   GOCKEY HALL
15 主网球场
   MAIN TENNIS HALL
16 训练场
   TRAINNING FIELDS
17 中华民俗园
   CHINESE ETHNIC CULTURE PARK

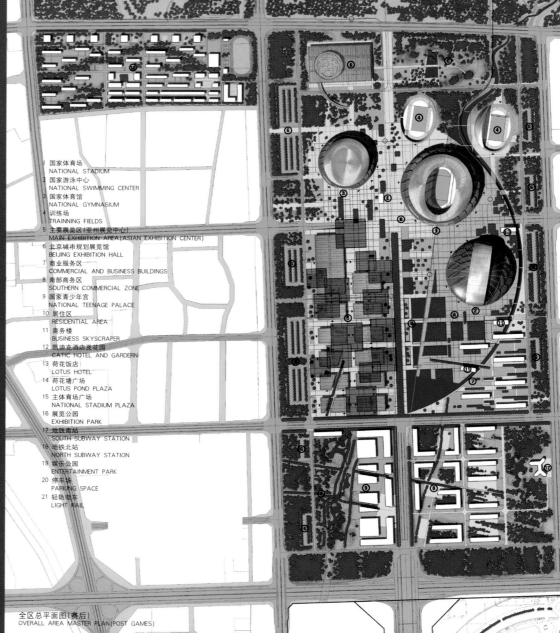

1 国家体育场
   NATIONAL STADIUM
2 国家游泳中心
   NATIONAL SWIMMING CENTER
3 国家体育馆
   NATIONAL GYMNASIUM
4 训练场
   TRAINNING FIELDS
5 主要展览区(亚州展览中心)
   MAIN EXHIBITION AREA (ASIAN EXHIBITION CENTER)
6 北京城市规划展览馆
   BEIJING EXHIBITION HALL
7 商业服务区
   COMMERCIAL AND BUSINESS BUILDINGS
8 南部商务区
   SOUTHERN COMMERCIAL ZONE
9 国家青少年宫
   NATIONAL TEENAGE PALACE
10 居住区
   RESIDENTIAL AREA
11 商务楼
   BUSINESS SKYSCRAPER
12 凯迪克酒店及花园
   CATIC HOTEL AND GARDERN
13 荷花饭店
   LOTUS HOTEL
14 荷花塘广场
   LOTUS POND PLAZA
15 主体育场广场
   NATIONAL STADIUM PLAZA
16 展览公园
   EXHIBITION PARK
17 地铁南站
   SOUTH SUBWAY STATION
18 地铁北站
   NORTH SUBWAY STATION
19 娱乐公园
   ENTERTAINMENT PARK
20 停车场
   PARKING SPACE
21 轻轨电车
   LIGHT RAIL

全区总平面图(赛后)
OVERALL AREA MASTER PLAN (POST GAMES)

Olympic

the axis is after the games also supported by Beijing Exhibition Hall that is designed in a spiral shape as continuous and endless movement of spiral captured in glass cube-a lotus fruit in water. By this solution, the Central Olympic Area presents a climax point of the axis in North. At last but not least, the turn of the main axis toward West is enclosing the city in a similar way as western and northern mountains are embracing the city of Beijing in larger scale.

Landscape design follows existing landscape and its characteristics: forest area in North, redesigned streams, tree patches and clumps, lotus ponds and rounds the lotus story. Ecological condition of the space is maintained and improved. Land-use in wider Olympic area includes forestry, grassland, wetlands (lotus ponds, marshes along the main stream and beside the Qinghe River), arable fields, and designed green open spaces in the central Olympic area.

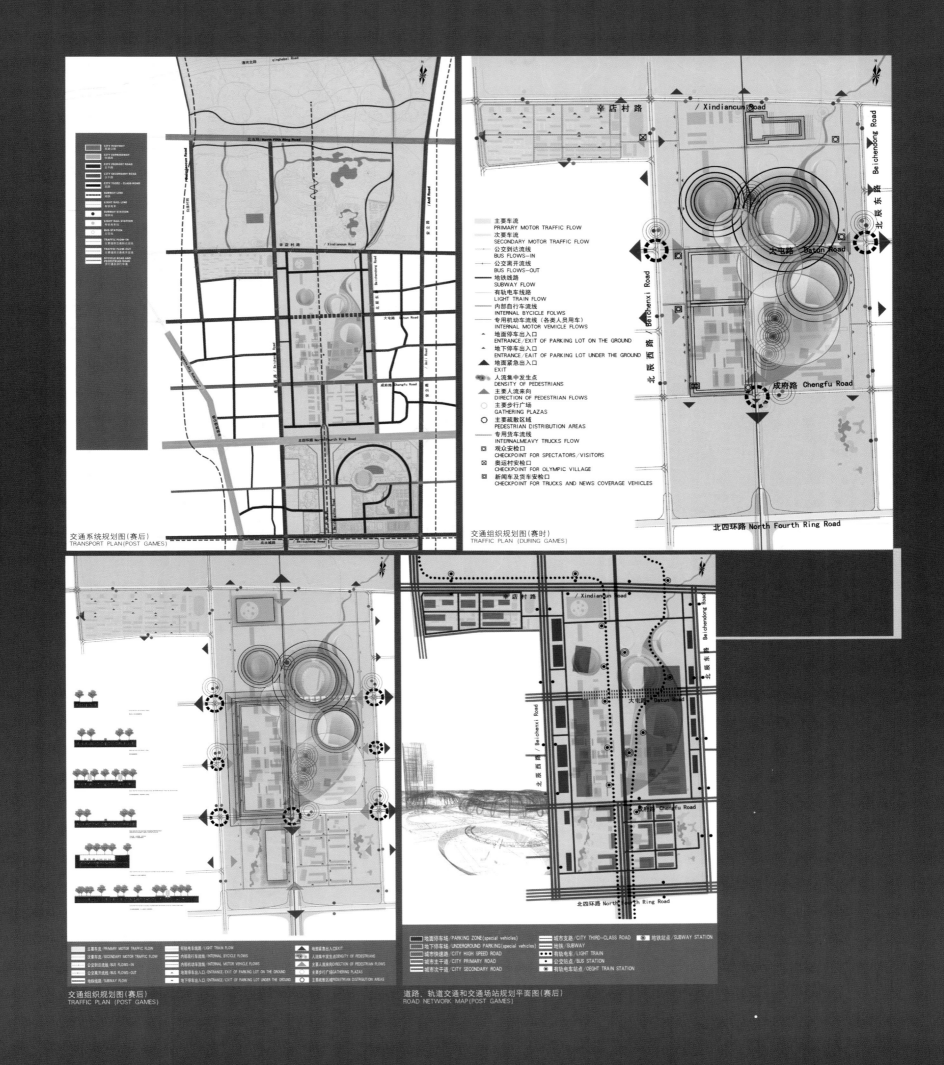

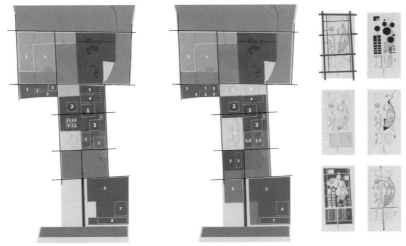

功能布局/分区图(赛时和赛后)
FUNCTIONAL LAYOUT/ZONNING MAP (DURING AND POST GAMES)

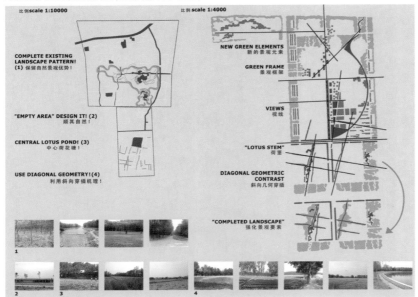

景观规划分析图
LANDSCAPE PLANNING ANALYSIS

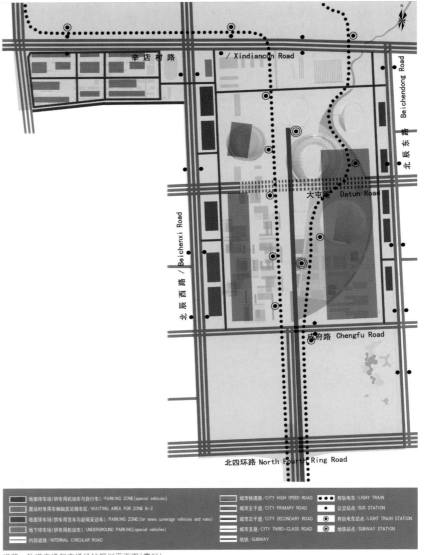

道路、轨道交通和交通场站规划平面图(赛时)
ROAD NETWORK MAP (DURING GAMES)

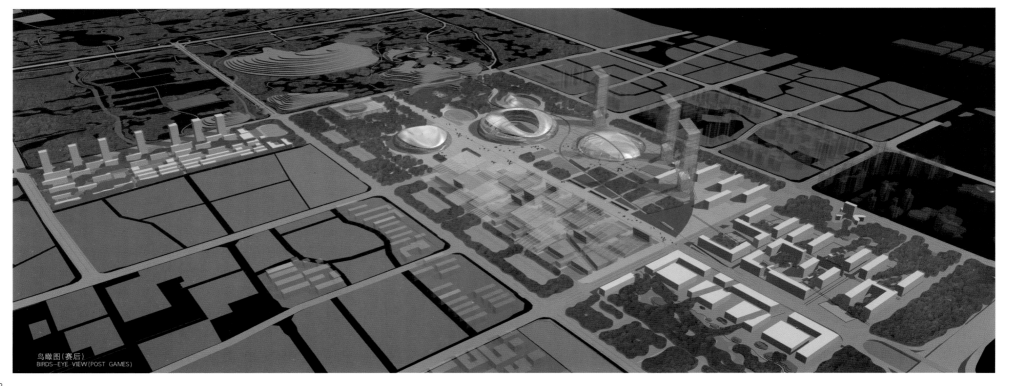

鸟瞰图(赛后)
BIRDS-EYE VIEW (POST GAMES)

Model pictures

模型
MODEL

北京奥林匹克公园设计方案 Beijing Olympic Green

参赛作品 Works

■ 规划设计构思

龙之谷——奥林匹克公园规划方案

本方案的九大特点：

尊重场地精神：阅读"洼里——荷田——龙王"的故事，遵从斜向肌理，保留林带与水系。

"龙之谷"意象：奥运内涵、民族之魂与场地精神的和谐。

区域景观生态战略：结合现有绿地系统和区域水系治理、雨水收集系统、南水北调工程，重建地表水系，建立生态基础设施，成为人行与自行车绿色通道和康体休闲走廊，文化遗产廊道和多样生物的栖息地。

从红墙禁宫走入森林童话：中轴线，一朵绽开的百合花，一条时代进步的轨迹。过成府路以北，中轴的交通功能消失，分解为一曲一斜的"V"字形内部道路，形成中部的龙之谷。

银龙盘轴，飞龙在天：三个主赛场作为一个整体，与中轴呈巨龙缠柱之势，取大量场馆建设之土方，堆就北端龙山。国家体育场南北端可开可合，延续中轴线。国家体育场、99m高的龙山及山上水火交融的"飞龙"，为这条古老的轴线画了一个感叹号。

潜龙在渊，莲池栈桥：核心区中部两叶建筑负阴抱阳，形成人性场所，营造生态景观。

五彩铺装导引系统：伸展奥运五环，导引五洲来客。

"三明治"式层圈结构与内外广场系统：明确的广场集散与安检系统。

森林公园：留下村庄的故事，重建湿地的精灵。

■ Planning and Design Concept

The Dragon Valley An Proposal for The Olympic Green

The Unique Features of This Proposal:

Respect for the genius loci: Read the story of "Marsh, lotus fields and dragon king" on the site; Follow the diagonal pattern and preserve the existing windbreaks and stream corridors.

The Dragon Valley Concept: Dragon is the right expression of Olympic spirits the soul of the Chinese culture and the genius loci.

The Regional landscape ecological solutions: An ecological infrastructure is designed by integrating the national water supply project that diverts water from the Yongtze River, and the local storm water collecting system to rebuild the greenways and blue ways.

From the Forbidden City to the Forest of Fairy Tales: After passing the Fourth Ring Rd., the volume of the floating traffic was largely diverted, and the transportation corridor is split into two local streets, making a big "V". The Dragon Valley was formed in between the enclosing arms of buildings, where lotus is in blossom. The Central Axis is treated as a blossom lily, a metaphor of progress, and an exclamation mark.

The Silver Dragon, the Flying Dragon and the Movable Stadium: The three main sports facilities are connected like the Silver Dragon embracing the Axis. At the north, the Dragon Hill, was built from the earth bulk during the construction. The stadium is composed of movable stands at north and south sides, so that they can be open or closed as a way of extending the Central Axis. The National Stadium, together with the 99 meters Dragon Hill and the Flying Dragon (made of fire on cascade), ends up the Axis in the "Flying Dragon".

Wetland, ponds, bridges and lotus flowers: Ecological landscape and people's places are created in the middle between the arms of the main sports facilities.

Beijing Olympic Green 2008

The orientation rainbow: Paths in Olympic colors extend along the Central Axis into the Dragon Valley in the north, to orient visitors and make the Dragon Valley lively.

The sandwich concept for inner and outer zones of open spaces: A clear open space and security checking system.

The Forest Park: With stories remain from the former villages and wetland rebuilt.

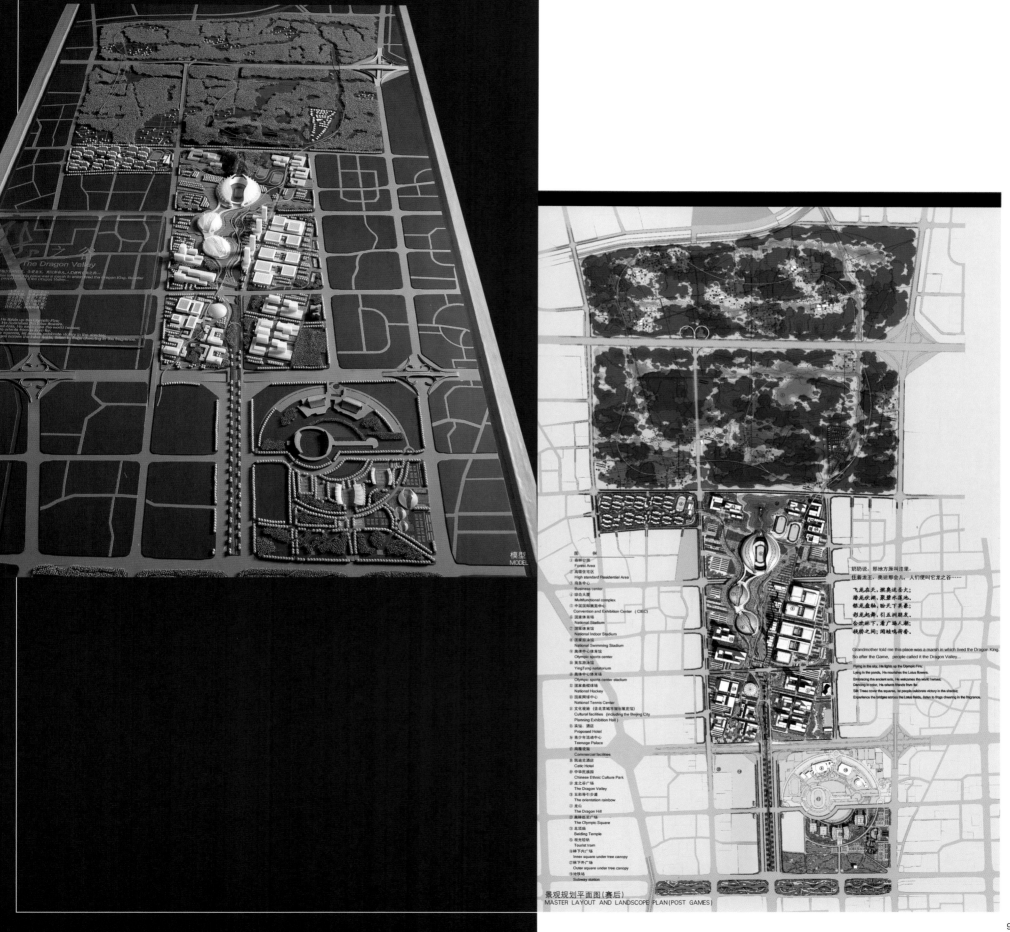

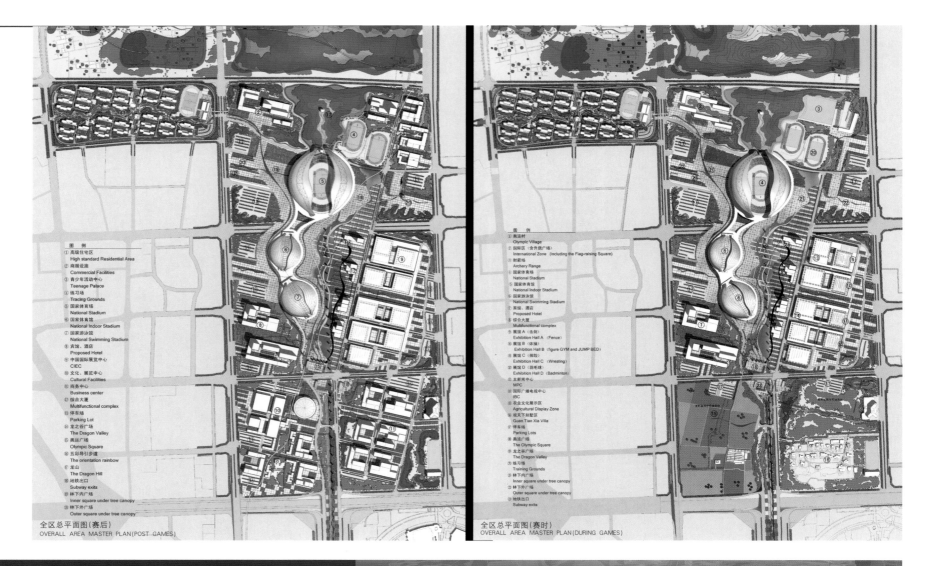

全区总平面图(赛后)
OVERALL AREA MASTER PLAN (POST GAMES)

全区总平面图(赛时)
OVERALL AREA MASTER PLAN (DURING GAMES)

鸟瞰图(赛后)
BIRDS-EYE VIEW (POST GAMES)

2008

北京奥林匹克公园设计方案 Beijing Olympic Green

参赛作品 Works

■规划设计构思

城市规划、建筑与生态环境相结合使北京21世纪在中轴线上有一个高潮终结。

在北京城中心的紫禁城基本定型于明清时代，这也是北京城市规划最基本、传统的布局。北京奥林匹克公园规划是作为21世纪城市发展的进一步回答。

奥林匹克公园作为城市中轴线的高潮和终结渐渐向自然环境过渡，其规模、含义和重要性具有和紫禁城相同的意义。在设计的风格上保留、延续了布局严谨、整齐对称、平缓开阔的城市空间形态和生动活泼的园林水系。在水系设计方面采用了与紫禁城相反的手法，即：从原来的护城河围绕长方形的紫禁城变换成建筑围绕长方形的奥林匹克湖。在城规设计这一点上，奥林匹克公园将作为21世纪北京城市建设的标志，其强化了中轴线在北京城市规划建设中的地位，即保护和延续历史。

从城市的南北轴线的引申形成了长方形的栈桥（也同时是奥林匹克广场），所有奥运会的体育建筑物都围绕这一广场。北侧生态小岛自然嵌入长方形的奥林匹克湖中，岛上的建筑与自然融成一体，奥林匹克湖的两侧设计具有很高居住质量的高层建筑，在北侧绿色植物呈森林结构向水面方向渐渐呈现草地，水泽地转化。通过先进水系管理系统为奥林匹克湖提供了充足、清洁的水源，另外使地下水位得以复苏并且还能防洪排涝。

2008年的北京奥运会提供了一个很好的机会，向世界展现北京既作为传统的历史古城又反映北京21世纪新的时代精神和发展建设的新成就。

■ Planning and Design Concept

Synthesis of Landscape and Architecture: Climax and Ending of Beijing's Central Axis in the 21st Century

The historical ground plan of the Forbidden City located in the center of the central axis represents a symbol of the imperial period (Ming and Qing dynasty) and is therefore the most prominent and central element in the ground plan of the city of Beijing. The Olympic Green as a symbol for the urban development of Beijing in the 21st century will also mark a strong sign in the city's ground plan.

欧博迈亚工程设计咨询公司（德国）
OBERMEYER PLANEN(GERMANY)

园林建筑师施密特教授（德国）
BERATEN RAINER SCHMIDT LANDSCHAFTSAR CHITEKTEN(GERMANY)

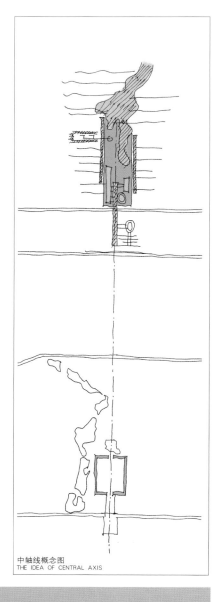

中轴线概念图
THE IDEA OF CENTRAL AXIS

Beijing Olympic Green 2008

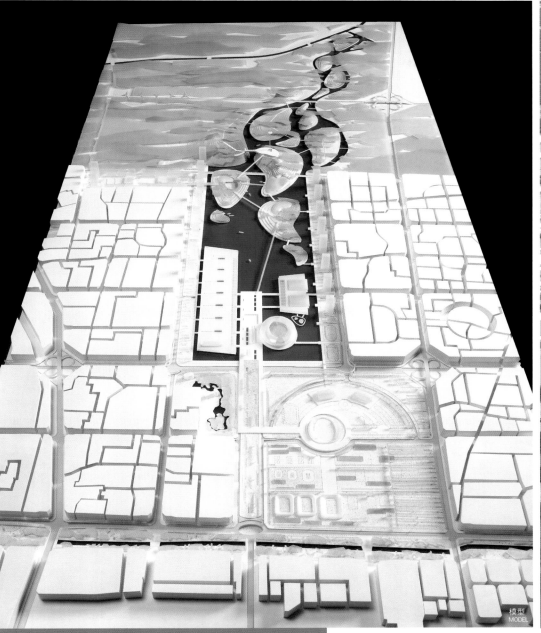
模型
MODEL

从南侧远眺的鸟瞰图
PERSPECTIVE VIEW FROM SOUTH

As the northern climax and final point of the central axis merging into the landscape, it marks a sign which can be compared in its dimension and succinctness with the Forbidden City. The design uses its historical, rectangular ground plan and the element of water in a different way opposite to the rectangular plan area surrounded by a water channel. it uses the form of a rectangular water basin.

From the south a rectangular pier (the Olympic Plaza) extending the urban axis leads out into basin. All Olympic buildings are concentrated here docking on this pier like large geometrically shaped ships. From the north organically shaped islands are moving from the curved water body into the rectangular Olympic basin. The buildings on these islands integrate perfectly with the landscape. The basin is bordered by two rows of high-rise-buildings. The landscape in the northern part is linked to the

Green Belt and dominated by a forest area. Its density decreases from the edges towards the water area and the landscape turns more and more from a woodland into an open landscape of swamps and meadows on the water edges. Despite scarce resources a cleverly thought out water system provides clean water and additionally makes a contribution to renewing the groundwater and perfecting the region against flood.

The design for the Olympic Green thus sets a new prominent sign for Beijing with respect to its urban development in the 21st century. It emphasizes the importance of the modern, northern central axis which leads on the traditional central axis and merges into the adjacent landscape. The Olympic Games of 2008 will offer the possibility to present both to the public, the historical and contemporary achivements of Beijing with respect to its urban development.

Beijing Olympic Green 2008

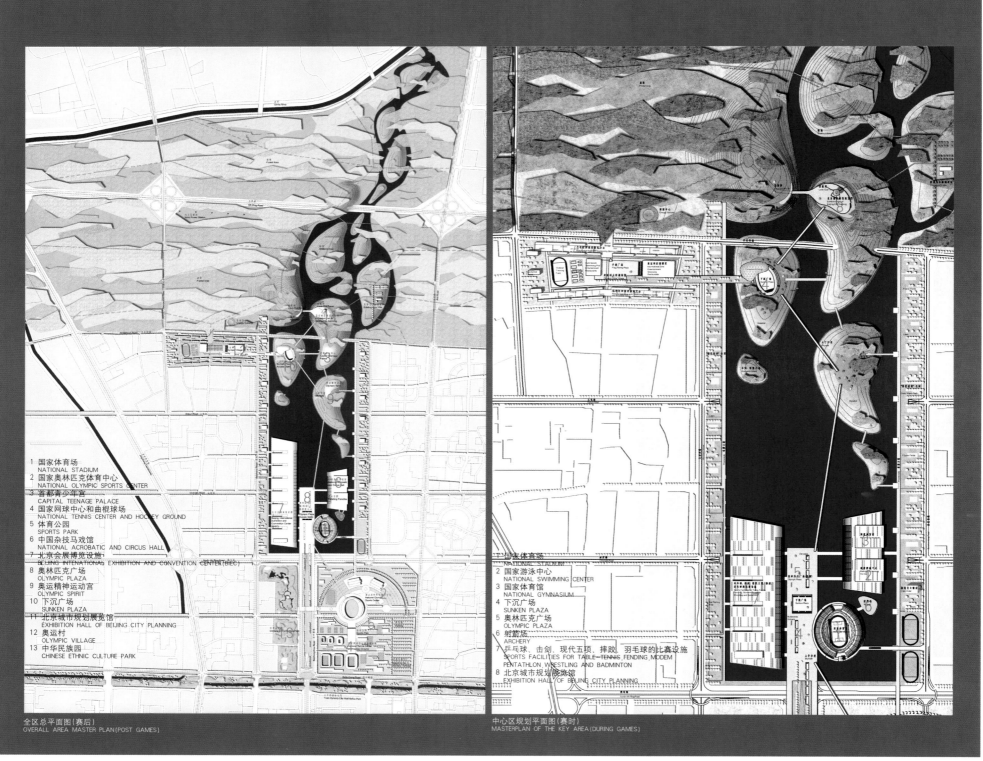

全区总平面图(赛后)
OVERALL AREA MASTER PLAN (POST GAMES)

中心区规划平面图(赛时)
MASTERPLAN OF THE KEY AREA (DURING GAMES)

构思的轴测分析图
ISOMETRIC DRAWINGS OF DESIGN IDEA

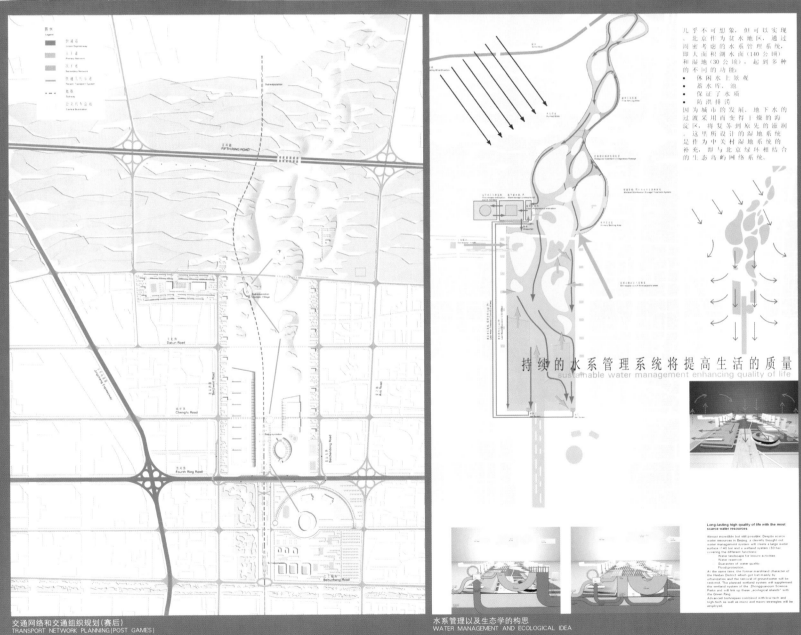

交通网络和交通组织规划(赛后)
TRANSPORT NETWORK PLANNING (POST GAMES)

水系管理以及生态学的构思
WATER MANAGEMENT AND ECOLOGICAL IDEA

持续的水系管理系统将提高生活的质量
sustainable water management enhancing quality of life

■ 规划设计构思

北京在21世纪的国际大都市中必将成为世界级的文化中心城市。北中轴线应建成城市的文化轴,强化北京城市文化的特色和国家文化中心的地位,增强中国文化对世界的影响力,体现世界文化中心城市应有的风采。

规划方案将B区和C区的部分功能加以置换,结合森林公园,在东西两侧分别安排会展中心区和奥林匹克场馆区,以保持中轴线的开敞性,充分利用五环路以优化赛场区的交通条件。

B区在奥运会期间将整体保持为开敞的城市公共绿地和城市广场,为奥运会期间举办大型集会活动和大规模人流疏散、临时停车留有充足的空间,以大面积的绿地空间突出"绿色奥运"的主题。赛时将安排大量临时性文化设施,用以展示和传播中国灿烂丰富的文化,组织各种文化活动,突出"人文奥运"的主题。赛后将对B区进行重新规划和整体开发,形成北京市的文化商务中心区。

洼里森林公园在赛后将依托奥林匹克设施建成集体育、文化、科学、旅游为一体的系列化主题公园。

■ Planning and Design Concept

Overall Scheme for Layout

Beijing aims to be one of the most important cultural central cities among all the International metropolises in the 21 century. The north Central Axis of Beijing should be the city's cultural axis to consolidate the characteristic of Beijing's urban culture and act as a national cultural center as well as infuence power of Chinese culture upon the whole world.

Permuting some functions of District B and District C.Based upon the forest park,exhibition section and Olympic fields and gymnasiums shall be placed on east and west sides respectively.Thus maintain the openness of the Central Axis and make full use of the Fifth Ring Road to optimize the traffic condition along the game area.

District B,as a whole,shall remain to be an open public activities, people evacuation and temporary parking during the Games.There will be a great number of temporary cultural canopies in District B for provincial and regional presentations to reveal and diffuse splendid Chinese culture,to organize diverse cultural exhibition meetings. This aims at highlighting the theme of "Cultural Olympic".This district will become a key quarter of post-game redevelopment to be Beijing's cultural and business section.

After the game,Wali Forest Park will be turned into an important cultural tourist park of Beijing city depending on the facilities of the Olympic Games .Among them,the Olympic fields and gymnasiums will be preserved and combined with the forest park to form an Olympic sports park with a theme of recreation,sports and well-being.

Works

北京清华城市规划设计研究院(中国)
URBAN PLANNING AND DESIGN INSTITUTE
TSINGHUA UNIVERSITY(CHINA)

Beijing Olympic Green 2008

鸟瞰图 BIRDS-EYE VIEW

模型 MODEL

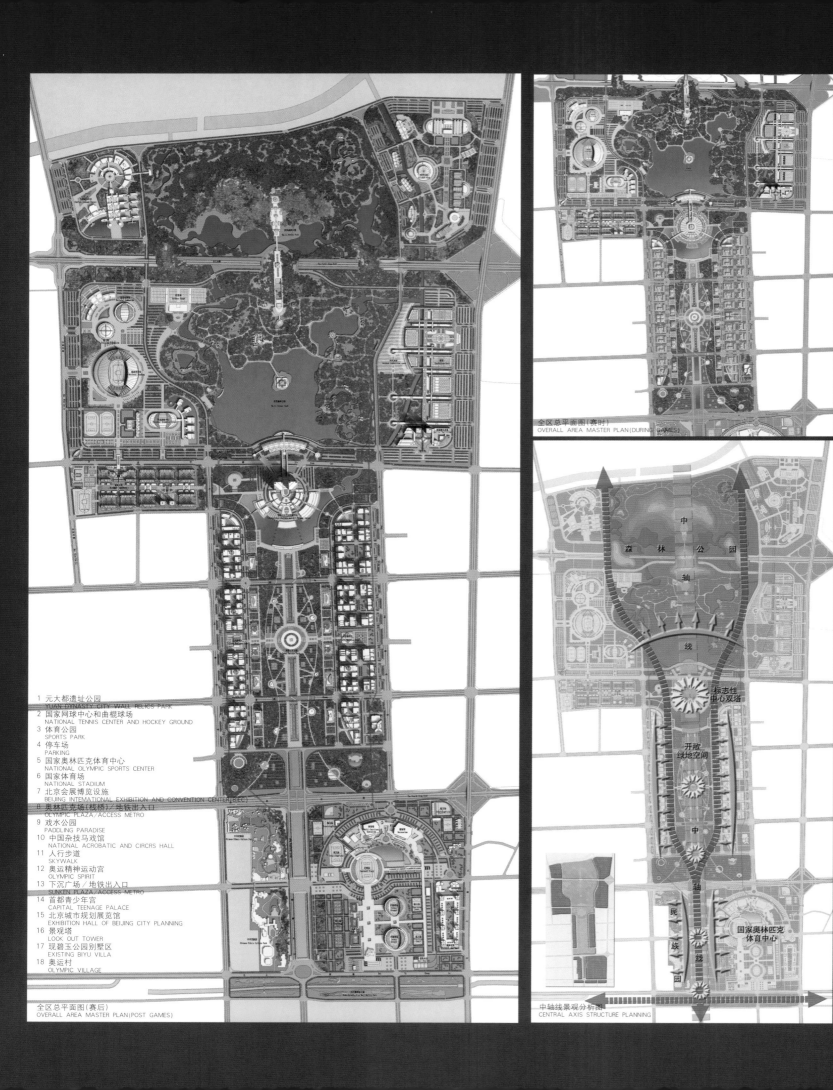

■ 规划设计构思

整个方案是建立在一个理念之上的。这个理念就是与历史上的皇城形式形成对比,在北京的南北中轴线的北面,用一系列的公园、运动和娱乐设施来完美结束这个轴线。

奥林匹克公园将有一系列大大小小的人们相聚的场所,而且在比赛期间将为人民提供愉快的娱乐空间,在赛后也将变成繁华的居住社区和旅游场所。生活和居住在这个新城市里的人们将享受高水平的城市生活,同时还享有国际著名城市的丰富性和多样化。

设计目标是建设一个奥林匹克公园,充分利用被动的和主动的环境技术,并且最大限度地使用最新的数码通信技术和知识工业。

设计将解决以下问题:

延长和光大北京的南北中轴线
强化北四环路和高科技区
将奥林匹克公园与周围环境相结合
建立永久性的国际联系
建立一个吸引人的绿化和园林网络
建设国际标准的居住区以完善奥运小区的设施
建设具有国际标准的水上娱乐和活动中心
引入生态的,能源的,水上的,绿色的,交通的概念来展示新的、绿色的技术
在奥运设施的建设上使用富有想像力的投资战略

■ Planning and Design Concept

Our proposal is built on the concept of a new public realm including a combination of formal and natural parks and significant sports and recreational uses.It's the perfect way to terminate Beiiing's Noth-South axis. in contrast to the imperial formality of the historic city.

The Olympic Green will have a network of large and small meeting places that will provide delightful recreation spaces during the Games, and will become a thriving community and tourist sites in the years to follow. The people living and working in this new city will enjoy the highest standards of urban life with the diversity and sophisticatian of internationally celebrated cities.

We have set out to build an Olympic Green that demonstrates the best use of passive and active environmental technology and is also optimised for the use of the latest digital communications and knowledge based industries.

Our concept proposes to integrate a new exciting lifestyle habitat mode which will:

Extend and celebrate Beijing North South axis
Consolidate North Fouth Ring Road into a Hi Technology Spine
Integrate the proposed Olympic Green with the surrounding areas
Establish lasting international links
Establish an extensive network of attractive parks and gardens
Develop an advanced residential model of international appeal to compliment Olympic precinct
Establish internationally competitive water based recreation and entertainment centres.
Introduce ecological energy water parks, transport concepts to showcase new green technologies
Support development of Olympic facilities through an innovative investment strategy.

Beijing Olympic Green 2008

北京市市政工程设计研究总院(中国)
BEIJING GENERAL MUNICIPAL ENGINEERING & RESEARCH INSTITUTE(CHINA)

UKY 公司(澳大利亚)
URBIS KEYS YOUNG PTY LTD(AUSTRALIA)

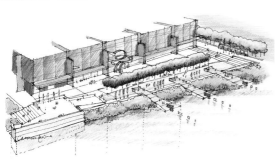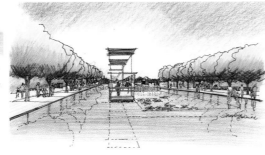

风景园林景观
LANDSCAPE VIEW

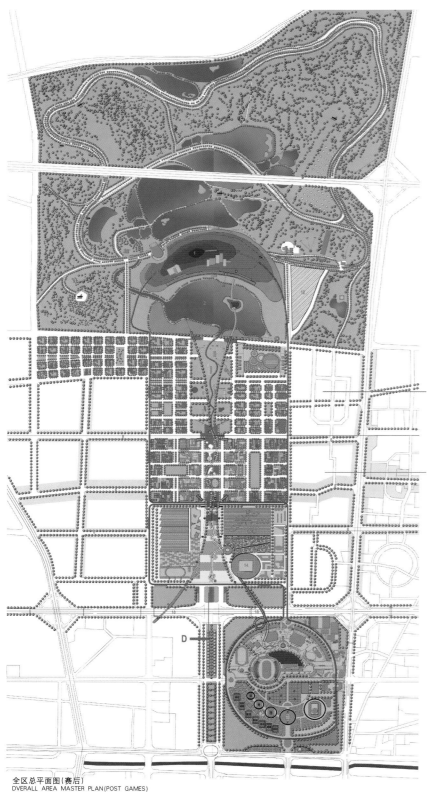

全区总平面图(赛后)
DVERALL AREA MASTER PLAN(POST GAMES)

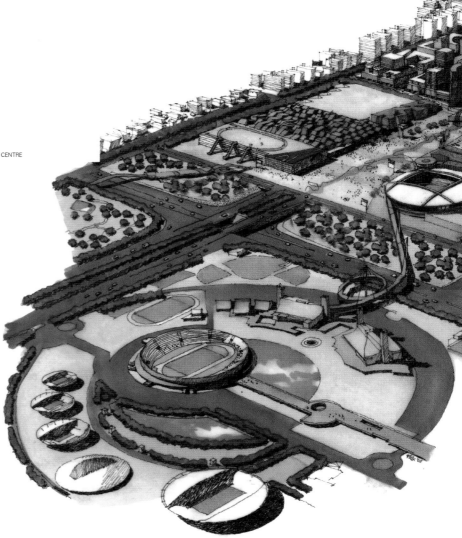

鸟瞰图(赛后)
BIRDS-EYE VIEW (POST GAMES)

1	生态山	ECO MOUNTAIN
2	人民湖	THE PIOOLE'S LAKE
3	混合居住	MIX USE RESIDENTIAL
4	生态园林区	ECO PARK FOREST AREA
5	奥林匹克城	OLYMPIC CITY
6	中央公园	CENTRAL PARK
7	城市广场	CITY PLAZA
8	商业带	COMMERCIAL ZONE
9	奥林匹克广场	OLYMPIC PLAZA
10	人行道/桥	PEDESTRIAN WALKWAYS/BRIDGES
11	宾馆	CATIC HOTEL
12	白玉别墅	BAIYU VILLAGE
13	国家奥林匹克中心	NATIONAL OLYMPIC CENTRE
14	国家体育场	NATIONAL STADIUM
15	国家展览中心	NATIONAL EXHIBITION & CONVENTION CENTRE

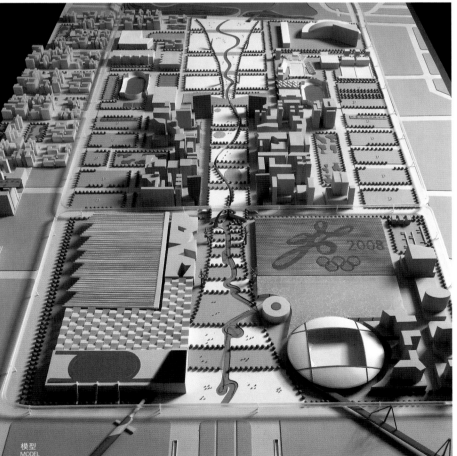

模型
MODEL

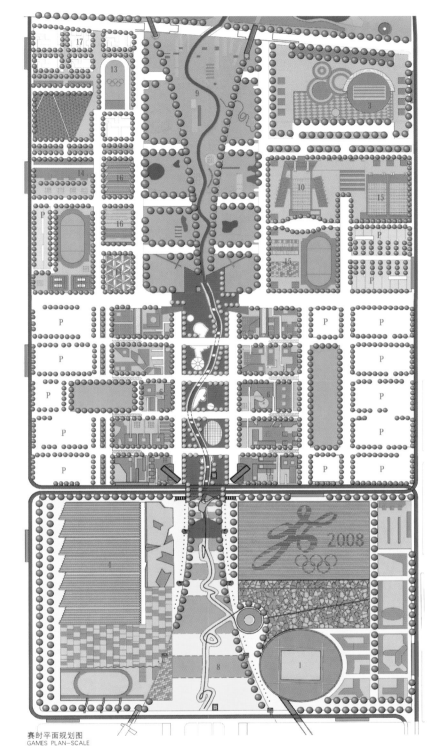

赛时平面规划图
GAMES PLAN-SCALE

1 国家体育场
　NATIONAL STADIUM
2 国家体育馆
　NATIONAL GYMNASIUM
3 国家游泳中心
　NATIONAL SWIMMING CENTER
4 国际展览中心
　INTERNATIONAL EXHIBITION CENTER
5 世界贸易中心
　WORLD TRADE CENTER
6 奥林匹克城
　OLYMPIC CITY
7 国家文化中心、中国青年文化宫、城市广场
　NATIONAL CULTURAL CENTER, CHINESE TEENAGE PALACE, CITY PLAZA
8 奥林匹克广场
　OLYMPIC PLAZA
9 中央广场
　CENTRAL PLAZA
10 射箭场
　ARCHERY (TEMPORARY)
11 国际奥委会办公室
　IOC OFFICES
12 奥运村
　OLYMPIC VILLAGE
13 升旗仪式和运动员集合处
　FLAG RAISING AND ATHLETES ASSEMBLY
14 后勤和保安
　LOGISTICS AND SECURITY
15 训练设施
　TRAINING FACILITIES
16 赞助商帐篷
　SPONSORS TENTS AND CULTURAL DISPLAYS
17 国际村
　INTERNATIONAL VILLAGE

107

北京奥林匹克公园设计方案
Beijing Olympic Green

参赛作品 Works

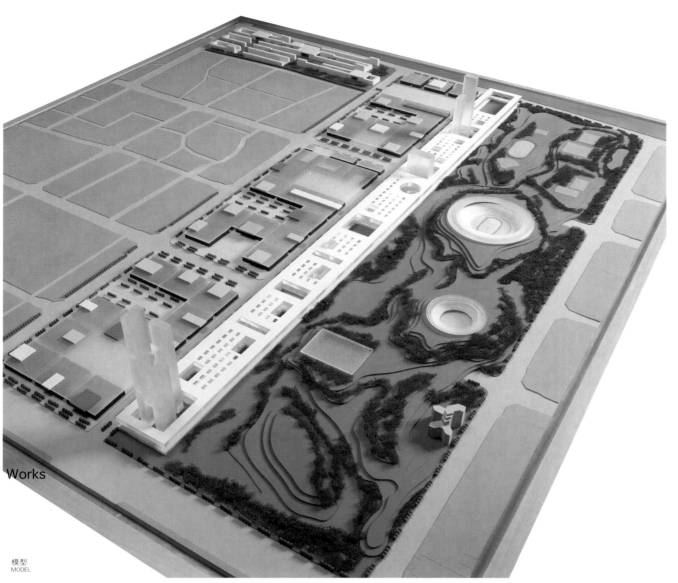

模型
MODEL

Beijing Olympic Green 2008

奥托·斯泰德勒（德国）
STEIDLE+PARTNER ARCHITEKTEN(GERMANY)
方略建筑设计有限责任公司（中国）
STRATEGY ARCHITECTURAL DESIGN CO.,LTD.(CHINA)

■ 规划设计构思

传统

亚洲的建筑艺术具有强烈的特色，它能给现代城市规划以巨大的想像空间，它同样适合2008年奥运会。亚洲的建筑和园林是和自然融合在一起，而不是单纯的几何形设计，它充分利用自然的山石、水、树等元素，用序列的排布方法，绚丽的色彩，院落和常见的大屋顶等，使建筑和自然达到和谐统一。

这些空间元素的特征同样运用在城市规划中，四合院成为城市的基本单元，从而形成了城市的肌理。中国历史上那些纪念性的重要建筑，它们很高大，但不是像现在的一些新的建筑像怪兽一样，而是具有鲜明的特点和很强的感染力。

轴线

奥林匹克公园的轴线是沿着紫禁城、鼓楼向北延伸的，它展示出一种新的、不同一般的内涵，具有强烈的东方特色，表现出高度的和谐统一。

景观

宏大的建筑与它周围的自然环境相统一，景观不是轴线的结束，而是原有北京古城建筑的延伸。奥林匹克公园的轴心是由一个完整的长条形的建筑群构成，它容纳了所有奥运会的设施，我们叫它"文化长轴"。

奥林匹克山

2008年奥运会是展示中国传统文化与自然景观完美结合的一个机会，体育馆坐落在山丘下面，人们只能看见它的屋顶，混凝土的结构构件都隐蔽在山体之中，游览者登上奥林匹克山，看到的是下沉式的体育场馆，它完美地与自然环境结合在一起。

文化长轴

延续北京建筑的中轴线，我们是用"文化长轴"的方式来体现的，在奥运会期间，它将成为为奥运所有项目服务的中心轴，展示轴，文化轴和商业轴。它位于很重要的中轴线上，它虽然不是纪念性的建筑的开始，但它却用新的形象成为与四环交汇的一个新的起点。三个塔式建筑，包括了很多辅助功能，如旅馆、会议中心、文化商务设施等，不仅将成为城市的地标，而且丰富了城市的轮廓线，烘托出奥林匹克公园的路口，使其更加突出。

第二座塔楼位于文化长轴的北端，邻近体育馆区域，它将容纳奥运会重要的服务内容，如行政管理和通讯设施，同样还将是展览中心的行政管理部门。

总体规划

赛时及赛后不同阶段建筑总体布局的不同主要体现在展览空间的使用上。奥运会期间，北部主要用于奥运村的国际区和公交车（临时为区间服务的短程汽车）的集中泊位及集中绿地，赛后则对外开放并作为国际展览中心的预留地，赛时作为临时场馆和IBC和MPC的展厅恢复原本的展览功能。

在赛后，文化长轴的基座层可包含不同的功能，如作为商场的扩充及娱乐场所和停车场。奥运村赛后将变为住宅区，A区的曲棍球中心和网球中心则对外开放。迅速有效地疏散及运送观众。四个地铁站沿长轴分布并通过电梯及扶梯与之相连，临时的短程公交观览汽车在长轴西侧的0.00层运转，主要的起始站邻近国际区，从此观众可到达通向所有体育场馆及展厅的主要出入口。

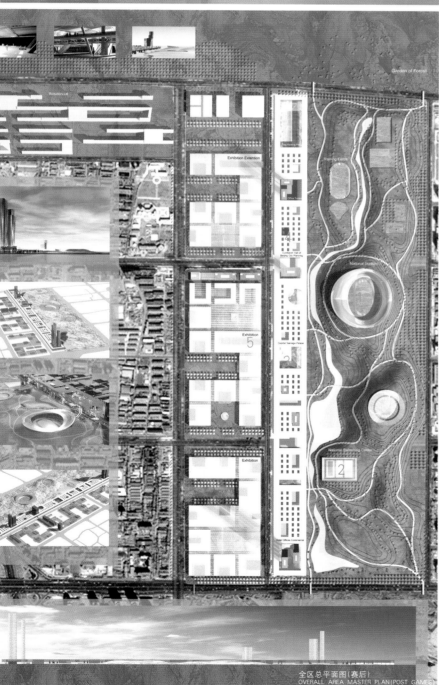

全区总平面图（赛后）
OVERALL AREA MASTER PLAN (POST GAMES)

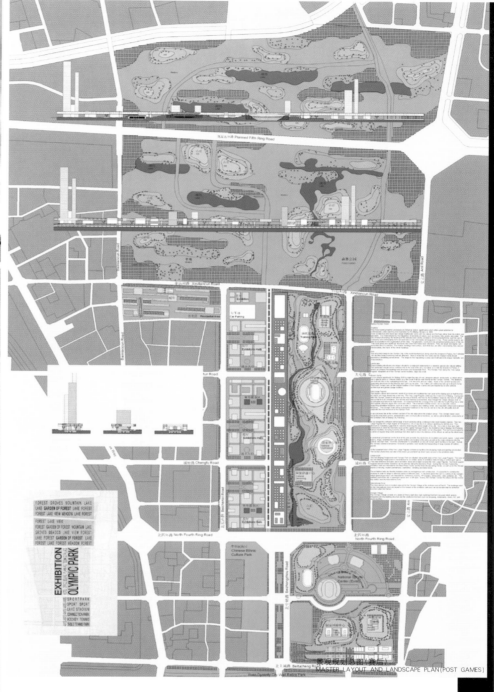

景观规划总图（赛后）
MASTER LAYOUT AND LANDSCAPE PLAN (POST GAMES)

1 国家体育场
  NATIONAL STADIUM
2 国家游泳中心
  NATIONAL SWIMMING CENTER
3 首都青少年宫
  CAPITAL TEENAGE PALACE
4 北京城市规划展览馆
  EXHIBITION HALL OF BEIJING CITY PLANNING
5 北京会展博览设施
  BEIJING INTERNATIONAL EXHIBITION AND CONVENTION CENTER (BIEC)
6 运动场
  TRAINING FIELD

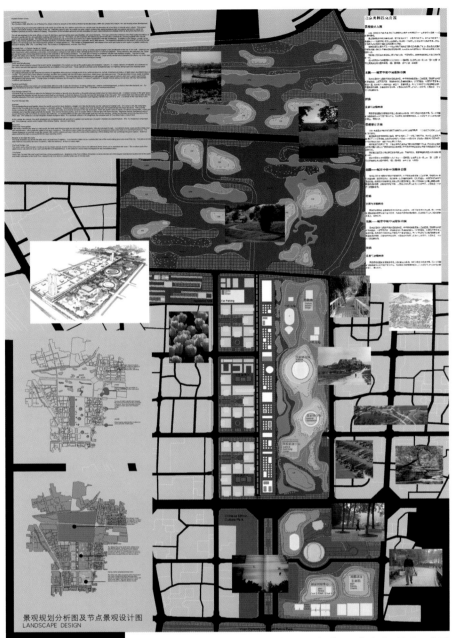

景观规划分析图及节点景观设计图
LANDSCAPE DESIGN

奥运村及其中的国际区的主入口设在B区的北端，从这里专用的汽车会把运动员安全地送到赛场。

Planning and Design Concept

Tradition

The Asian building tradition possesses an immense historic significance which offers great potential for contemporary urban planning and inspiration for the Beijing Olympics in 2008.

Asian architecture and landscape design traditionally works with the natural and the fluid, rather than the orderly and geometric. It is characterized by the integration of elements of nature, such as stones, hills, plants, and water. The use of linear structuring, the layering of the splendid and colorful, avoidance of monumental forms, and the use of courtyards and overhanging roofs are best seen in the colonnades and pagodas in Asia. These spatial aspects are also incorporated in the organization of the cities. The geometric courtyard is an essential building block of the city. The most grand and significant structures also draw from these basic elements. They are larger and higher, but avoid that which is ordinary. The historic buildings have pathos, and while they are often monumental, they are not monstrosities, like many of the newer buildings.

Axis

The axis which leads to the Olympic City is the most important axis along which the Emperor's Palace, the Forbidden City and the Great Drumming Hall are attached. Here is where the new world and the western world should experience something extraordinary. The stage for this happening is asian and it will lend it its unique mentality and characteristics.

Landscape

Grand building structrures are always situated in a balanced relationship to a similarly spectacular natural setting. The landscape should not be confined only to the end of the axis, but rather as is with the Emperor's Palace, it should accompany the buildings and structures and incorporate them. The Olympic Park along the "Very Large Pagoda" integrates all of the Olympic sports facilities; large and small.

Mount Olympic

It is an unique opportunity for Beijing 2008 to make the idea of vast, designed natural landscapes, a culture which has a special place in Chinese history (e.g. the Great Wall) its main theme for the Olympic Games. The stadiums are built into hills in the modulating landscape. Only the roofs are visible. None of the cement facades and supporting structures for the tribunes and roofs are seen. The visitors can walk to the hills up to Mount Olympic. The Asian Games carry on the original tradition of the Olympic Games and are inspired by its own chinese architecture and garden design tradition.

Very Large Pagoda

The largest cultural and economic event of our times will constitute the next pearl of the Bejing axis so that this line of culture can forge ahead into a next era. The Very Large Pagoda continues the Great Axis of Beijing. During and after the Olympic Games it will serve as the Axis of Events, the Axis for the Exhibition Grounds, the Axis of Culture and as the Axis of Commerce. It is not first and foremost a monument, even though the "Urban Olympic Axis" is accentuated and braced by high towers. It is important that the crossing of the axis with the Fourth Ring has a clear, and spatially effective signal and conclusion effect. The ensemble of three towers can incorporate various supporting exhibition functions, such as a hotel, a congress and meeting center, and important cultural and business functions.. This Olympic Gate will be a prominent landmark in the Axis as well as in the city silhouette and will celebrate the main entrance to the Olympic Park.

A second tower lies in the northern section of the site adjacent to the stadium areas. This Olympic tower could house the important service, administrative and communications functions, as well as administration components of the Exhibition Hall of Beijing City Planning.

Planning in Masterplanning

The phases before and after the Games are mostly visible in the area of the exhibition spaces. During the Games, the northern part of the site is used for functions of the International Zone and central bus parking, as well as for green park zones. After the Games this area is open to public use and in the long term the expansion of the Beijing Fairgrounds will be realized in this area. The Exhibition Halls used for Olympic sporting events, IBC and MPC will be transformed into normal exhibition spaces.

The base of the Very Large Pagoda can serve different functions after the Games: enlarged shopping opportunities, entertainment and parking. The Olympic Village will be transformed into a residential area. The hockey rinks and tennis courts in Part A will open for public use.

■ 规划设计构思

从故宫天安门"广场"到北京奥林匹克"桥场"

桥——时间
传承之桥——呈现"时间"走廊
"中轴"——时间之"桥"——传承过去、现在和将来
→ 故宫天安门广场 → 奥林匹克桥场 → 未来

桥——空间
友谊之桥——跨越"空间"阻隔
"中轴"——空间之"桥"——连接五大洲
中国——奥林匹克桥场——五大洲

桥——人间
胜利之桥——塑造"人间"辉煌
"中轴"——奥运精神：更快、更高、更强
V形桥场：胜利！
零起点、逐渐升高的V形桥面及铺地：一步一个脚印，从胜利走向胜利！

桥——之间
实与虚"之间"
中轴线上构成"奥林匹克桥场"的三座大型实体建筑国家体育场、国家体育馆和国家游泳馆准确地呈现出奥林匹克公园主题，1700m长、30m宽、坡度为1∶28的V形桥场步道本身则形成从昨天到今天、走向光辉灿烂明天的独特景域。从位于地面的北京猿人广场起始，渐次升高的桥场展现出一幅中华民族灿烂文明的雄伟长卷，尽端广袤无垠的星空则预示着充满希望和机遇的未来，中轴线"实轴"与"虚轴"合二为一。

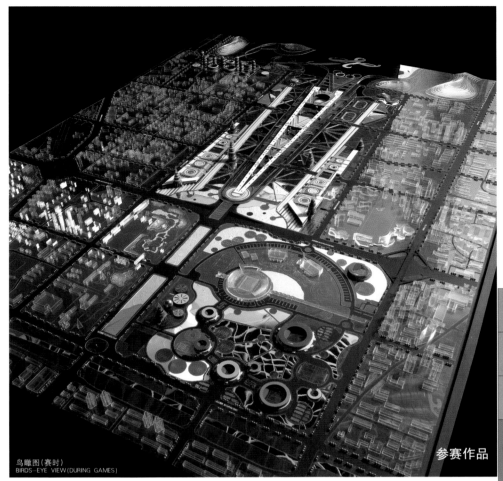

鸟瞰图（赛时）
BIRDS-EYE VIEW (DURING GAMES)

Beijing Olympic Green 2008

清华大学建筑设计研究院（中国）
ARCHITECTURAL DESIGN AND RESEARCH INSTITUTE OF TSINGHUA UNIVERSITY(CHINA)

继往与开来"之间"
对称布局，转译中国古代建筑之精华——故宫建筑群平面，故宫9999间 VS 奥林匹克桥场2 330 000m²；
层层叠翠，再现中国古代高台建筑营建。

五环与五色"之间"
奥运五环——绿、黑、蓝、黄、红代表世界五大洲，与中国古代万物之源五行说"五色"：绿（地）、黑（铁）、蓝（水）、黄（木）、红（火）相对应，奥林匹克公园建筑及桥场、广场大量采用五色，寓示古代与现代、中国与世界之间的交流对话。

山与水"之间"
挖土堆山，形成大片水面，既土方平衡，又呈现山水意象，同时串联北京现有及规划水系，使奥林匹克森林公园成为一个亮点。

水与火"之间"
北京西南的山顶洞人点燃了新石器时代人类之火，呈火炬状的北京2008年奥林匹克公园平面则预示着人类迈向未来的又一火种。

桥场与广场"之间"
桥场与广场相互交错，形成独一无二的水平视野空中景观及活动走廊，最大限度地创造市民易接近的立体休闲空间。

高与低"之间"
100～200m高塔（写字楼、宾馆等）如雨后春笋，拔地而起，形成与桥场、广场之间空间上的强烈对比。

竞技与休闲"之间"
将大型竞技体育场馆和部分娱乐商业实施置于中轴线上，与空中桥场结合，既体现奥林匹克公园的主题，又形成兼顾竞技与休闲的公共活动所；将会展中心（赛中为场馆或新闻中心）、写字楼、宾馆等建筑安排在两侧，有利于赛后地块的可持续发展，更利于形成与现状街道建筑相呼应的宜人尺度。

赛中与赛后"之间"
在满足奥运会赛中及赛后竞技体育的前提下，更多地考虑赛后地块发展，形成市民和游客流连忘返的、人气旺盛的大型公共休闲场所。

■ Planning and Design Concept
FROM FORBIDDEN CITY / TIAN'ANMEN SQUARE TO BEIJING OLYMPIC V-BRIDGE PLAZA

Bridge-Time
Bridge of continuum-form a corridor in time
"Axis"-a bridge of time-a continuum connecting the past, the present and the future
The Forbidden City-The Olympic Green-the futureBridge-Space

Bridge to friendship-span across the space

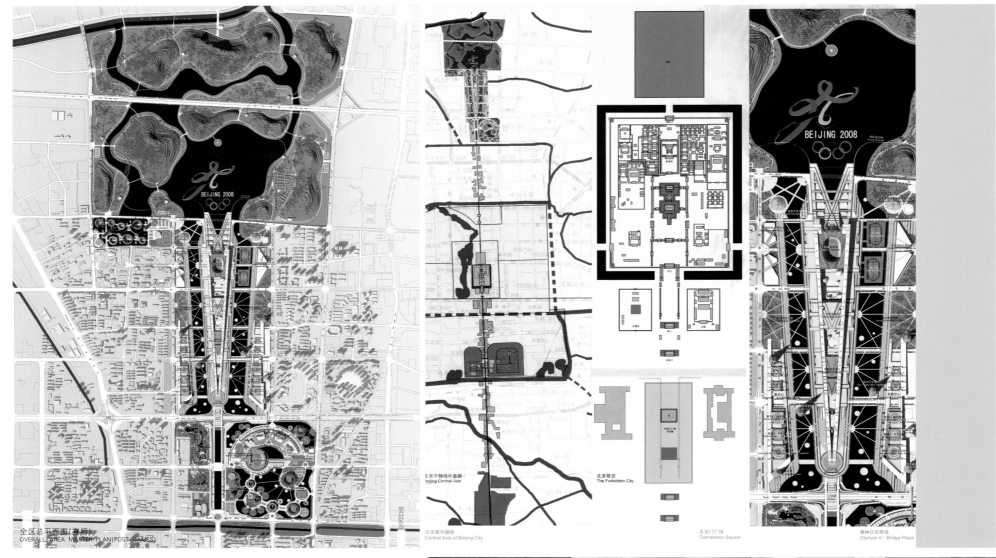

全区总平面图(赛后)
OVERALL AREA MASTER PLAN (POST GAMES)

北京城中轴线
Central Axis of Beijing City

天安门广场
Tian'anmen Square

奥林区V桥广场
Olympic V - Bridge Plaza

1 国家体育场
  NATIONAL STADIUM
2 国家体育馆
  NATIONAL GYMNASIUM
3 国家奥林匹克体育中心
  NATIONAL OLYMPIC SPORTS CENTER
4 首都青少年宫A区
  CAPITAL TEENAGE PALACE A
5 首都青少年宫B区
  CAPITAL TEENAGE PALACE B
6 北京城市规划展览馆A区
  EXHIBITION HALL OF BEIJING CITY PLANNING A
7 北京城市规划展览馆B区
  EXHIBITION HALL OF BEIJING CITY PLANNING B
8 国家曲棍球场
  NATIONAL GOCKEY FIELD
9 体育场
  SPORTS CENTER STADIUM
10 垒球场
  SOFTBALL FIELD
11 英东游泳馆
  YING TONG NATATORIUM
12 练习场
  TRAINING FIELDS
13 会展中心
  CONVENTION & EXHIBITION CENTER
14 奥运村
  OLYMPIC VILLAGE
15 中华民族园
  CHINESE ETHNIC CULTURE PARK

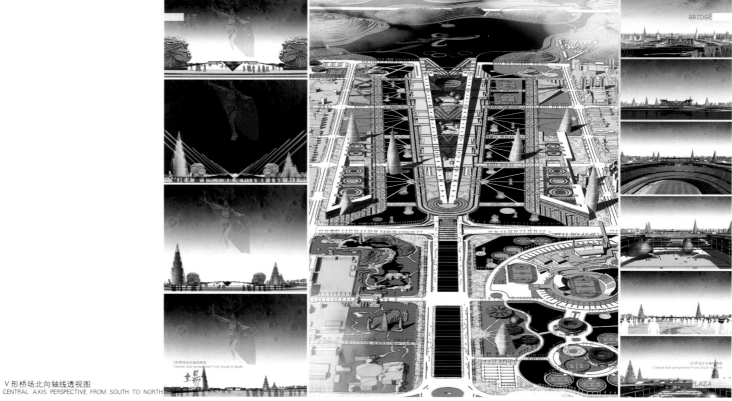

V形桥场北向轴线透视图
CENTRAL AXIS PERSPECTIVE FROM SOUTH TO NORTH

112

"Axis"-a bridge of space-connecting the five continents
China-The Olympic Green-five continents

Bridge-man
Bridge to victory-build up the glory of life
"Axis" -The Olympic Spirit: to be quicker, higher and stronger
The V-shaped bridge: Victory! -the V-shaped bridge and the plaza rise from the ground embodying the meaning of shifting from one victory to another step by step.

Bridge-in-between

Between "solid" and "void" axis
The axis of the bridge forms a special image, which spans over the past, the present and leads to the bright future, while the open-ending of the bridge signifies a future which is full of hope and chance. All these come up with a harmony between solid and void.

Between the past and the future
The structure of tierring up represents the art of building hathpaces in ancient China; while the symmetric structure interprets the essence of ancient Chinese architecture-the layout of The Forbidden City.with its 9, 999 rooms which cover an area of 2, 330, 000 square meters.

Between Five Rings and five colors
The Olympic Five Rings of green, black, blue, yellow and red, which signify the five continents, is in parallel with the Chinese concept of "Wu Se", which refers to the five ultimate elements of the world-with "green" referring to earth, "black" to iron, "blue" to water, "yellow" to wood, and "red" to fire. These five colors are widely employed in the design of the buildings, bridges and squares, which symbolizes the exchanges and conversation between the past and the present, and that between China and the world.

Between mountains and waters
By piling up mountains and digging out large areas of water, the balance of earthwork is maintained and the image of mountains and waters is formed. At the same time, it connects the existing water system of Beijing and makes the Olympic Park of Forestry an attractive spot.

Between water and fire
In the Neolithic Age, Upper Cave Man, who once lived in the southwest of Beijing, lit up the fire of man. The layout of the Olympic Park of 2008, Beijing, which takes the shape of an torch, symbolizes the kindling that will lead the human man to the future.

Between the bridge and the plaza
The bridge and the square intertwine with each other, forming up a unique horizontal view, a special scenery and a corridor to provide the most convenient space of leisure for the residents.

Between the high and the low
In strong contrast to the bridge and the square, high buildings like office buildings and hotels, whose height range from 100 m to 200 m, rise from the ground like bamboos.

Between competition and entertainment
Large stadiums and some buildings of entertainment and commerce, that are built on the axis, together with the bridge provide large areas for public entertainment, which carter for both the need of competition and that of entertainment. In order to make better of the land after the Olympic Games, and keep in harmony with the present streets and buildings, buildings like the Conference Center (used as the stadium or press center during the Olympics), office buildings and hotels are arranged on the two sides.

Between "during" and "after" the games
In the precondition that there will be enough stadiums during and after the Olympic Games, the designer puts more emphasis on the development of the land after the Games, so as to engender large popular place for public entertainment, that will put the citizens and tourists on the scoop.

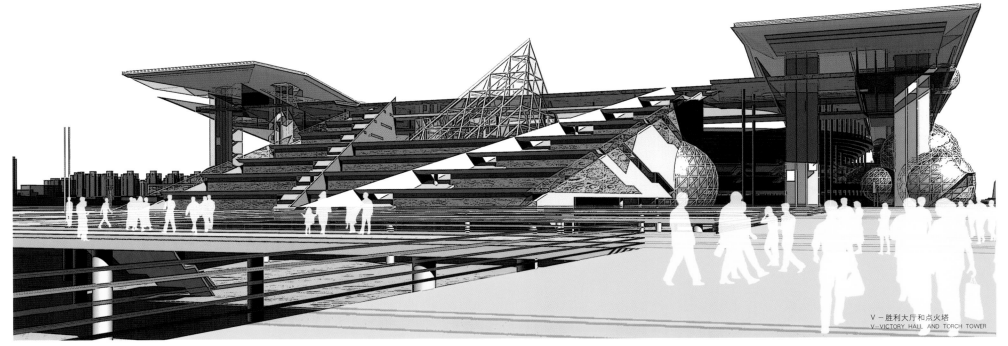

V－胜利大厅和点火塔
V-VICTORY HALL AND TORCH TOWER

北京奥林匹克公园设计方案
Beijing Olympic Green

参赛作品 Works

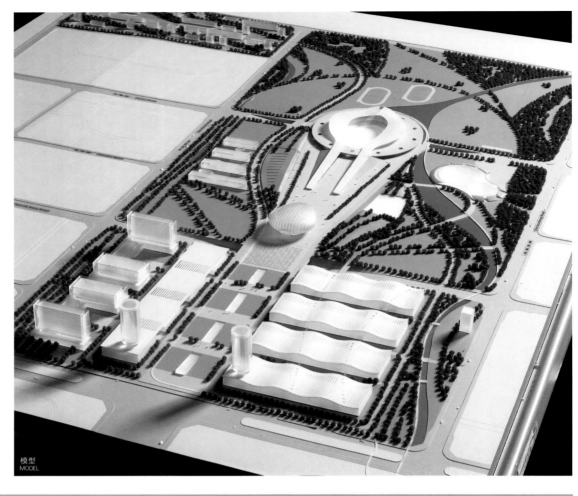
模型
MODEL

ARUP 合伙人有限公司（英国）
ARUP ASSOCIATES LIMITED(ENGLAND)
深圳市陈世民建筑师事务所（中国）
SHENZHEN CHEN SHIMIN ARCHITECTS ASSOCIATES(CHINA)

Beijing Olympic Green 2008

■ 规划设计构思

作为创立总体规划的基本框架的三个概念是：

历史性同时又具现代性，建立"科技奥运"的形象。

在生态环境方面渐进地建立多样化的自然生态，减少能量消耗，达到"绿色奥运"的要求。

建立多用途的完满活力的北京新城区，供市民游客长期享用的广场、公园，实现"人文奥运"的目标。

以体育场作为北中轴线的终点

北京的城市中轴线现在以森林区域为终点，但没有任何标志性的建筑。我们的方案以体育场标志中轴性的终点，体育场构图就像中国著名的二龙戏珠图案，并将其一分为二，以释放中轴线所蕴藏的能量，拥有传统与现代、东西文化结合的内涵。依照中国传统，体育场坐北向南，并且以一个同样坐落在中轴线上的较小一些的国家体育馆作为它的大门。这个表现方法既强调了中轴线的历史意义，同时又突出了21世纪的形象。

■ Planning and Design Concept

These three concepts are the framework that will create a master plan that is:

Historically sensitive but also very modern, an image for a High-Tech Olympics.

Ecologically and environmentally progressive creating diverse natural habitats and reducing energy consumption, a Green Olympics.

Creating a vibrant new area of Beijing with a good mix of uses, new squares and parks for the long term benefit of visitors and residents of Beijing, a People's Olympics.

Key conceptual ideas:

Continuation of the Central Axis of Beijing City

The proposed Beijing Olympic Green is set on a central axis. This provides a continuation to the central axis of Beijing City and enhances its significance. Along the middle section of the central axis is the Forbidden City which represents the historical Chinese culture. The Olympic Green extends the reform and modernisation of China. It is proposed that the city central axis will be extended to the south and terminated at the planned World Village-a symbol of international culture.

Termination of the North Central Axis

The existing Forest Area terminates the axis but is not marked by any building. Our proposal uses the stadium to mark the end of the axis, but is split into two halves allowing the energy of the axis to pass through; this resembles the famous Chinese image of two dragons and a pearl. This combines the traditional and the modern, the east and the west culture. In line with Chinese tradition the stadium faces south and has the smaller indoor stadium as a 'gate way' in front of it also on axis. This method addresses the historical significance of the axis but at the same time is a modern 21st century icon.

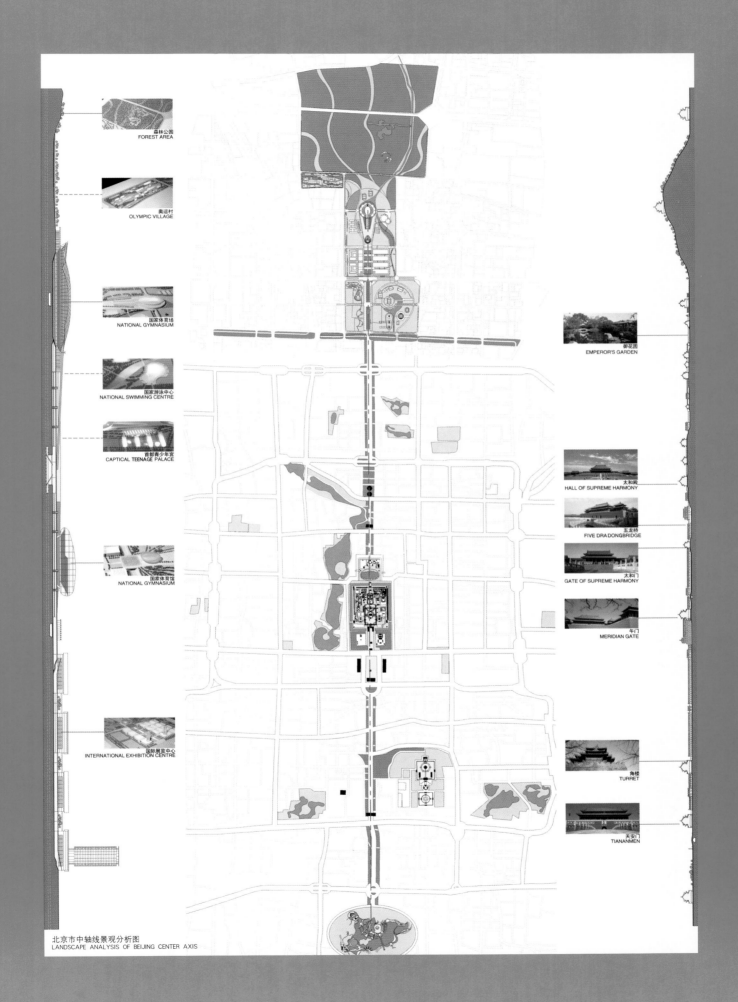

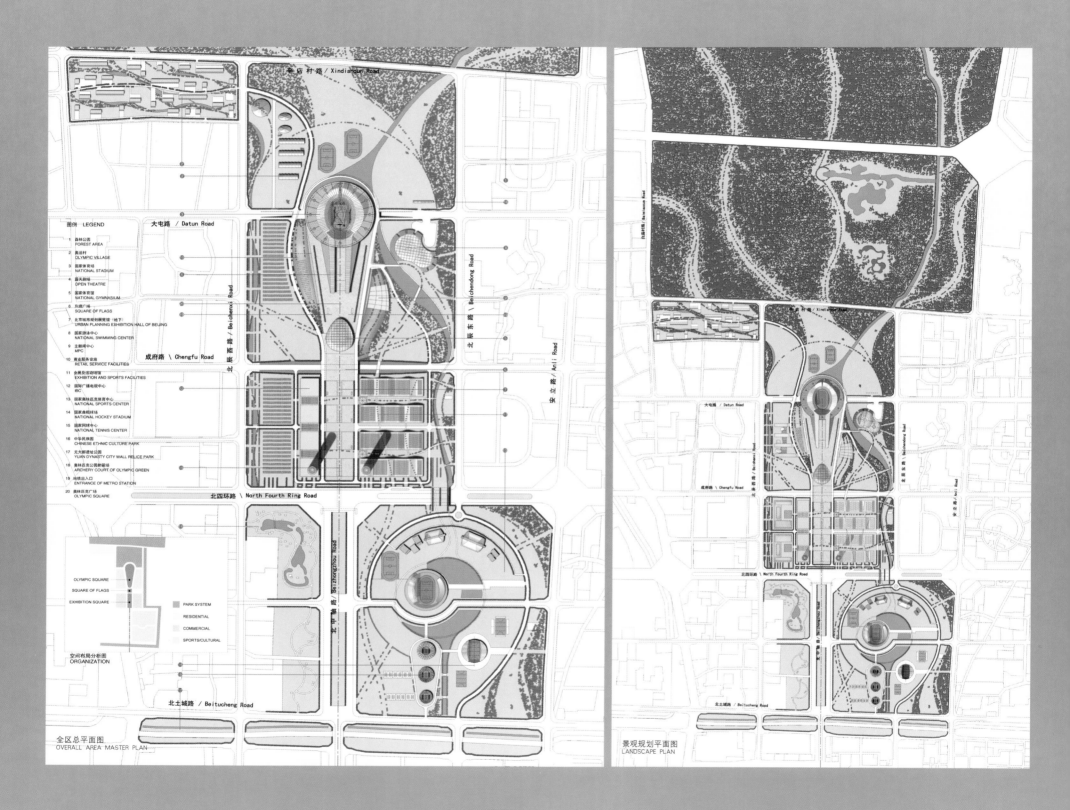

模型 MODEL

■ 规划设计构思

奥林匹克公园位于中轴路的特殊位置为设计提供了丰富的思路和设计灵感。

以中轴线构思整个景观及建筑,保持中轴路的视觉特征和延伸性。

空间布局中充分考虑到功能的合理,主体育场、体育馆等建筑与原有体育设施相结合,肌理保持"龙"的韵律感。重要单体组合如国家体育馆把具有中国文化特征的太极图与具有希腊文明特征的爱奥尼柱巧妙结合,强化了奥林匹克运动与东方文化的结合。

创造性、参与性、功能性、历史性与资源的多元结合,如森林公园内的未来奥林匹克博物馆、奥林匹克山的创造,别具特色,将使奥林匹克公园成为第29届奥运会的纪念性设施。

■ Planning and Design Concept

Our design is to integrate the Eastern culture into the human beings' civilization, to integrate the Olympic spirit into the modern technology. On base of the fact that Olympic Park is located on the Axis Road of Beijing, which provides us with so many inspirations, we started to move.

All the landscape is designed and constructed on base of the Axis. The Olympic Park will be the axis of Beijing, geographically in landscape and historically in the mind of human beings.

To make full use of the current resources we have, and design different function areas reasonably, as National Stadium which combines the Chinese "Tai Ji" Map with Greek pillar enhancing the combination of Olympic and Eastern culture.

To be creative is another important design principle. We try our best to make the Olympic Park a milestone of human beings' creativity, and a milestone of sports, culture and art.

参赛作品 Works

北京奥林匹克公园设计方案 Beijing Olympic Green

Beijing Olympic Green 2008

九源建筑设计有限公司(中国)
JIUYUAN ARCHITECTURAL DESIGN CO.,LTD(CHINA)
北京达沃斯景观规划设计中心(中国)
BEIJING DAVOS LANDSCAPE PLANNING AND DESIGN INSTITUTE(CHINA)

B区总体规划平面图(赛后)
AREA B GENERAL DESIGN PLAN (POST GAMES)

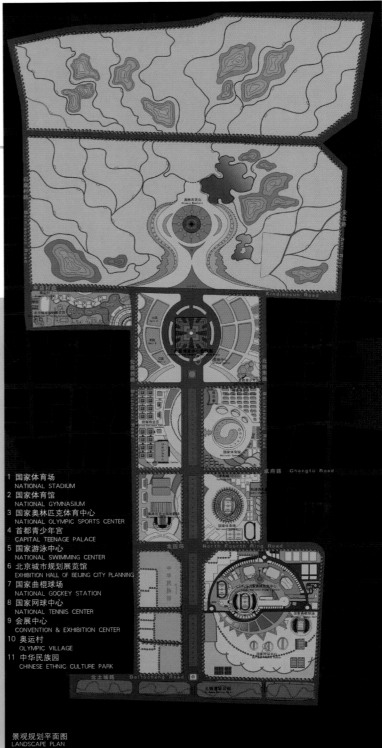

1 国家体育场
  NATIONAL STADIUM
2 国家体育馆
  NATIONAL GYMNASIUM
3 国家奥林匹克体育中心
  NATIONAL OLYMPIC SPORTS CENTER
4 首都青少年宫
  CAPITAL TEENAGE PALACE
5 国家游泳中心
  NATIONAL SWIMMING CENTER
6 北京城市规划展览馆
  EXHIBITION HALL OF BEIJING CITY PLANNING
7 国家曲棍球场
  NATIONAL GOCKEY STATION
8 国家网球中心
  NATIONAL TENNIS CENTER
9 会展中心
  CONVENTION & EXHIBITION CENTER
10 奥运村
   OLYMPIC VILLAGE
11 中华民族园
   CHINESE ETHNIC CULTURE PARK

景观规划平面图
LANDSCAPE PLAN

B区鸟瞰图（赛后）
AREA B BIRD'S EYE VIEW (POST GAMES)

北京奥林匹克公园设计方案 Beijing Olympic Green

参赛作品 Works

模型 MODEL

■ 规划设计构思

方案中的南北轴线将是一条标志着北京历史文化的城墙，它蜿蜒向北，起始于一系列的雕塑、绿化地带和中央水流。这条水流一直流入森林公园中的瀑布，从一座很高的假山上飞流直下。站在山顶，可以遥望紫禁城。

占地广阔的体育场馆看上去就像"大弹坑"。地像受到了挤压而凹进去。与此相似的是，一个椭圆形的广场（集会广场）也低于平均地面海平线，满足城市系统的需求，改变地形来适应大众需要，提供商业场所等等。

在中央地区，楼房集中在4对高层建筑群内。这些高层建筑群分别面对东部和西部的郊区，其中，2对坐落于大屯路轴线上，另外2对坐落于成府路轴线上。

十字路口，如辛店村路和北四环路的交叉路口，低于周围的绿化地带，但高于地铁8号线。地铁的出口连接在周边的停车场，为地下大规模的商业区和娱乐区的建设提供了方便。

Beijing Olympic Green 2008

PCA INT PICA恰马拉R.L设计事务所（意大利）
PCA INT PICA CIAMARRA ASSOCIATES.R.L(ITALY)

■ Planning and Design Concept

The ideal north-south axis which marks the historic structure of Beijing lying beyond the city walls, opens up northwards in a sequence of soil mouldings, and spaces of greenery and water, which flows into the Forest Area and into the waterfall which cascades down from the artificial hill. From the top of this hill, the view intersects the Flag Raising Square and reaches the Forbidden City.

The extensive sports facilities, which occupy the area, are interpreted as "craters", land swellings and soil depressions. Similarly, the ellipsoidal square (the meeting Place) 7 metres lower than the basic level, among the leading focus points of the new urban system, shapes the terrain to create a public space close to the trade fixtures provided for.

In the central zone building is concentrated into four pairs of high-rise buildings, towards the eastern and western boundaries of the area, two on the Datun Road axis, while the other two on the Chengfu Road axis.

These crossways are just like Xindiancun Road and the North Fourth Ring Road, passing below the Olympic Green but above line 8 of the Olympic Subway: the exits from the stations of this underground are linked to perimeter car parks and facilitate the implementation of large concentrations of underground spaces for trade fixtures and entertainment.

On the Western front of the area, along the Belchenxl Road, appear spaces of the convention and exhibition facilities and the Capital Teenage Palace in direct relationship with the ellipsoidal Square. In the south-east corner of Central Green - near to the intersection between North Fourth Ring Road and the Belchendong Road - is located the Exhibition Hall of Beijing City Planning, linked to underground entertainment fixtures.

鸟瞰图
BIRDS-EYE VIEW

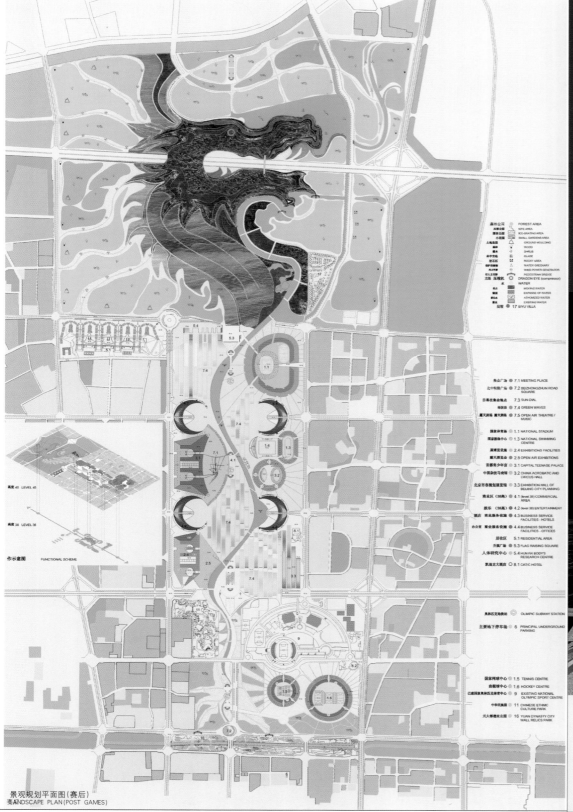

景观规划平面图（赛后）
LANDSCAPE PLAN (POST GAMES)

Beijing Olympic Green 2008

景观规划分析图及重要节点的景观设计图（赛后）
LANDSCAPE PLANNING (POST GAMES)

■ 规划设计构思

"山水奥运"作为主要概念，不仅包含了"人与地球"、"人与人"相融与共的生存观念，而且更广义地将人与人的竞技比赛演变为人与人和平相处、人与自然和平相处的崇高境界。

创造壮美轴线

贯穿南北的中央绿色通道在结合西方广场和中国街道空间的基础上导入生态理念，形成绿野空间并以"鱼骨"、"鱼刺"的方式有机渗透场馆区，提高环境质量。国家体育场位于轴线尽端，成为壮美轴线高潮之所在，圆形体育场形如玉璧，成为中国传统文化中"宇宙并提，时空一体"的意境空间。

创造三组互动的奥运组团

体育馆、游泳馆、展览馆是在化整为零，又化零为整中形成连续的"大地工程"，从而产生了"无限"的意境。

构成与北京城相适宜的轮廓线

"冰山"轮廓和"水波"轮廓的节奏变化多种多样，"匀速"中见高潮，"急速"中见安详……古典的、抒情的、摇滚的、间断的节奏都将成为可能。

以单元组团形成节奏变化的时空形态

山水单元并不是互相平等的被随意排列在一起的，而是通过功能组织形成时空秩序。在这里，时间、运动是空间时间化的主要因素。走建筑、读建筑成为主题，"叙述性"增加，园林的"步移景迁"得到广泛应用。

■ Planning and Design Concept

As the main principle of design, "landscape Olympics" not only contains the living conception that the humankind and the earthsmoke the calumet together and people get on quite well with each other. Furthermore, in a broader sense, it evolves the athletic contest among people to a realm of lofty thought that people live together peacefully.

Create a magnificent and great axle

The axle line passes through the fourth ring road. The central green road leading to the north and the south takes into account the design of the western square and Chinese street. It also introduces the tenet of ecology. It forms a green space with the "fish bone" and "fish thorn" organically interpenetrating with the whole stadium. It greatly improves the environmental quality. The National Sports Center locates at the end of the axle and it fully demonstrates the magnificence of the stadium. The round stadium looks like a "jade wall" described in the Chinese historical book Zhou Li. It not only embodies the spirit of the Olympic Games, but also further develops the traditional Chinese culture.

Create Three Interactive Olympic Zones

The stadiums, natatorium and exhibition hall are consistent "land project". They are not designed lonely but closely connected with each other. It creates the idea of infinity.

Constitute the Contour Line That Suits Beijing City

Iceberg and wave contour are interpenetrated with multiple changes. Visitors can not only experience quiet and peace but also fast and quickness---classical, poetic, rock-and-roll, discontinuous rhythm are possible.

Form the Space Structure with Separate Units

The unit landscape is not simply collocated at will. They are arranged in conformity with the spatial-temporal sequence according to different functions. Time and sports are the main factors contributing to the timing of the space. The theme is to understand and know more about the architecture by looking and walking. The distinguished feature of the park, i.e., scenery changes by walking, is widely applied.

Beijing Olympic Green 2008

中国科学院北京建筑设计研究院有限责任公司（中国）
INSTITUTE OF ARCHITECTURE DESIGN & RESEARCH, CHINESE SCIENCE ACADEMY (CHINA)

国家体育场
NATIONAL STADIUM

模型
MODEL

创造壮美轴线
Create a magnificent and great axes

创造三组互动的奥运组团
Create the three interactive Olympic zones

构成与北京城市相适宜的轮廓线
Constitute the contour line that suits Beijing city

以单元组团形成节奏的时空形态
Form the space structure with separate units

南端位于土城北路的双塔犹如门阙，标志着奥运公园开端，它的地标特征和元大都遗址公园的文化属性成为重要的城市节点。
The National Olympic Sports Center organically integrates with the new stadium in a helical structure. It extends clockwise and points to the axle line. The round external form is similar to that of the Heaven Temple.

景观规划平面图(赛后)
MASTER LAYOUT AND LANDSCAPE PLAN(POST GAMES)

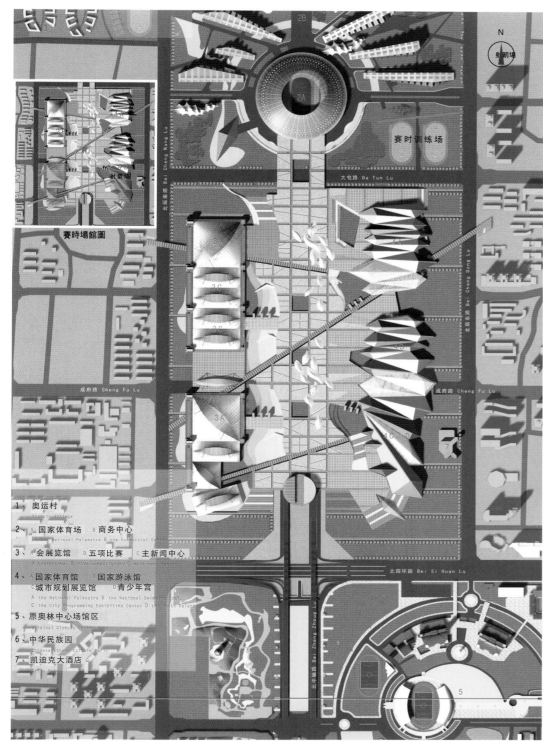

全区总平面图(赛后)
OVERALL AREA MASTER PLAN(POST GAMES)

1、奥运村
  Olympic Village
2、A 国家体育场 B 商务中心
  A the National Palaestra B the Commercial Center
3、A 会展览馆 B 五项比赛 C 主新闻中心
  A Exhibition B five competitions C main news center
4、A 国家体育馆 B 国家游泳馆
  C 城市规划展览馆 D 青少年宫
  A the National Palaestra B the National Swimming Pool
  C the City Programming Exhibition Center D the Youth Palace
5、原奥林中心场馆区
  Original Olympic Centre Area
6、中华民族园
  Chinese Ethnic Culture Park
7、凯迪克大酒店
  Caidike Hotel

中心区
CENTRAL AREA

主展览馆
MAIN EXIHIBITION

125

■ 规划设计构思

交通的畅通安全并与环保配套，是2008年北京奥运能否成功的第一决定因素。因此，我们首先从"科学的和环保的奥运"来解决最头疼的交通问题。我们的交通专家有40年的在国际特大城市及最大项目的实战经验。我们用国际上最先进的环保系统工程，规划设计好了这块基地的交通。同时，本规划设计配合使用功能的核心构思真正做到 "以人为本及以经营为前提"的理念，因此，我们才能做到以下"三赢"：

一赢：在奥运比赛及展览会议期间，使用人（包括选手）都很方便、快捷而非常安全地往来于（有环保设计的）环境幽美的住地和场馆之间。

二赢：比赛后，选手村将作为商品房出售。现在这些住宅区的附近已有了这么好的体育设施和美好环境，房地产商肯定能够以很好的价格出售这些商品房。

三赢：体育设施的经营管理者使住宅区的居民很方便就能到场馆活动，所以他们的营支经济效益肯定也很好。

■ Planning and Design Concept

The core idea of planning and design is the philosophy of the "three party winners":

Whether or not that,the traffic and transportation (and the security) could be properly handled(with the environmental protection as one package)is the top vital factor that could decide the success or failure of the 2008 Beijing Olympic Games.Therefore,we commence our task of this planning from"Olympic of Science& Technology and Environmental Protection"and trying to solve the most headache traffic and transportation problem.Our traffic and transportation expert has had 40 years thought practical experiences in international cosmopolitans, supper large scale projects.we have used the most advanced international traffic and transportation system engineering to plan and design our traffic and transportation system on this project site.To meet the human beings' requirements as the core object as well as the management and operation is the primary concern for our planning and design idea.

The first winner-the user:during the exhibition,conference and games,it will be very convenient for the users including the athletes.They would be able to travel in a very pleasant environment while back and forth between the halls and buildings and the places they are staying.

The second winner-the new home owners:after the games,the athletes apartment will be sold on the commercial market. Environment and sports facilities already,all the units will be sold at a good price.The new home owners will be happy to live in such surroundings.

The third winner-sport man and women:with sports facilities close by ,it will be convenient for local residents who enjoy sports to use them.

北京奥林匹克公园设计方案
Beijing Olympic Green

参赛作品 Works

Beijing Olympic Green 2008

沈祖海建筑师事务所（中国台湾）
HAIGO SHEN & PARTNERS ARCHITECTS AND ENGINEERS(TAIWAN CHINA)
大地建筑事务所（国际）（中国）
GREAT EARTH ARCHITECTS & ENGINEERS INTERNATIONAL(CHINA)
美商迪斯唐工程顾问股份有限公司台湾分公司（中国台湾）
DESHAZO,STAREK & TANG,INC(TAIWAN CHINA)
卡瑞尔合伙人工程顾问有限公司（美国）
CAROL R. JOHNSON ASSOCIATES INC(USA)
异空建筑师事务所（韩国）
BEYOND SPACE ARCHITECTS (KOREA)

景观景点构想
LANDSCAPE PERSPCCTIVE VIEW

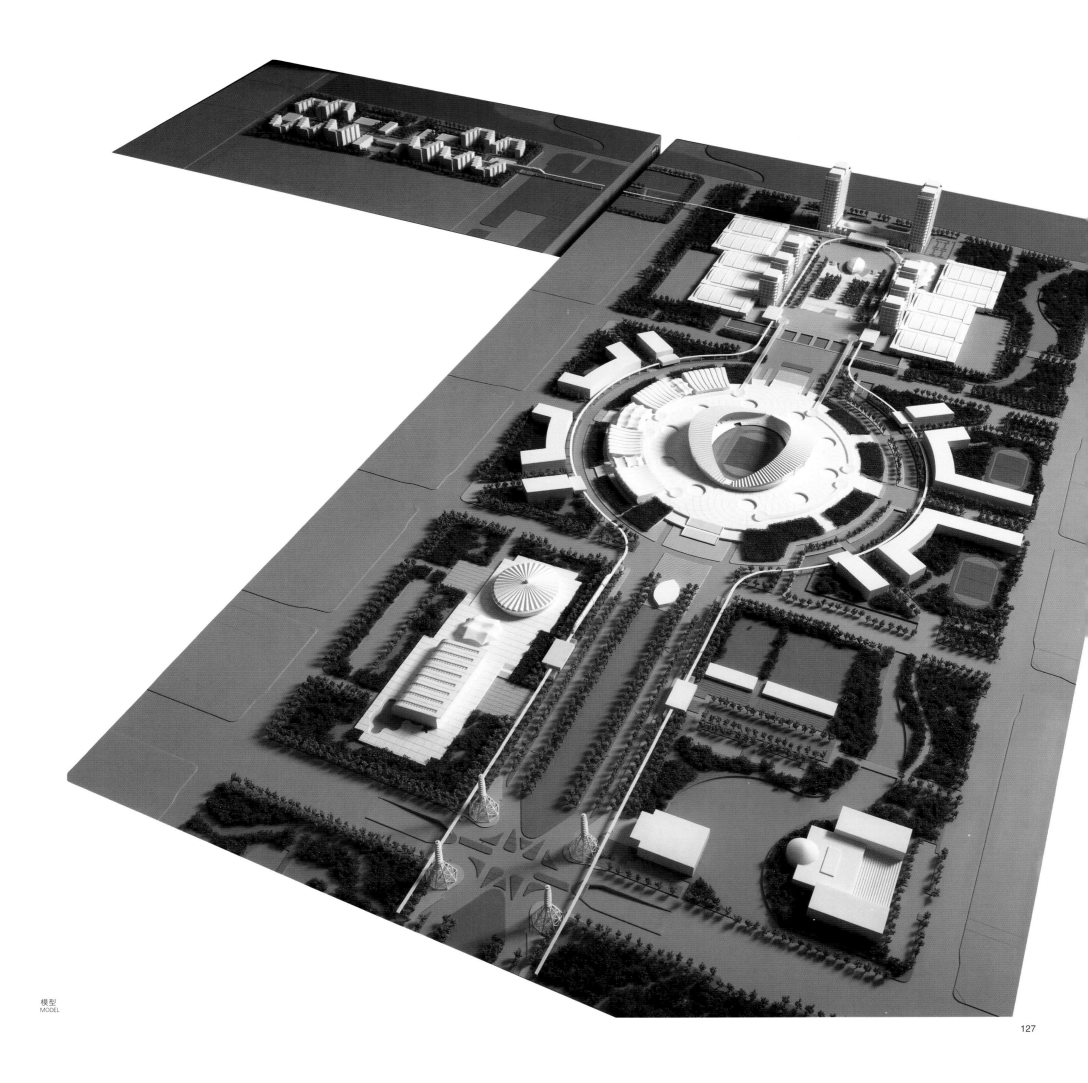

模型
MODEL

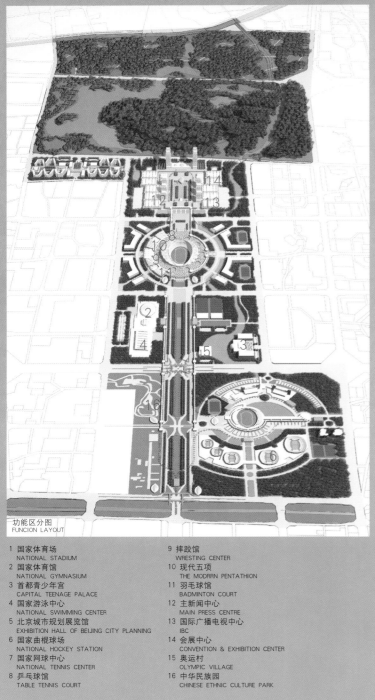
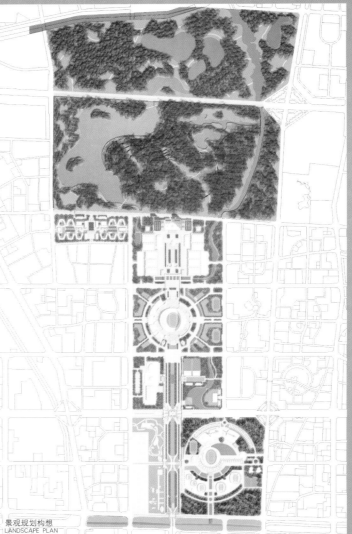
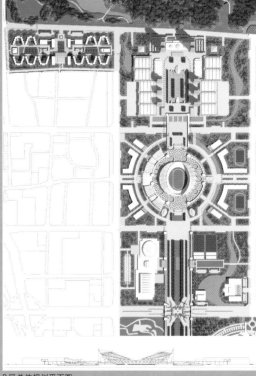

功能区分图
FUNCION LAYOUT

景观规划构想
LANDSCAPE PLAN

B区总体规划平面图
SITE B MASTER LAYOUT PLAN

1 国家体育场 NATIONAL STADIUM
2 国家体育馆 NATIONAL GYMNASIUM
3 首都青少年宫 CAPITAL TEENAGE PALACE
4 国家游泳中心 NATIONAL SWIMMING CENTER
5 北京城市规划展览馆 EXHIBITION HALL OF BEIJING CITY PLANNING
6 国家曲棍球场 NATIONAL HOCKEY STATION
7 国家网球中心 NATIONAL TENNIS CENTER
8 乒乓球馆 TABLE TENNIS COURT
9 摔跤馆 WRESTING CENTER
10 现代五项 THE MODRRN PENTATHION
11 羽毛球馆 BADMINTON COURT
12 主新闻中心 MAIN PRESS CENTRE
13 国际广播电视中心 IBC
14 会展中心 CONVENTION & EXHIBITION CENTER
15 奥运村 OLYMPIC VILLAGE
16 中华民族园 CHINESE ETHNIC CULTURE PARK

Beijing Olympic Green 2008

中元国际工程设计研究院（中国）
IPPR ENGINEERING INTERNATIONAL(CHINA)

BCL 国际公司（香港）
BCL INTERNATIONAL(BCLI)(HANGKONG)

■ 规划设计构思

规划概念是以真正实物的方式去推崇轴线，而不仅是一个简单的视觉。以组织和连接所有北京奥运公园站点的步行大道作为中央轴。

步行大道的南部将从北土城路开始，这是一个位于北中轴路的行车路之间的正式美化环境空间，并在北四环路上形成一条96m宽的步行大道，直达（和超越）洼里森林公园的中央。

比赛场地主要被组织于步行大道的东面，于是各场馆之间的人流将被改善。步行大道的西边将有一片草地，把元朝城墙遗址公园、中华民族文化公园和洼里森林公园等空间连贯起来。

惟一的永久例外是于奥运公园中大屯路轴线上占据一个奥运体育馆，这是出于对体育馆的象征性和最终对历史的重要性方面的尊敬。

总计划将把更多重点放于对软性风景的设计上。轴西面大部分的地方，将归于真正的奥运公园。

这个拥有草坪和林阴的美化环境空间，将从元朝城墙遗址南部延至洼里森林公园的顶端，成为新的"市肺"。

■ Planning and Design Concept

Our master plan concept promotes the celebration of the Axis in a real and physical manner - not just a simple visual cue. We are promoting the central Axis as the principal pedestrian mall organizing and linking all of the sites of the Beijing Olympic Green.

The mall will commence as far south as BeiTuCheng Road where it is principally a formal landscaped space between the carriage ways of Bei Zhong Zhou Road and becoming a 96m wide pedestrian promenade above Bei Si Huan Road which extends all the way to the center, (and beyond) of Wa Li Forest Park.

The Games venues are organized principally on the Eastern side of the Promenade in such a manner that the convenience of pedestrian circulation between venues is enhanced. The western side of the Promenade is reserved for parkland establishing an urban scale link between the spaces of the Yuan Dynasty City Wall Relics Park, the Chinese Ethnic Culture Park and, ultimately, Wa Li Forest Park.

The only permanent exception is the Olympic Stadium which occupies a place on the axis of Da Tun Road within the linear Olympic park. We have done this in homage to the symbolic and eventual historic importance of the Stadium.

Our master plan will place more emphasis on soft landscape design than perhaps has ever been made before. The majority of the Western side of the Axis will be dedicated to the true Olympic 'Green'. This pattern of rich landscaped spaces of lawns and thick trees will extend from the southern most edge at the Yuan Dynasty City Wall Relics Park to the very top of the Wa Li Forest Park - becoming new 'lungs' for the City.

鸟瞰图(赛后)
BIRDS-EYE VIEW (POST GAMES)

模型
MODEL

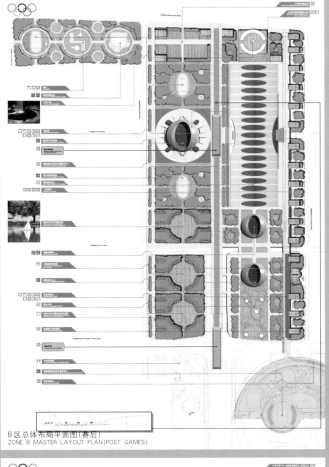
ZONE B MASTER LAYOUT PLAN (POST GAMES)

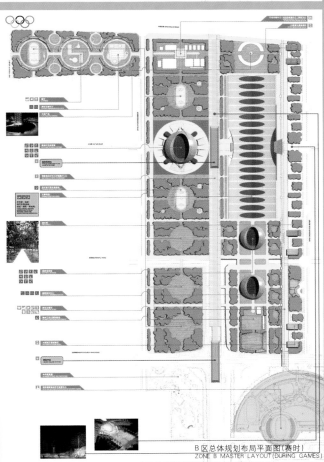
ZONE B MASTER LAYOUT (DURING GAMES)

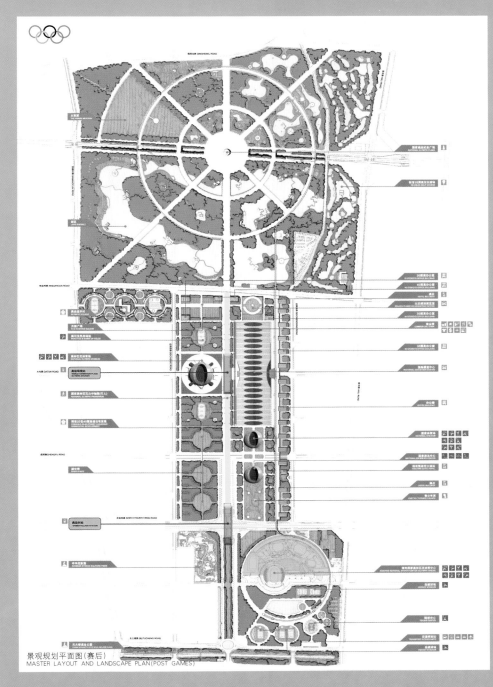
MASTER LAYOUT AND LANDSCAPE PLAN (POST GAMES)

Beijing Olympic Green 2008

北京奥林匹克公园设计方案 Beijing Olympic Green

■ 规划设计构思

规划提议在规划区内建设寓意着生命的源泉的水池——"天池"和辽阔葱绿的山丘——"地林"来作为北京市的未来城市的象征。

象征着这片土地的水和绿将北京市重重围绕，将现有的公园和水池网联起来，形成动植物相共生的"生态走廊"，展现出与自然相共存的崭新的城市形象。

在阳光照射下璀璨辉煌的"天池"宽800m，长2300m，她将与古老的万里长城并肩成为中国向全世界·全宇宙宣传"环境都市北京"的纪念碑。

利用建设剩土堆建起来的"地林"，模仿老北京城的景山，形成能够饱览新北京市全景的观景点。

"天池"与"地林"在面临高度的发展与开发的北京城里，将成为给予城市生活以滋润与恬静的宝贵的都市开放空间。正如现今的天安门广场与曼哈顿的中心公园一样，"无为而为"，亦是本规划的设计理念之一。

■ Planning and Design Concept

As symbols of the future urban development of Beijing, this plan proposes the construction on this site of a reservoir, called the Heavenly Pond, in praise of the water that is the source of all life, and an Earthly Woods, a rich expanse of green hills.

The water and greenery symbolized by this location will envelop the city of Beijing, forming a network with existing parks and ponds and an "ecological corridor" for animal transit, and creating the model for a new type of city based on coexistence with nature.

The gleaming "Heavenly Pond" will reflect sunlight and will measure 800 m x 2,300 m. It will stand with the Great Wall as a monument to China's commitment, made to the world and the entire universe, to establish Beijing an environmentally friendly city.

The Earthly Woods, on a hill created using the earth left over from construction, together with the former Beijing Castle and Prospect Hill will become a new scenic view location providing a panoramic view of the city of Beijing.

This Heavenly Pond and Earthly Woods will provide the city of Beijing, which is undoubtedly about to experience a high level of growth and development, with valuable open space that will provide the people of Beijing with a place for rest and relaxation in their urban lifestyles. As exemplified by the existing Tian'anman Square in the city, and by Central Park in Manhattan, the value of "not building" is another principle of this plan.

Beijing Olympic Green 2008

株式会社日建设计（日本）
NIKKEN SEKKEI (JAPAN)

参赛作品 Works

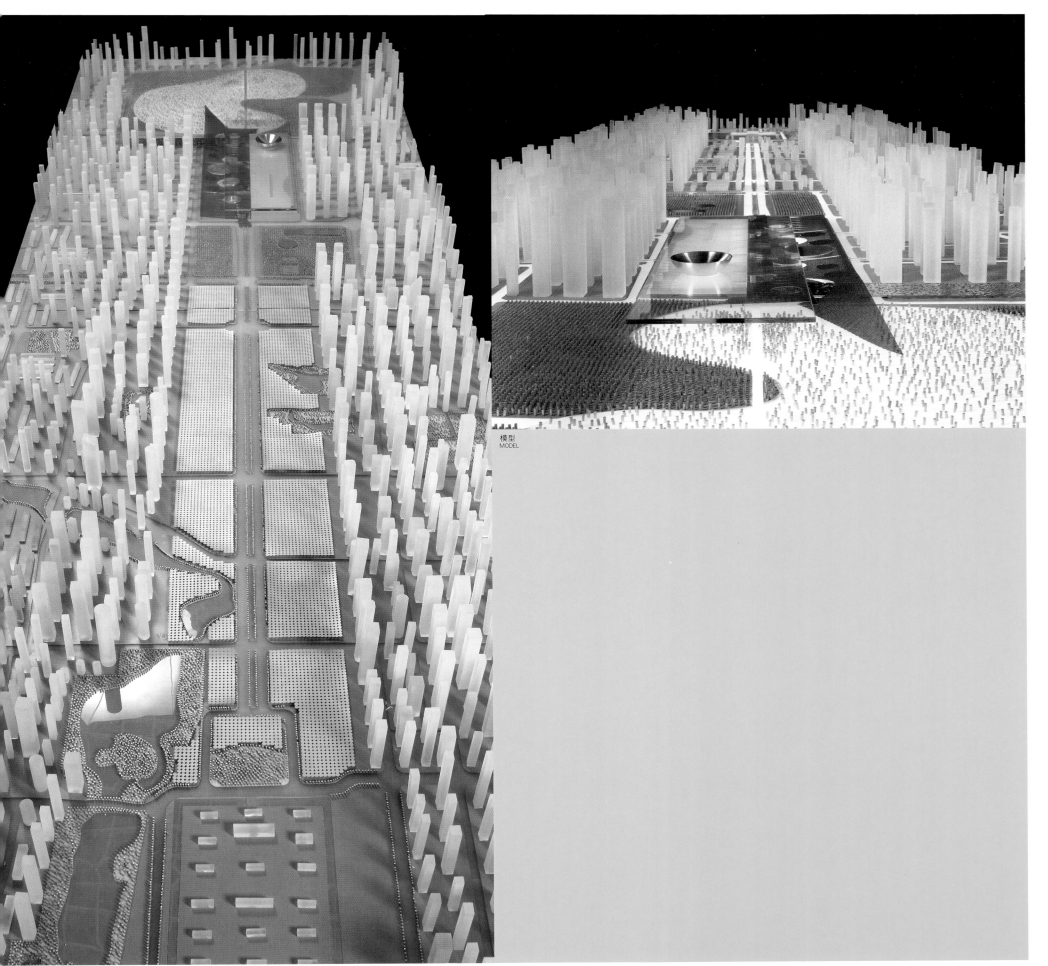

模型
MODEL

市政规划平面图
PLAN OF MUNICIPAL DESIGN

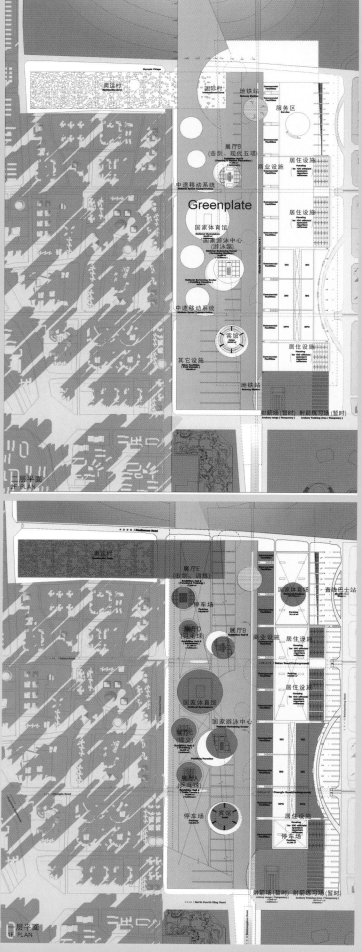

北京奥林匹克公园设计方案 Beijing Olympic Green

■规划设计构思

总体结构

绿色生态的导入

一系列奥运体育设施在中轴线东侧连续展开，坐落在连绵起伏的草坡之上，形成带状体育公园，与南端的国家体育中心连成一体，并将森林公园通过体育公园与绿色中轴，将绿色生态环境导入城区。

北京中轴线的延续

中轴线空间是奥运期间的主要人流集散空间，也是重要的观景走廊。体育场馆、商业建筑及会展建筑群，如同展品般在两侧展开。中轴上的下沉广场，连接着地下商业街与地铁交通，中轴线上的"金、木、水、火、土"的系列主题景观设计，延续了北京中轴线的历史人文气息。

城市公共活动中心的塑造

将会展建筑群布置于四环路北侧的中轴线两侧，整体性的建筑群体形象，成为中轴北端重要的标志性景观，将商业娱乐设施集中布置形成街区，有利于营造富有活力的城市生活气氛，沿北辰西路与北辰东路带状办公与综合商业服务建筑群，在完善公共活动中心功能的同时，为丰富城市生活提供便利。

功能布局

国家体育场、国家体育馆、国家游泳中心，在中轴一侧带状展开，利用中轴空间作为赛时主要的人流集散场所，并成为中轴东侧优美的城市景观，强化了中轴的景观。场馆之间以起伏的绿化平台相互沟通，并与森林公园的水面相连，使体育、绿化与城市有机结合，浑然一体。

文化建筑周围的城市开放空间更强调休闲、融合的氛围，是城市公共活动中心区的调和空间。将首都青少年宫、中国杂技马戏馆、北京城市规划展览馆分散布局在中心区内，与公园、体育、会展等设施相邻或结合布置，将在城市中增添一系列有吸引力的场所，增强城市活力和文化氛围。

会展中心多重的角色和功能的多样化，决定了她应成为一个巨型的综合体，城市的多功能厅。

商业空间是保证中心区健康运转和活力的重要元素，将商业呈带状集中布置于中轴线地下两侧，形成地面、地下乃至空中不同层

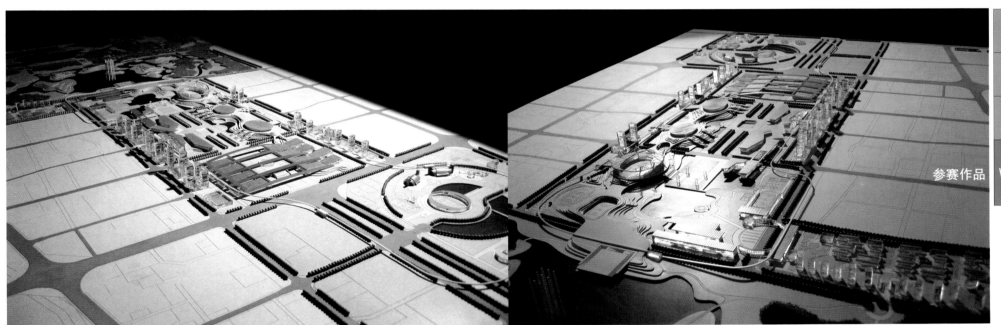

参赛作品 Works

Beijing Olympic Green 2008

上海同济城市规划设计研究院（中国）
SHANGHAI TONGJI URBAN PLANNING & DESIGN INSTITUTE (CHINA)

面的步行街，将商业空间与城市开放空间紧密结合。同时使之成为联系其他城市功能区的通道，将城市公共空间串联起来。商务区位于中心区两侧，以两条高层带强化中心区外围界面，使之具有完整的城市形象。

城市景观
以中轴线串联起一系列景观节点，构成序列，由南向北逐步将城市景观推向高潮。

五行系列广场
以金、木、水、火、土为主题的系列广场，构成人文主题的景观系列。

水之广场
200m高的喷泉以一柱冲天的形象作为中轴线终点的景观，在水面的"七大洲"环绕下，充分体现世界各地人们聚集在奥林匹克大家庭中的融洽氛围。

交通组织
倡导性流线——公交、自行车绿色流线

适当限制流线——私家机动车流线

优先满足交通流线——运动员、官员、媒体及补给救护等

■ Planning and Design Concept

Overall Structure
Promoting a green, ecological environment - Situated in rolling green meadows a series of Olympic facilities will form a belt shaped sports park and connect with the National Sports Center situated at the southern end. A green ecological environment will be promoted by linking the Forest Area with the sports park thus creating a green city central axis.

Extending the Beijing city central axis - The area around the city central axis will become the primary spectator distribution space while the Games are in progress. The landscaping design, seen in a series of the five traditional themes of gold, wood, water, fire and earth extending along the city central axis, will reflect the historical, humanitarian culture of Beijing.

Creating a city central area - The exhibition buildings will be arranged along the two sides of the city central axis and they will become landmark buildings at the northern end of the city central axis. Commercial and entertainment facilities will be grouped together to form a block, creating a vibrant city atmosphere. Office and commercial buildings will be arranged in a belt shape to enrich urban life as well as improve the functions of the central area.

Function Layout
The National Stadium, National Gymnasium and National Swimming Center will be extended along one side of the city central axis. These buildings will become an image in the city landscape along the eastern side of the city central axis. Event venues will be connected by rolling green areas, water and forest thus integrating sports, greenery and city. The capital teenage palace, Chinese Circus and Acrobatic Hall and the Exhibition Hall of Beijing City Planning will disbute

central area adjacent to parks, sports and exhibition facilities. This will add a series of organic, attractive venues to the city to enhance the vitality and cultural atmosphere of the city. The Convention and exhibition facilities will combine many roles in the city central area; the diversity of its functions will create a large scale complex as well as a multi-function hall. Commercial facilities will be arranged in a belt shape centered along the two sides under the city central axis, forming a pedestrian street with different levels, This closely connected commercial space to the city open space will become a pathway to the other functional areas within the development. The business district will be located at the two sides of the central avenue flanked by two high rise building belts to create an integrated city image.

City Landscape
A series of landscape features connected by the city central axis will push the landscape of the city to a climax from south to north..

"Five theme"Squares – A series of squares with the themes of gold, wood, water, fire and earth will be strung along the city central axis like pearls, creating a series of human scale landscapes.

Water square – A huge 200m high fountain, dominates the water square located at the northern end of the axis and will become a key landscape feature. Surrounded by "Seven Continents" in the lake, this feature fully expresses the harmony of people all over the world coming together in the Olympic family.

Transportation
Primary Flow - Bus and bicycle (green flow).

Relatively Restricted Flow - Private Vehicle Flow.

Priority Flow - Athletes, officials, media, supplies and emergency vehicles.

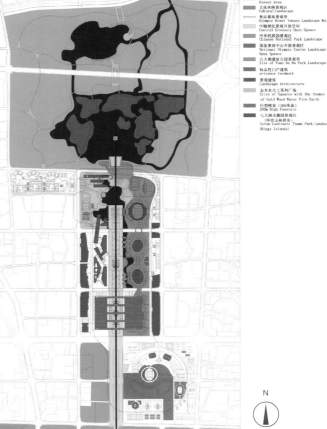

北京奥林匹克公园设计方案 Beijing Olympic Green

参赛作品 Works

全区总平面图二期工程
ZONE MASTER LAYOUT PLAN PHASE 2

■ 规划设计构思

本设计方案以北京南北中轴线为出发点提出了一个新的城规设计理念。

空间上,该轴线作为一个大型的具有规则几何形的"水景公园"继续延伸,一直向北伸展至森林公园的自然景观与主交通道相互分离。以这样的布局不仅产生了一个大型"奥林匹克公园",而且在运动会之后亦可作为休闲的公园,即作为城市的"绿色之肺"而发挥作用。它同国家体育场一起形成通达四方的中心,在该中心周围排列着2008年奥林匹克项目的各个功能区域。

在南部耸立着两个雕塑般具有有机造型的约250m高的塔楼群组,它们作为风格独特的"迎客大门"而构成整个区域的标志。

这个大门与"世贸大厦"亦即商务中心同时构成绿色奥林匹克公园入口和前广场,可以把它看作是沿整个奥林匹克公园周边而建的功能设施在建筑上的序曲。整个公园的重点建筑是国家体育场,它作为拥有升旗广场的奥林匹克运动场位于水景公园的中轴线上。

国家体育场位于中心位置,这使该建筑在城市空间中心占有一个突出的地位,其意义永久而重大。在北京的南北中轴线上,它是由北京城向外延伸的中轴线建筑的终端,同时又构成森林公园通向城市轴线上的开端。

水景公园和国家体育场的东西侧平行布置建筑用地,建筑用地分别划分为功能区域,用于体育馆和服务设施等功能,并以建筑产生城市的轴心空间。

"奥林匹克村"设置在规定的B—1区里,在规划上与中心区域衔接。东南部现存的体育设施区域(A区)在形式上和功能上均得到补充并以其圆形的设计而得以完善。

■ Planning and Design Concept

The design transforms the important north-south axis of Beijing into a new model for urban design. The axis will extend and project as a large geometric ‚Water Park into the free-plan Wa Li Forest landscape and will be free of through-traffic. In this way, a large Olympic Parkland is created, called the Olympic Green, which will be a green lung' recreation area for the city after the games. The National stadium surrounded by the functions of the Olympic Project 2008 will make the link.

To the south, two clusters of organically shaped, 250-m high-rises articulate a welcoming landmark ‚city gateway situation.

Together with the Business Center and the Market Plact, this forms the access point and plaza in front ofthe Olympic Green and is the architectural opening beat of the orchestration of functions flanking the Olympic grounds.The design focus of the project area is the National Stadium, i.e. the Olympic Stadium with a large open forecourt and flagging plaza. This is sited on the axis that leads up to the Water Park.

Beijing Olympic Green 2008

GMP 建筑师事务所(德国)
ARCHITECTS VON GERKAN, MARG UND PARTNER(GERMANY)

景观规划和主要景点设计
LANDSCAPE PLANNING ANALYSIS AND KEY AREAS

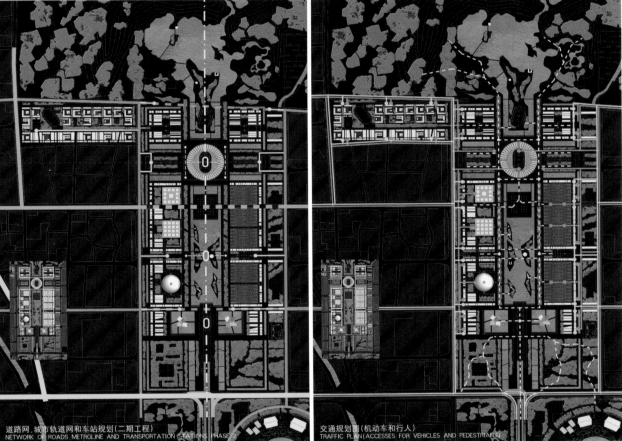

道路网,城市轨道网和车站规划(二期工程)
NETWORK OF ROADS, METROLINE AND TRANSPORTATION STATIONS PHASE 2

交通规划图(机动车和行人)
TRAFFIC PLAN (ACCESSES FOR VEHICLES AND PEDESTRIANS)

The central position of the stadium makes it the hub of this urban area, its value always obvious as the architectural termination of the central Beijing tranic axis in both directions: from and to the city, from and to Wa Li Forest Park.

East and west of Water Park and Stadium, a number of development areas are aligned in parallel rows. Each area defines five functians, including sports halls and sevice facilities, and forms the architectural and urban design frame for the entire length of the axis area.

The Olympic Village is sited in area B-1 and connected to the central area in terms of urban planning. The existing sports complex (area A) in the south-east will be added to both formally and functionally and its circular footprint filled in.

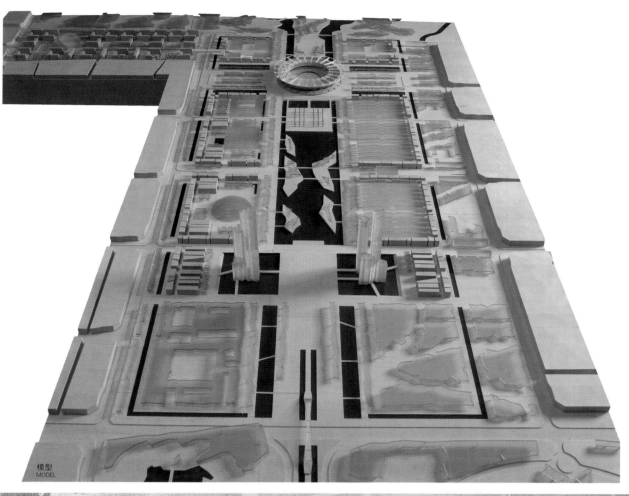
模型
MODEL

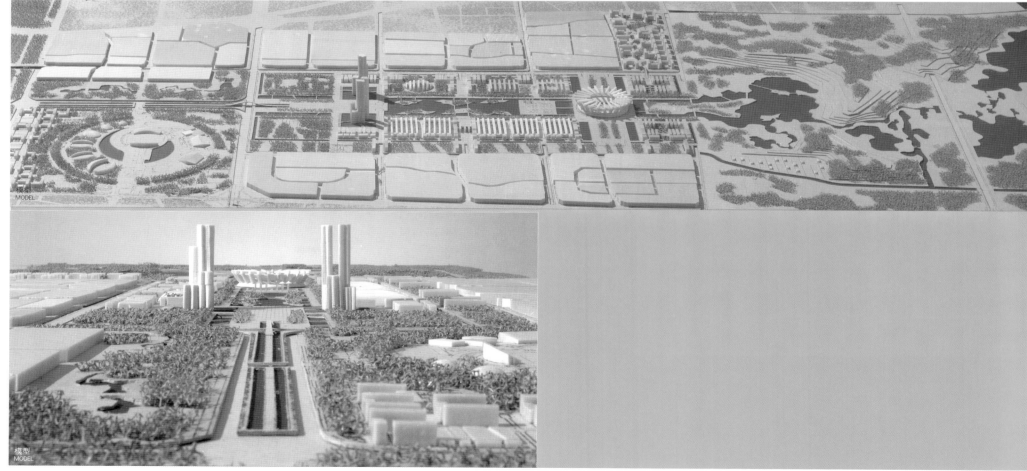
模型
MODEL

北京奥林匹克公园设计方案 Beijing Olympic Green

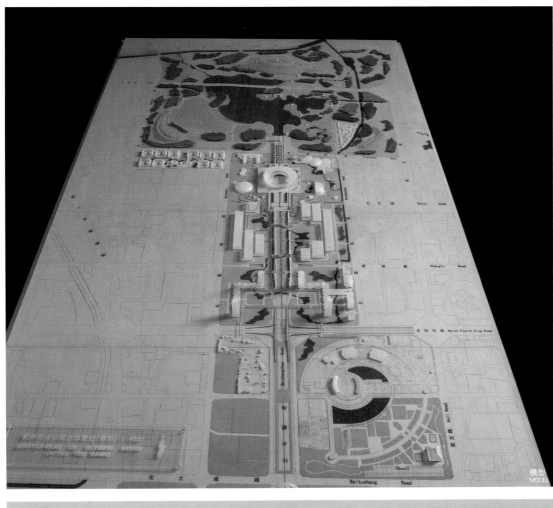

模型 MODEL

■ 规划设计构思

奥运会的场馆结合北京的城市发展考虑

利用奥运会的契机，规划都市的基本骨架（包括交通，市政设施，场馆以及都市景观等），充分挖掘地域的活动潜力，为今后的发展提供可能。

市民活动中心的设计以整体为目标

充分考虑参与性，力求成为具有丰富内容的新的活动中心，以及能为市民的日常生活提供多样有益活动的新场所。

创造具有可持续发展的新的都市环境典范

利用丰富的水面及绿地为市民提供宁静、休闲的理想的环境的同时，也能让市民重新认识环境。

营造能进一步促进发展的城市结构

在设计中融合各种最新的设计方法和技术的同时，考虑了地区的经济性，为实现可持续的分期建设提供可能。

利用已有的文化遗产作为北京市发展的象征

延续作为历史遗产的城市中轴的概念的同时，把其作为开拓新时代精神的表象空间化，具象化。

综合四个系统的规划设计：整体规划上的水、绿、人的活动，象征的轴等四个系统互相重叠，表现整体上的形态。

水的系统：在设计中把全长5km的水面以龙的形态形化，作为中国的精神象征。

绿的系统：创造宁静祥和的环境，在北中轴的山顶上能感受到绿是环境的主角。

人的活动：利用廊的系统为地区内的人群提供行走，交流以及赏景的场所。

象征的轴：让人们能体会到贯穿故宫的北京市的象征中轴的真实感。

参赛作品 Works

Beijing Olympic Green 2008

环境设计研究所（日本）
ENVIRONMENT DESIGN INSTITUTE(JAPAN)

上海市地下建筑设计研究院（中国）
SHANGHAI UNDER(CHINA)

佐藤尚已建筑研究所（日本）
NAOMI SATO ARCHITECTS(JAPAN)

鸟瞰图 BIRDS-EYE VIEW

■ Planning and Design Concept

Combine with the urban development of Beijing to consider the fields and gymnasia for Olympic Games

Utilize the turning points of Olympic Games, plan the basic framework of the metropolis including traffic, municipal facilities, fields and gymnasia, and urban landscapes, etc., fully unearth the active potential in the region, and provide possibilities for the future development.

The design of Civil Active Center is directed toward the integrity

Fully consider the visitors and participation of people from in and out of Beijing, and strive to become the new active center with rich contents and the new place which can provide multiple beneficial activities for the citizens' daily life.

Set a new example of metropolitan environment with sustainable development

Utilize the abundant water surfaces and greenbelts to provide a quiet, leisurely and ideal environment, which can also let citizens refresh themselves, and recognize and learn about the environment at the same time.

Build the urban framework which can further promote development

At the same time of blending various up-to-date methods and techniques in the design, consider the economical efficiency of the region and provide possibilities for the realization of sustainable construction by stages.

Utilize the existing cultural heritages as the symbol of the urban development of BeijingAt the same time of extending the concept of the city as the central axis of the historical heritages, civilize and concretize the representation which everybody can understand and pass down from generation to generation as the spirit of exploiting a new era.

Design of integration of the four systems

(Explanation of perspective drawing) The four systems, namely, the activities of the water, green and human as well as the axis of symbol, overlap each other in the integrated planning and represent the form as a whole.

System of water: visualize the water surface with an overall length of 5km according to the form of the dragon as the spiritual symbol of China in the design.

System of green: create a quiet and serene environment and experience the feeling that the green is the leading role of the environment at the mountaintop of the north central axis.

Activity of human: utilize the corridor system to provide the recreational places for the crowds in the region to walk, communicate and enjoy the scenes.

Axis of symbol: Let the people have cognizance of the new and true feeling of the symbolized central axis of Beijing through the Imperial Palace.

System of water
System of green
Activity of human
Axis of symbol

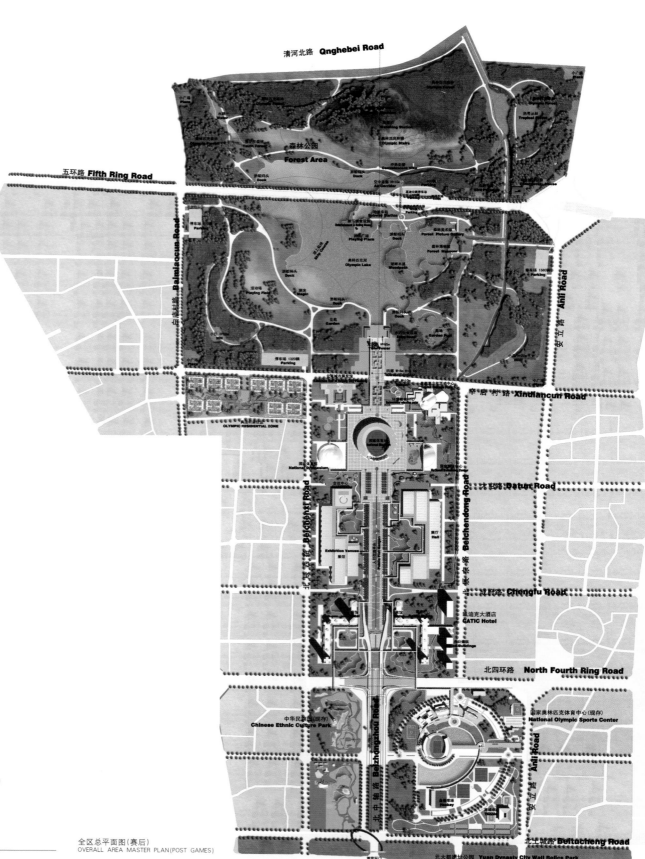

OVERALL AREA MASTER PLAN (POST GAMES)

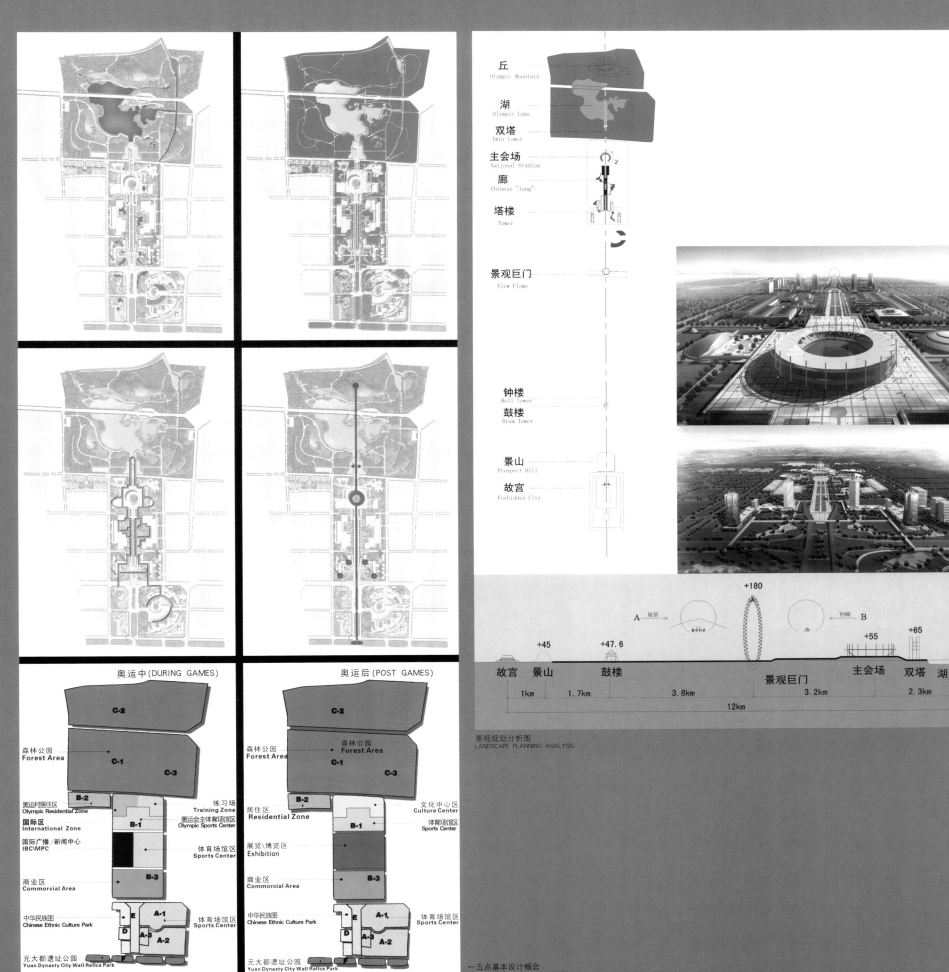

北京奥林匹克公园设计方案 Beijing Olympic Green

参赛作品 Works

■ 规划设计构思

建设一个体育、娱乐、商业区的战略

体育、娱乐、商业区战略是指在社区附近配置主要的公共集会设施和景点,并辅有高级、有影响力的娱乐场所。该战略可以促进北京的旅游业的发展,提高对外来人员的接待水平,并可通过以下战略推动经济发展。

在一个单独核心区内整合公共集会设施,运营商可以大大降低设施及基础设施的成本,有效地利用资产,便于设施的使用,提高经营利润。

通过促进旅游业、创业和私人投资,推动经济发展,创造就业机会。

支持已有商业中心的保存及扩建,从而与地方和地区企业建立经济联系。

提高现有和将来的周边住户的生活质量。

为该地区树立新形象,增强地位,使其成为区域中心和聚集目的地。

对体育、娱乐、商业区的公共和私人投资必须以潜在的投资回报期待为前提——市盈率、直接收入、辅助收入及对经济发展的贡献。该战略的基础是:通过一级市场,以及通过设立完成目标(基于切实的市场潜力评估来设立)来增加旅游收入。

奥运体育馆、奥运运动场、会议中心以及会议宾馆的主要坐落位置有利于协同发展,这是在北京奥林匹克公园内建立一个生机勃勃的体育、娱乐、商业区的基石。

■ Planning and Design Concept

Creating a Sports/Entertainment/Business District Strategy

Sports/Entertainment/Business District Strategy refers to the placement, in relatively close proximity within a community, of major public assembly facilities and attractors supplemented with high profile/high impact entertainment venues. This approach can advance the development of tourism and hospitality in a city like Beijing, thus promoting its economic growth through the following strategies:

Consolidation of public assembly facilities in a single core district: Consolidation of facilities and operators offers significant reductions in facilities and infrastructure costs, efficient utilization of assets, ease of access, and improved operational profits.

Spurring economic growth and job creation through increased tourism, entrepreneurship and private investment.

Supporting the retention and expansion of the existing business centers, thereby creating economic links to local and regional businesses.

Enhancing the quality of life for current and future residents in the surrounding area.

Fostering a new and enhanced identity for the area as a regional center and destination place.

The public and private investments in a Sports/Entertainment/Business District must be premised on an expectation of the potential return based on marketplace yield, direct revenues, secondary revenues, and related contributions to economic growth. This strategy builds upon increasing tourism revenues by drawing on the primary market and by establishing "performance goals" based on a realistic assessment of market potential.

The prime location of the Olympic Gymnasium, the Olympic Stadium, the convention center complex, and the convention hotels lends itself to the kind of synergistic development that is the cornerstone of the creation of a vibrant Sports/Entertainment/Business District within the Beijing Olympic Green.

中国建筑设计研究院(中国)
CHINA ARCHITECTURE DESIGN INSTITUTE(CHINA)
KPF 设计公司(美国)
KPF ASSOCIATES(USA)
泛亚易道公司(香港)
HONGKONG FANYAYIDAO INC.(HANGKONG)

Beijing Olympic Green 2008

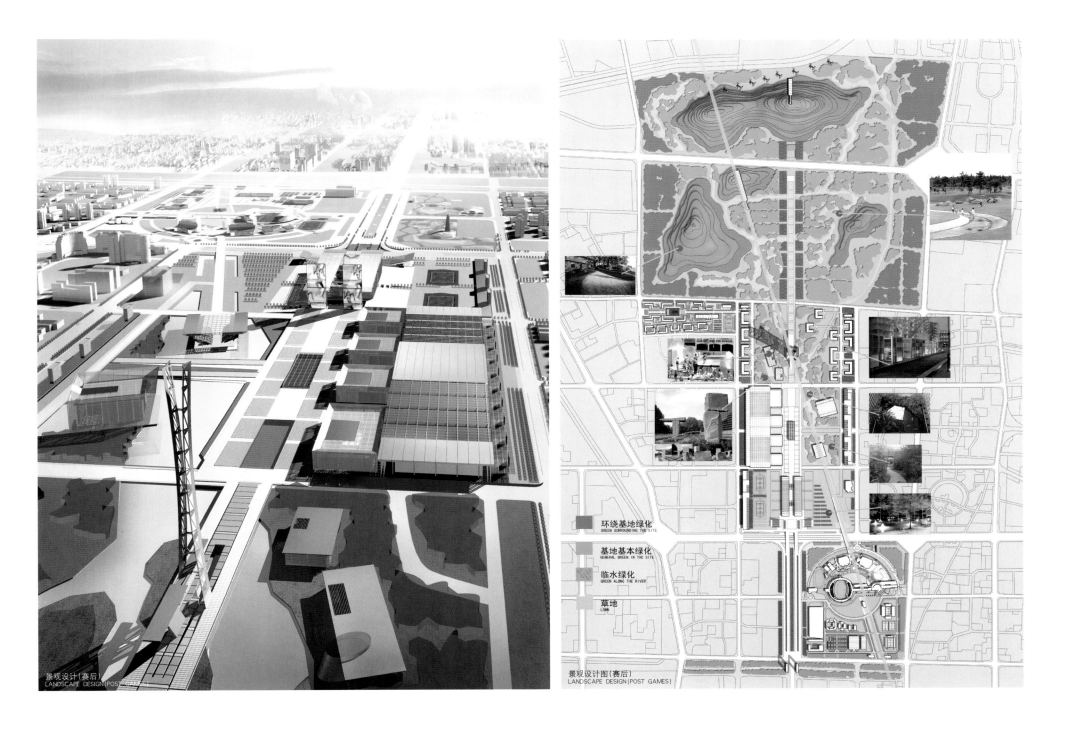

景观设计(赛后)
LANDSCAPE DESIGN (POST GAMES)

景观设计图(赛后)
LANDSCAPE DESIGN (POST GAMES)

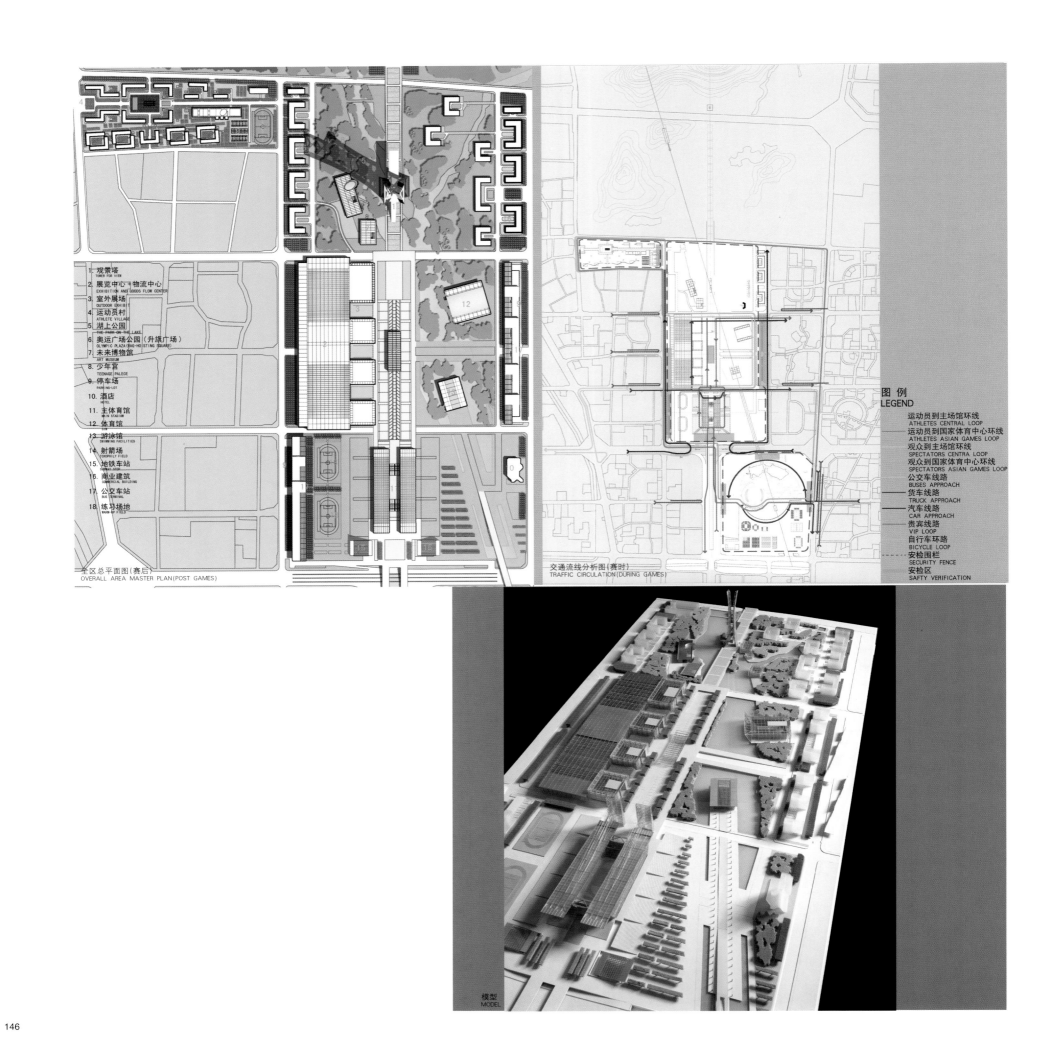

■规划设计构思

1.规划设计原则

(1)城市发展的延续性

本方案将中国5000年历史与2008年奥运会及以后长久发展统一起来认识,认为城市未来5000年的发展重于奥运场馆的具体规划。

(2)传统与现代

奥运短暂,未来漫长,而奥运也终将成为历史。所以,对历史的尊重使我们在现代设计中充分考虑中国的人文传统,并力图反映在设计当中。

(3)理想与现实

2008年奥运会将记载全体中国人的美好理想,我们仔细分析了北京现有条件,试图将现实条件与理想主义追求协调起来。

(4)情感与形式

对历史的尊重,结合设计构想的现代化,将超越建筑的局限。

(5)人的价值观

我们希望自己的作品在反映人们物质条件提高的同时提升人们的精神境界。我们希望展现在世界舞台前的是一个焕发活力的文明古都。

(6)开发与自然环境的关系

即做到自然与城市的共生,其他生物与人类的共生。

(7)对高技术的慎重采用

(8)经济性

规划方案要有经济上的可实行性,城市的开发需要阶段性的规划,不可能一蹴而就。

(9)象征性

2.设计要点

(1)设计之源

承载了5000年的历史与文明,在这个欣欣向荣的国度里,我们发现源是中国。

(2)城市之轴

这也是一条城市发展的生命线,它的延长就是历史的延长,是5000年的延长。我们认为在这条轴线上的规划设计,应注意理念上的象征,意义上的追求。

(3)奥林匹克

毫无疑问,2008北京奥运将是奥运历史上最重大的事件之一,同时,我们希望看到该地区的持续良性发展。

■ Planning and Design Concept

1.Principles of Planning and Design

(1) The sustainability for urban development. Our project tries to integrate the cognition on the relationship between the 5000-

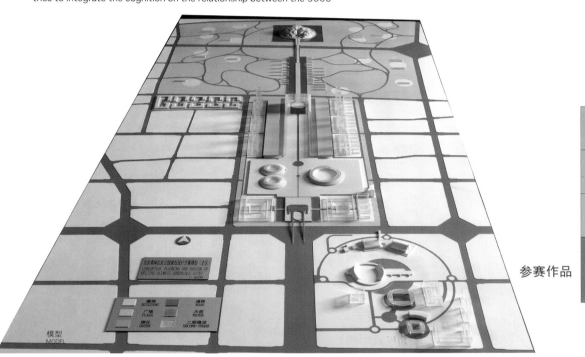

Beijing Olympic Green 2008

(株)川口卫建筑构造设计事务所(日本)
KAWAGUCHI & ENGINEERS(JAPAN)

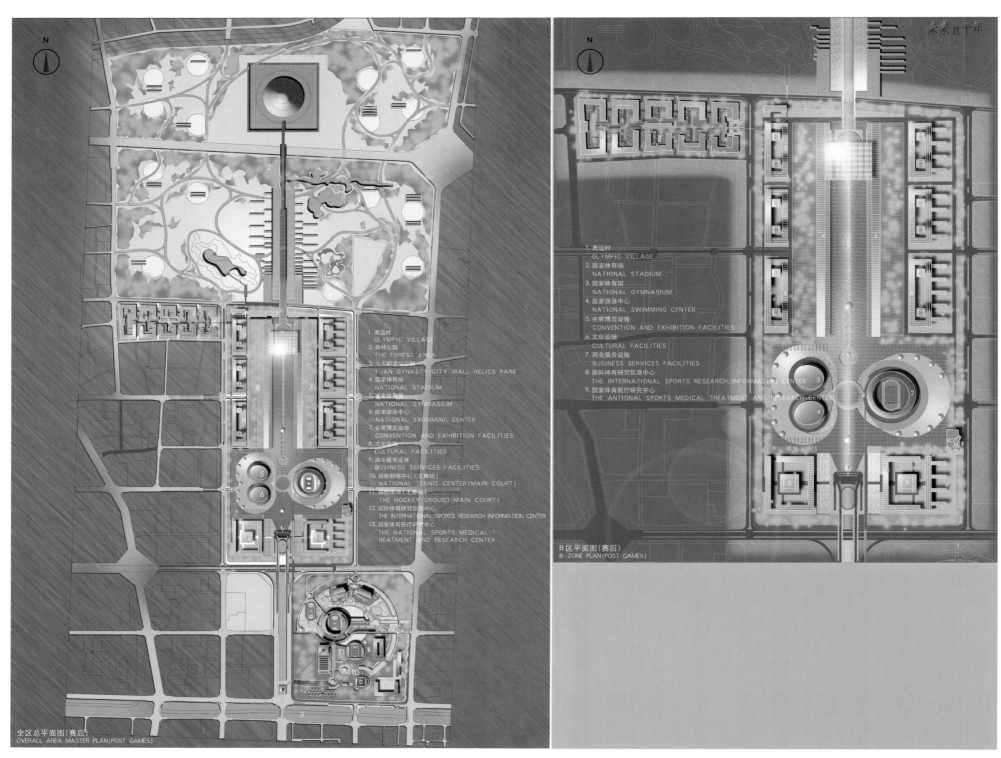

year-long history of China and the 2008 Olympic Games

(2) Tradition and modernization. The transience of the holding for the Olympic Games will be accentuated by the elapse of the future years, which will put the Olympic Games into distant memory, too. For that reason, the history needs respect. Further more, in our project, the Chinese traditional context has been fully considered and reflected.

(3) Ideality and reality. The 2008 Olympic Games will record the Chinese people's honors and dreams. Here, we have carefully analyzed the present conditions of Beijing, trying to harmonize the ideality with reality.

(4) Emotion and formalization. That is, the idea including both the respect for our tradition and the recognition for modernization of the design united, can transcend the limitation of the single building itself.

(5) The point of view on value. we are making efforts in our design to promote people's spirit standards, while their living conditions keep improving. We hope the city revealed on the stage is a flourishing metropolis with both long tradition and bright prospect.

(6) The relationship between the exploitation and the nature. In other words, that means the accretion between the cities and the nature, as well as human beings and other creatures.

(7) Cautious utilization on High-tech.

(8) Economy. The growth of the cities depends on the cycling of the social resources. So, the urban planning should be available and economical for coming into reality. In fact, the exploitation to cities needs step-by-step urban planning. It is unrealistic to make a terminal planning for once.

(9) Symbolization.

2.Key points in planning and design

(1) The source of our design. Beijing loads the 5000-year-long history and civilization as well. Here, in this flourishing country we have found that, the source of our design is, China herself.

(2) The axis of our city. It is conducted in the north-south direction, symbolizing a connection between the city and the universe. When extended, the axis, the artery for Beijing's development, also extends Beijing's 5000-year-long history, as well. Thus, we suppose that the layout on this axis should pay much attention to the symbolization of its concept and the pursuit of its content.

(3) Olympics. Undoubtedly, the 2008 Olympic Games in Beijing will turn out to be one of the greatest events in the Olympics history. We are looking forward to the well sustainable development in this area even after the Olympic Games.

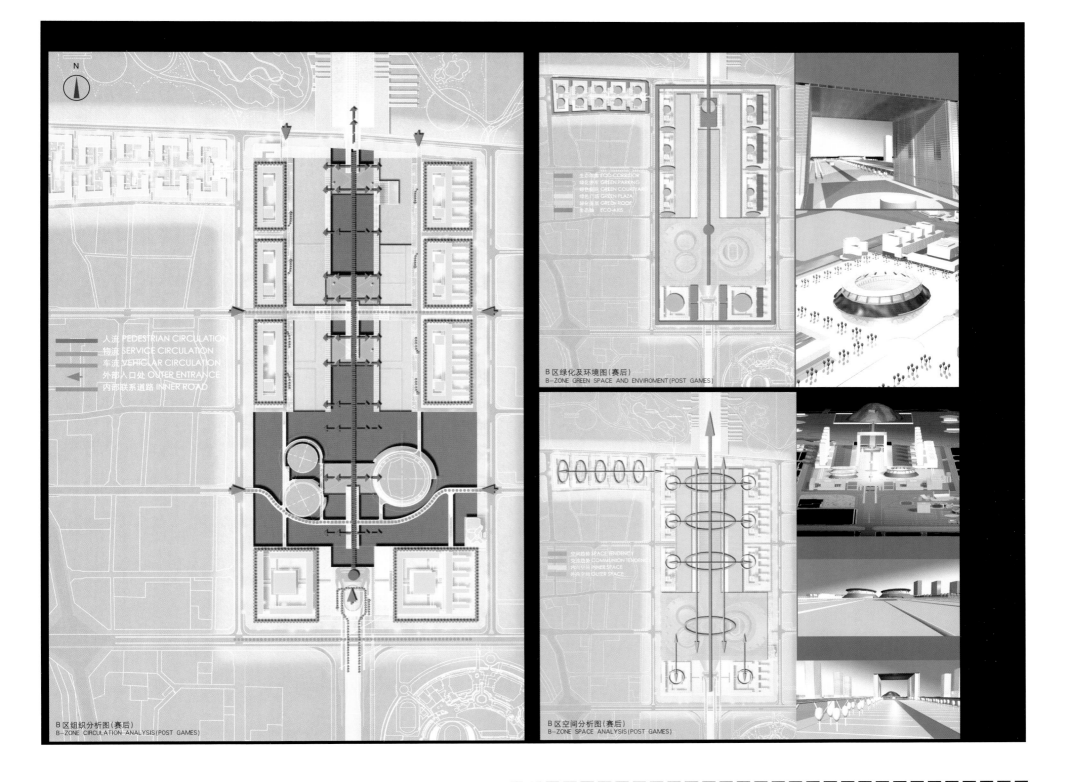

Beijing Olympic Green 2008

DECATHLON S.A. 规划及工程公司（希腊）
DECATHLON S.A. PROJECT PLANNING & ENGINEERING CONSULTANTS(GREECE)

■规划设计构思

方案不仅在设计概念上，而且在部分工程上，比如：交通、绿化、基础设施等方面，突出以下重点：

尽管举办奥运会是临时性的，但设施的外观及运行功能将是完善的（美观与运行功能的持续性）。

楼群、空地等的设计，尽量做到稍经改动，就能适应奥运会期间的需求（节约）。

设计的灵活性。不仅做到整体而且局部也能适应可能的改动（灵活性）。

给予这些设施醒目的特征，使之成为北京城市历史上的一个现代的新起点（穿时性）。

方案视整个设计为一个有机结合的整体，比如：基础设施、交通、地面、高架行人区、楼群。也就是一个基础坚实的设计方案，不仅允许也能接受未来的用途。

设计的重点主要是公众使用区域，而不十分注重楼群的外表改建。

预期目标如下：

景观部分，在实质上将是城市相关区域的延续。

通过营建一个动（人行道）、静（广场）态高架线状活动区域，不仅与各楼群，而且还和市区相接，作为城市中心区域的继续发展与完善。这一设计方案既突出了轴心区的作用，同时亦使其成了市区各区间的活动中心。增强了市区的观感及优越性。高架轴心区对研究领域来说是一个真正的里程碑，对北京城市发展也是一个里程碑。

■ Planning and Design Concept

In terms of technical concept,as well as in terms of the different specific designs/strdies (traffic,green spaces,utilities etc.),the proposal places emphasis on the following:

Despite the fact that the Olympic Games period represents a temporary phase, the look and the functionality of the space should be complete (Aesthetic And Functional Continuity).

The design of buildings,open spaces etc. should follow princeples allowing them to function with the minimum adaptations during the Games period (Economy).

Flexibility should be built into both the overall result and its individual components can respond to changing requirements (Flexibility).

Assignment to the area of a particular identity that will constitute a contemporary starting point for the urban history of the city of Beijing(Timelessness).

At the planning level,the propsal has been perceived as a system of distinct successive "Status Layers",such as utilities,traffic, ground level,raised pedestrian spaces,buildings.In other words,the proposal constitutes a well-structured system which not only allows or is susceptible of future uses but also induces such uses.

Emphasis is mainly placed on the planning of the Public Communal Space and -to a lesser extent- to the morphology of the individual buildings.

The following are proposed:

Ground level:A -mainly landscaped- open space, constituting in effect a continuation of the corresponding urban tissue.

Implementation of continuity and extension of the main axis of the city through the establishment of a set of linear movenents and stationary locations on a raised level that provides access to all buildings and maintains the connection to the urban tissue. This layout choice symbolises the central axis while at the same time organises this axis as a tool for the arrangement of movements and connections between the various areas of the Village, in other words, it strengthens the inteoigibility and accessibility of the Urban Space. The raised central axis becomes the reallandmark for the area of the design,but also a landmark for the city of Beijing as a whole.

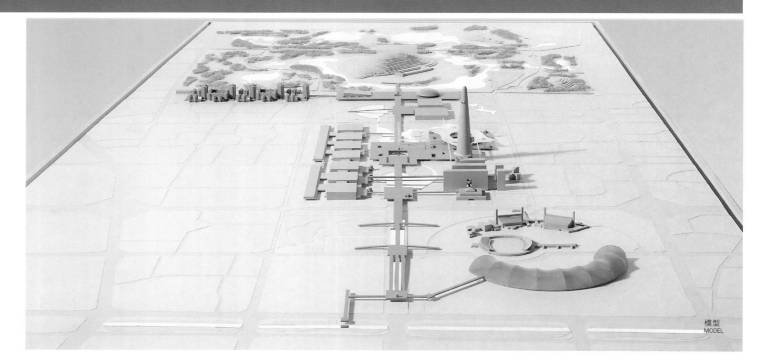

模型 MODEL

北京奥林匹克公园设计方案 Beijing Olympic Green

参赛作品 Works

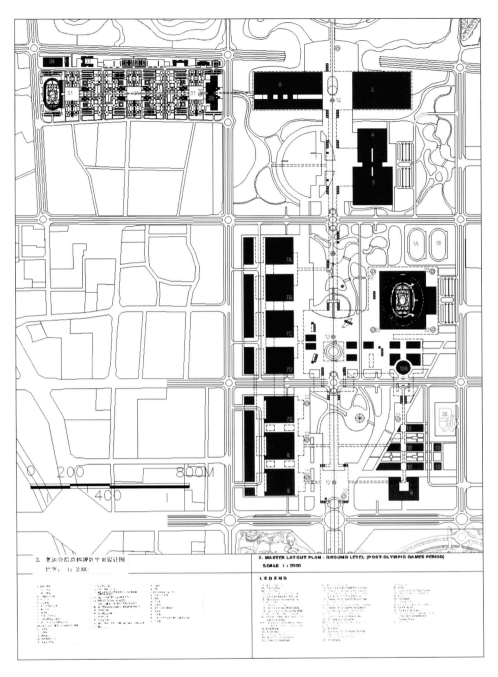

全区总平面图(赛后)
OVERALL AREA MASTER PLAN (POST GAMES)

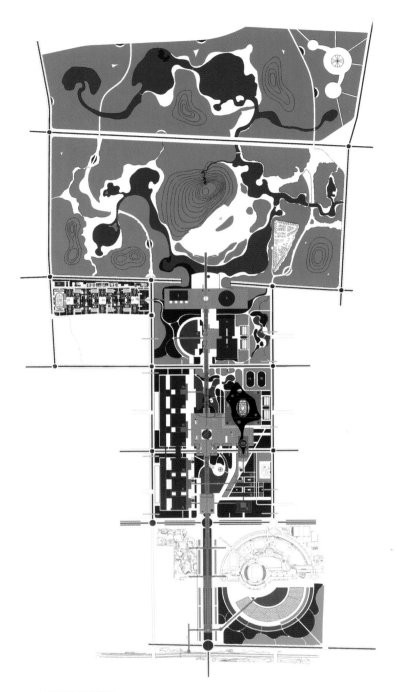

景观规划总图(赛后)
MASTER LAYOUT AND LANDSCAPE PLAN (POST GAMES)

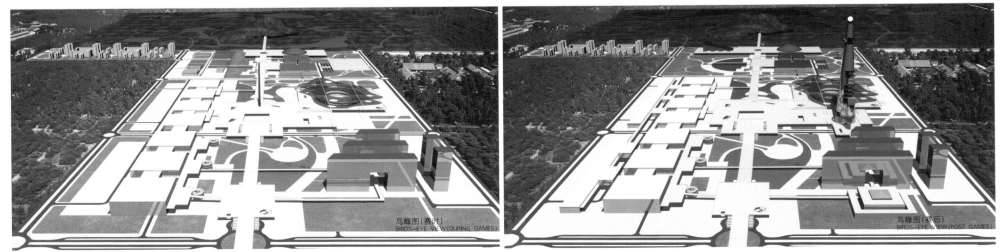

鸟瞰图(赛时)
BIRDS-EYE VIEW (DURING GAMES)

鸟瞰图(赛后)
BIRDS-EYE VIEW (POST GAMES)

北京奥林匹克公园设计方案 Beijing Olympic Green

参赛作品　Works

奥林匹克中心透视图
OLYMPIC CENTER PERSPECTIVE

■ 规划设计构思

规划的构思是做一个集中的建筑综合体或建筑群。将奥运设施相对集中布置，可缩短场馆之间及场馆与交通站之间的交通距离，并增加交流的机会。奥运设施不再是单个的建筑物，而是一个有机的城市空间，同时节省了用地，并为赛后的扩建提供了更为自由的空间。

用一个简洁的圆锥体来表达这个集中的综合体的形式，使得奥运场馆以一种前所未有的竖向集中的方式重叠在一起，同时也是对北京城市中轴线终点的一个解答。竖向交通在这里成为一种重要的交通形式。

景观设计结合了许多生态设计的理念，如雨水收集，植物净化水系统，生物气支持的水净化系统，自然水域被导入地下水等。能量及能量利用的概念在中轴线上以具体的形式得以诠释，使得中轴线成为一条可以被体验的"能量轴"。

■ Planning and Design Concept

Programming thought to make a compact building or building group. It is better if we build them together in one part of the site. In this way the transportation between the facilities would be shortened, as well as between the facilities and the transport hub. At the same time, the facilities will not be just separated buildings. Instead, they will form an organic city space. What's more, this will also save the land and offer more space for the post Olympic development.

Use a succinct cone as the form of this multifunctional building, which makes the Olympic stadiums piling up one onto another in such a way, that no one has done before, is also suitable for the end of the north axis in Beijing. The vertical transportation becomes very important.

Landscape design indicates many economical design concepts such as rain collection, water infiltration, natural water purification through aquatic plants, water purification plant in conjunction with a biogas system, etc. The concept of energy use is interpreted to a concrete form on the central axis, such as energy phenomena, street wave, wave biking. This axis is an "energy axis".

IFB Dr.布拉斯克尔AG公司（德国）
IFB DR.BRASCHEL AG CORPORATION(GERMANY)

Beijing Olympic Green 2008

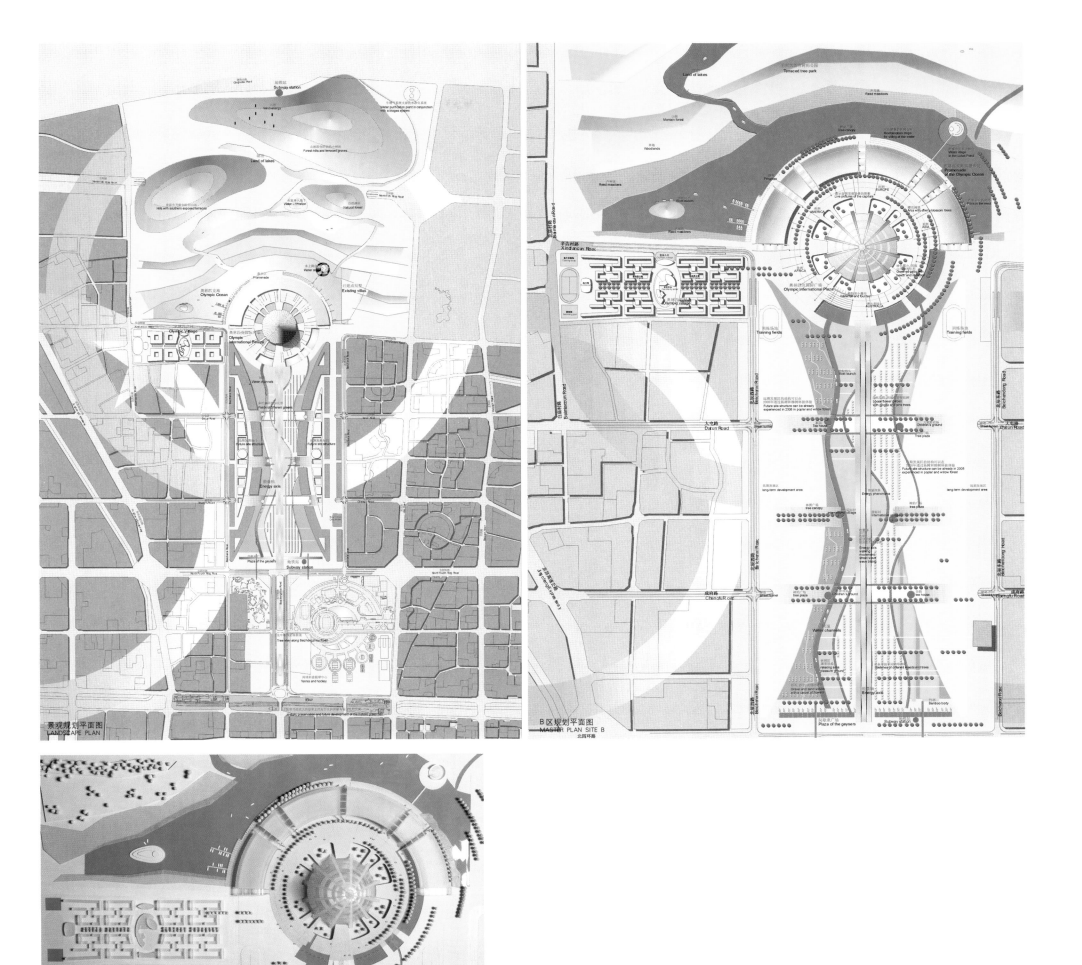

北京奥林匹克公园设计方案 Beijing Olympic Green

参赛作品 Works

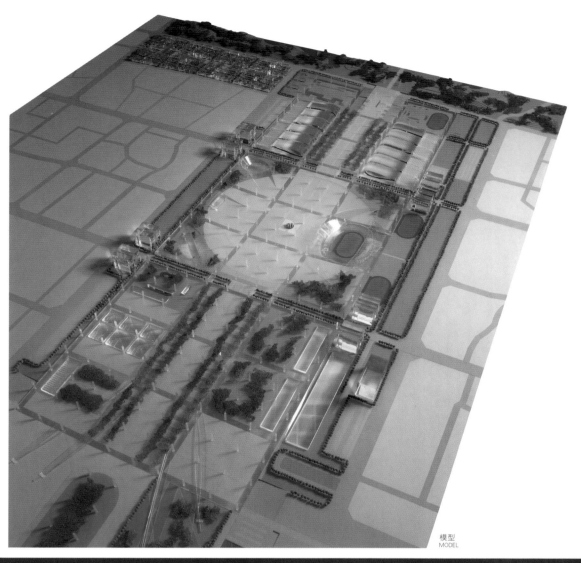

模型
MODEL

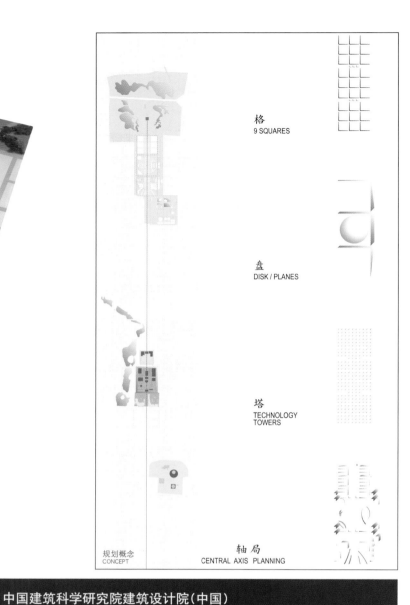

格
9 SQUARES

盘
DISK / PLANES

塔
TECHNOLOGY TOWERS

规划概念
CONCEPT

轴局
CENTRAL AXIS PLANNING

Beijing Olympic Green 2008

中国建筑科学研究院建筑设计院(中国)
INSTITUTE OF BUILDING DESIGN OF CHINA ACADEMY OF BUILDING RESEARCH(CHINA)
巴马丹拿集团(香港)
PALMER & TURNER INTERNATIONAL INC.(HONG KONG)
百和纽特公司(澳大利亚)
BLIGH VOLLER NIELD PTY LTD.(AUSTRALIA)

鸟瞰图(赛时)
BIRDS-EYE VIEW (DURING GAMES)

154

■规划设计构思

"格"——九宫格

"九宫格"的理念在中国城市规划文化中有着悠久历史,也成为奥林匹克公园规划布局构思的精髓。场地B被划分成三个地块,每个地块等分成9个约240m×240m的方格。"九宫格"使北京的城市中轴线在场地中得以延续和强化,更为其中的建筑体块提供适当的比例,使之井然有序。奥林匹克体育场占用其中1个方格,展览中心占用4个方格,赛后将扩展至6个方格。27个方格中,每个方格再分割成9个小方格,构成适合人类比例的体块,并加入照明灯光和绿化元素。

■ Planning and Design Concept

The use of 9 squares has a long and significant history in Chinese architecture and urban planning, and it is this principle which inspires the site order for Olympic Green. The site has been clearly divided into 3 zones, each made up of 9 squares of approximately 240m x 240m in size. These squares are used to create a powerful extension of the Beijing city axis through the site, and to order the buildings and public spaces. In the context of the major 9 square grid, the module provides an appropriate scale for the required buildings on the site. The Olympic Stadium occupies one square, and the Exhibition Centre occupies four squares, extending to six squares after the Olympic Games. Within each of the 27 squares, a further 9 square module provides an opportunity to structure elements at a human scale, including lighting and landscaping elements.

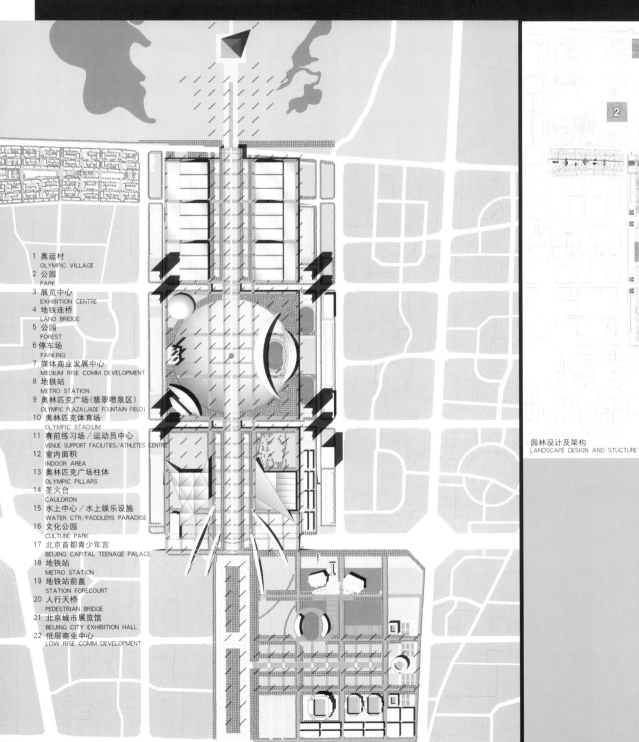

1 奥运村 OLYMPIC VILLAGE
2 公园 PARK
3 展览中心 EXHIBITION CENTRE
4 地铁连桥 LAND BRIDGE
5 公园 FOREST
6 停车场 PARKING
7 媒体商业发展中心 MEDIUM RISE COMM.DEVELOPMENT
8 地铁站 METRO STATION
9 奥林匹克广场(翡翠喷泉区) OLYMPIC PLAZA (JADE FOUNTAIN FIELD)
10 奥林匹克体育场 OLYMPIC STADIUM
11 赛前练习场/运动员中心 VENUE SUPPORT FACILITIES/ATHLETES CENTRE
12 室内面积 INDOOR AREA
13 奥林匹克广场柱体 OLYMPIC PILLARS
14 圣火台 CAULDRON
15 水上中心/水上娱乐设施 WATER CTR/PADDLERS PARADISE
16 文化公园 CULTURE PARK
17 北京首都青少年宫 BEIJING CAPITAL TEENAGE PALACE
18 地铁站 METRO STATION
19 地铁站前盖 STATION FORECOURT
20 人行天桥 PEDESTRIAN BRIDGE
21 北京城市展览馆 BEIJING CITY EXHIBITION HALL
22 低层商业中心 LOW RISE COMM DEVELOPMENT

全区总平面图(赛后) OVERALL AREA MASTER PLAN (POST GAMES)

园林设计及架构 LANDSCAPE DESIGN AND STUCTURE

1 森林公园步行径 TRAIL—FOREST PARK
2 森林公园教育中心 EDUCATION CENTRE—FOREST PARK
3 森林公园-农庄公园 BOTANICAL GARDEN FOREST PARK
4 森林公园绿化金字塔 PYRAMID—FOREST PARK
5 中央喷泉 OLYMPIC PLACE FOUNTAIN
6 绿化遮盖 GREEN ROOF
7 南北轴线景观 CENTRAL AXIS LANDSCAPE
8 长线形公园 LINEAR PARK

北京奥林匹克公园设计方案 Beijing Olympic Green

■ 规划设计构思

宏扬"绿色奥运、科技奥运、人文奥运",立足中轴线的延伸与发展,充分考虑城市建设的远、近分期规划,借鉴并演绎传统的空间思想理念,将传统的城市形态和自然山水引入奥林匹克公园是我们规划的主要思想与理念。

在B区南部商贸会展区,采用元大都所效仿的周礼"营国之制"方格的布局形式,形态上具有"城"的意味。在城市景观规划空间形态上,构筑三山、一水、一城的主导空间形态,符合中国传统造园手法"一池三山"分格,其喻意为人间仙境。

在森林公园堆山作为城市中轴线的底景,引清河之水入奥林匹克公园形成具有自然风格的奥运湖,湖水向南注入商贸区环城河内,再经A区和元大都遗址公园注入坝河,水景是串连奥林匹克公园整个空间形态的主要景观因素。湖岸两侧的会展与文化建筑,其退台不规则的形态如天然河岸的自由延伸,既是城市向自然的过渡,也是规整而富有理性走向自由感性的铺垫。

■ Planning and Design Concept

In order to advocate the view of Beijing Olympic Games, which is "Green, High-Tech and People", this proposal is based on the extension and development of the central axis and takes fully account of the present and the future plan of the city constructions. It learns from the traditional space concept to introduce the traditional city form and natural environment into the plan of Olympic Green.

At for the Site B' south business exhibition area, we follow the construction concept of Yuan Dynasty City which is crosshatches resemble the forms of "city". As for the urban landscape design, we use the guiding space concept of "three hills, one water system and one city". It follows Chinese traditional garden construction concept and creates "heaven on the earth".

The hills made in the Forest Park are the background of the central axis. The Olympic Lake's water is introduced from Qing River and goes into the south business area's river that encircles the city. The water also goes from Site A and Yuan Dynasty City Wall Relics Park into Bahe. Water landscape is the dominant element to link each space in Olympic Green. The exhibitions and the cultural buildings go with the natural extension of the riverbank, which are the transitional areas from city to nature as well as the preparation from reason to perception.

参赛作品 Works

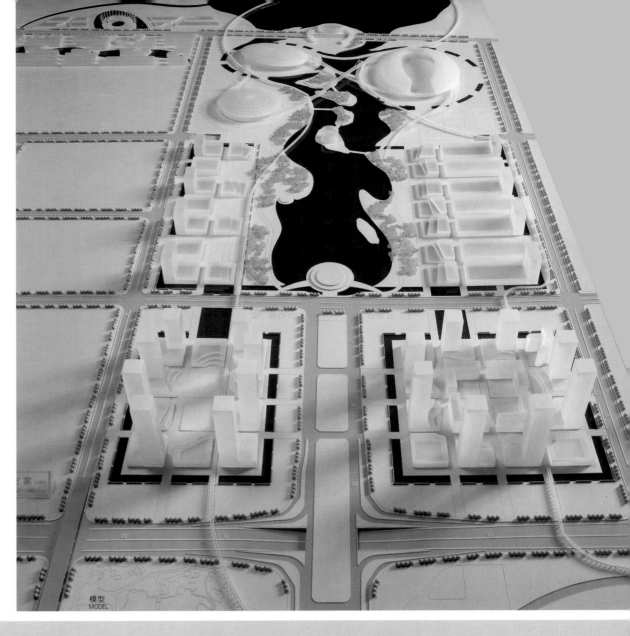
模型 MODEL

Beijing Olympic Green 2008

北京建筑工程学院建筑设计研究院(中国)
ARCHITECTURAL DESIGN AND RESEARCH INSTITUTE OF BEIJING INSTITUTE OF ARCHITECTURAL AND ENGINEERING (CHINA)

鸟瞰图(赛后) BIRDS-EYE VIEW (POST GAMES)

6	商贸区 BUSINESS AND TRADE DISTRICT
7	会展配套设施 CIEC SERVICE FACILITIES
8	会展中心 CIEC
9	环境教育中心 ENVIRENMENTAL EDUCATION CENTERS
10	市民绿化广场 CITY LANDSCAPE PLAZA
11	射箭场 ARCHERY RANGE
12	奥运湖 OLYMPIC LAKE
13	激光幻影台 LASER PLAT
14	国家体育馆 NATIONAL GYMNASIUM
15	国家游泳馆 NATIONAL SWIMMING CENTER
16	国家体育场 NATIONAL STADIUM
17	高架桥 OVERPASS
18	观景台 VIEWPOINT PLAT
19	运动员村 ATHLETE VILLAGE
20	配套设施 SERVICE FACILITIES
21	升旗广场 THE FLAG RSISING SQUARE
22	国际区配套设施 INTERNATIONAL ZONE SERVICE FACILITIES
23	中心湖 CENTER LAKE
24	碧玉公园别墅 BIYU PARK VILLAES
25	凯迪克酒店 CATIC HOTEL

中心区景观规划平面图(赛后)
MASTER LAYOUT AND LANDSCAPE PLAN (POST GAMES)

中轴线鸟瞰图
BIRDS-EYE VIEW ON CENTRAL AXIS

北京奥林匹克公园设计方案 Beijing Olympic Green

参赛作品 Works

模型
MODEL

■ 规划设计构思

规划将建设一个充满朝气的中心,让市民享受生活情趣的公共空间。此设计建议偏于把各种体育活动集中在连接上层公园、绿茵区、空地、娱乐、零售、商业区及公共基建的建筑群。

结合体育、文化、展览、休闲及旅游。中央建筑群任何地点提供连接零售、商业、娱乐、运输及体育设施的方便联系。核心设施和服务结合而成为一个公园及以空间为主的地区,为奥运的场地提供服务。

多用途的支援服务区

由公园空地往来的各商业、零售、娱乐及住宅区组成,将环绕中央建筑群。上述场地将分期地以渐进方式,而不是一蹴而成的方式开发,按照市场需求制定建筑的功能及用途,以创造建筑功能、用途的多元性。

空敞的奥运绿区

场地北面提供绿区,包括一个面积100ha的湖。此外,奥运绿区也沿着中央轴线中枢,向南伸展,连接奥运场地的各种设施,直到曲棍球和网球中心为止。

■ Planning and Design Concept

A centre of vibrant and popular public spaces for the enjoyment of citizens.

This design proposal concentrates sports activities into one cluster of buildings that are interconnected with the park layers above, areas of greenery, open space and entertainment, retail and commercial zones, and public infrastructure.

A combination of sports, cultural, exhibition, leisure and tourism.

Beijing Olympic Green 2008

汉沙杨建筑工程设计有限公司(马来西亚)
T.R. HAMZAH & YEANG SDN BHD(MALAYSIA)

Easy linkages to retail, commercial, entertainment, transport and sports facilities can be gained from any point within the Central Cluster of buildings. Core facilities and services will service the site post-Games through their integration as a Park and Open Space focused Precinct.

Multi-functional districts of supporting facilities

A mix of commercial, retail, entertainment and residential zonings, easily accessible through park space, will come to surround the central cluster. Phasing and allowing these sites to be developed incrementally rather than monumentally allows building use and function to be market oriented, as such a diversty of use and function will arise.

Spacious Olympic Green

Green space is provided to the north of the site, including an 100-hectare Lake.

Additionally, the Olympic Green extends southward along the Central Axis Spine, linking the various facilities of the Olympic Site and terminating at the Hockey and Tennis Centres.

模型
MODEL

全区总平面图(赛时)
OVERALL AREA MASTER PLAN (DURING GAMES)

1 国家体育馆
　NATIONAL GYMNASIUM
2 国家游泳馆
　NATIONAL SWIMMING CENTER
3 国家体育场
　NATIONAL STADIUM
4 北京会议中心
　BEIJING CONVENTION CENTER
5 广场
　PLAZA
6 少年宫
　TEENAGE PALACE
7 展览馆
　EXHIBITION

B区规划平面图
SITE B MASTER LAYOUT PLAN

1 公园和城市结合在一起
　INTEGRATION OF PARK INTO
　THE CITY AND THE CREATION
2 奥运设施
　CONCENTRATION OF OLYMPIC
　FACILITIES
3 南北轴线
　NORTH SOUTH AXIS
4 重新使用的场地
　AGAIN USE FIELDS
5 分期建造和开发的场地
　PHASING OF CONSTRUCTION AND
　DEVELOPMENT OF WHOLE SITE

Beijing Olympic Green 2008

北京市住宅建筑设计研究院（中国）
BEIJING RESIDENTIAL BUILDING DESIGN NETWORK(CHINA)

北京码维建筑设计咨询有限公司（中国）
BEIJING MAWEI ARCHITECTURAL DESIGN CONSULTATION CO.,LTD(CHINA)

■ 规划设计构思

规划方案围绕2008年北京奥运会的三个理念"人文奥运、绿色奥运、科技奥运"，紧密结合中国悠久的文化内涵、北京古都的城市文脉以及21世纪现代技术文明的高度发展，构造了一个体现奥运五环精神，由五层环形元素为触发点，由外向内，由低向高，逐层递进的现代奥运新城。

构图的核心为中国古老自然科学中的五个基本元素：金、木、水、火、土。方案引入上述元素，在城市规划过程中将其重新定义、定位，使其成为整体规划方案中功能明确的有机组成部分。

■ Planning and Design Concept

This infarastructure plan revolves around three concepts of 2008 Beijing Olympic Games, namely "the Peopl's"Olympics, Green Olympics, High-Tech Olympics". These concepts are incorporated with depths of long-thriving Chinese culture, amazing achievements of the ancient city Beijing, and have taken full advantage of highly-developed modern technology in the 21st century. This plan is to build a brand-new modern Olympic Green that corresponds central and east, and that expands out from interior. Based on the idea of fiverings of the Olympic Flag, the Green is designed to rise by five circular staircases to symbolize the height of Olympic spirits.

Five basic elements, derived from ideas of natural science of ancient China, have been introduced into the plan. They are gold, timber, water, fire and earth. The entire plan is a composite of these ideas that have been reformed to be function-specific in planning the Green.

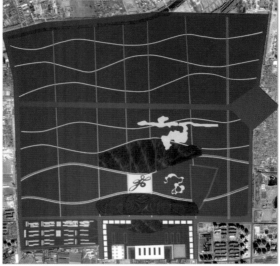

景观网格由软线条和硬线条组成
LANDSCAPING GRID OUT OF HARD LINES AND SOFT LINES

将网格中的软线条渗透到奥运村内
THE SOFT LINE OF THE GRID DIFFUSES INTO THE OLYMPIC VILLAGE AREA

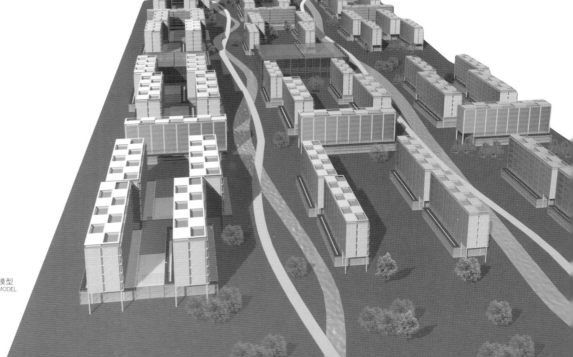

模型
MODEL

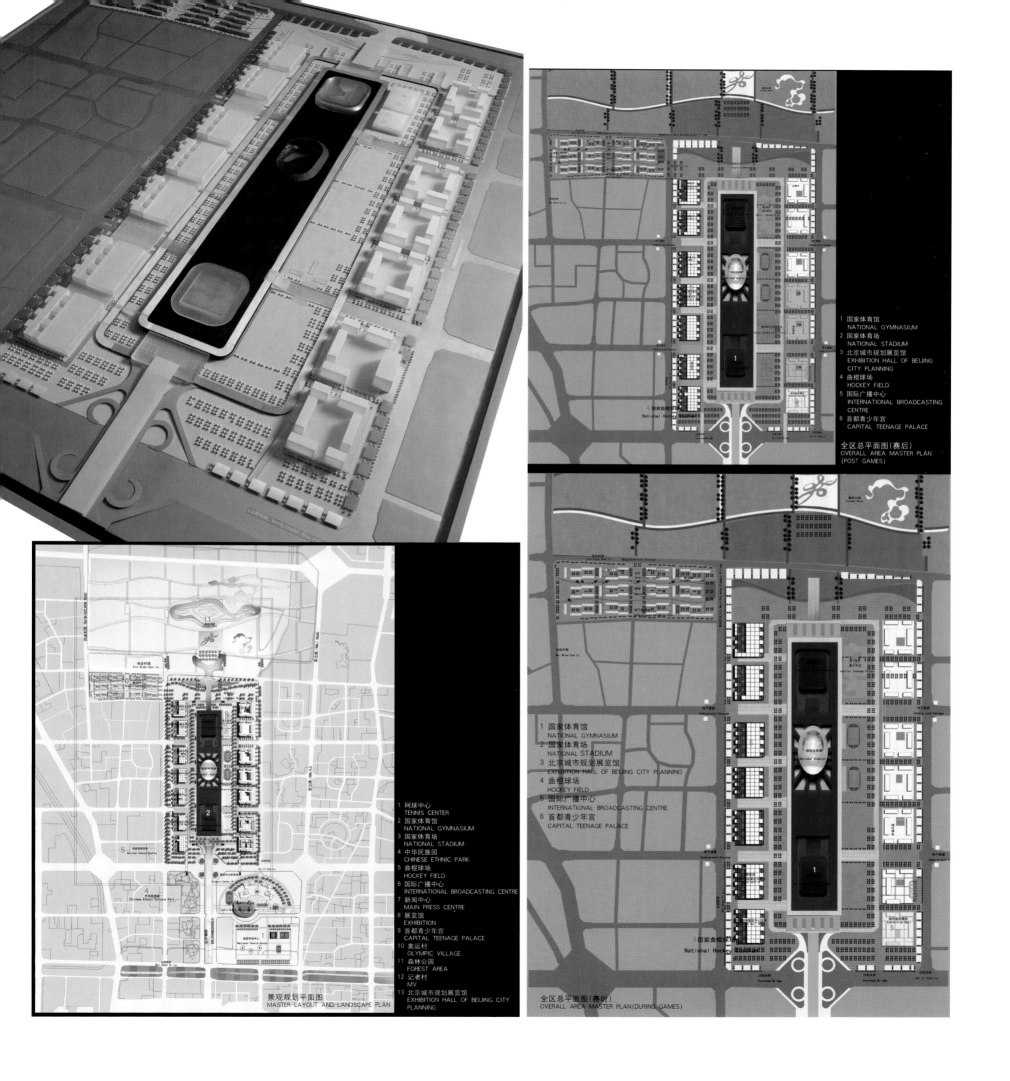

Beijing Olympic Green 2008

北京奥林匹克公园设计方案

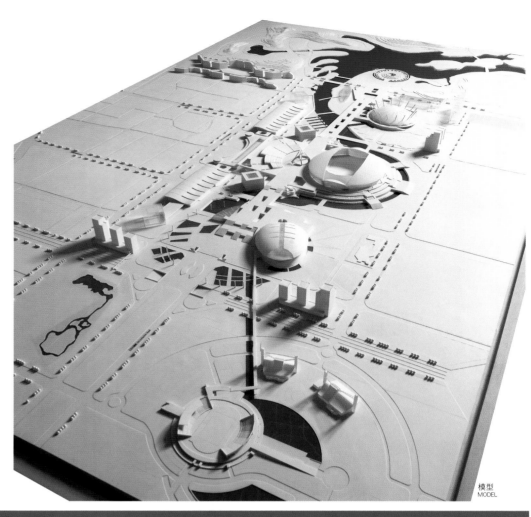
模型 MODEL

■ 规划设计构思

北京奥林匹克公园雄踞北京的城市中轴线北端。这条中轴线，有着太多与"龙"相关的渊源，它统领着整个城市的规划，是北京几千年城市历史精华所在。

将"绿色奥运、科技奥运、人文奥运"化做"龙的奥运"是方案构思的出发点。通过表现龙的精神，展示龙的形象，向世界展示自信自强的民族精神，兼容并蓄的传统文化，并在北京留下有中国特色的奥林匹克运动遗产，是该规划设计的目的。

希望能用有限的"形"，表现无限的"神"。让北京的龙脉延伸，让"龙的奥运"带领中国腾飞！

■ Planning and Design Concept

It is with this spirit, that we have chosen the Dragon as the theme of our design for the 2008 Olympic Green site. According to legend, the Dragon is of the water, an idea, which we have taken and expanded upon to create a Dragon shaped Lake/River - The Dragon Lake/River. The location of the park along the central axis of the city of Beijing is also appropriate, as the axis is believed to be the pulse of the dragon. We also believe that the Dragon best represents the motto of the Olympic Games for a People's Olympics, a Green Olympics as well as a High-Tech Olympics

Let the Dragon Olympics be the spring board for China, showing the world the best of Chinese culture, economy and spirit.

参赛作品 Works

亚太专业同盟有限公司（新加坡）
PROFESSIONAL ASIA-PAC ALLIANCE PTE LTD(SINGPORE)
三连庄建筑设计有限公司（中国）
3HP ARCHITECES PTE LTD(CHINA)
中国对外建设总公司设计研究院（中国）
INSTITUTE OF DESIGN & RESEARCH,CHINA CONSTRUCTION INTERNATIONAL CORPORATION(CHINA)

全区总平面图（赛后）
OVERALL AREA MASTER PLAN (POST GAMES)

1 国家体育馆
  NATIONAL GYMNASIUM
2 国家体育场
  NATIONAL STADIUM
3 北京城市规划展览馆
  EXHIBITION HALL OF BEIJING CITY PLANNING
4 首都青少年宫
  CAPITAL TEENAGE PALACE
5 五环广场
  FIVE RING PLAZA

1 国家体育馆
  NATIONAL GYMNASIUM
2 国家体育场
  NATIONAL STADIUM
3 国家游泳馆
  NATIONAL SWIMMING CENTER
4 射箭场
  ARCHERY RANGE
5 升旗广场
  FLAG RAISING SQUARE
6 国际广播中心
  INTERNATIONAL BROADCASTING CENTRE
7 新闻中心
  MAIN PRESS CENTRE
8 乒乓球
  TABLE TENNIS
9 击剑、现代五项
  FENCING HALL/MODERN PENTHATHLON
10 摔跤
   WRESTLING
11 击剑练习馆
   FENCING TRAINING HALL
12 羽毛球馆
   BADMINTON HALL

全区总平面图（赛时）
OVERALL AREA MASTER PLAN (DURING GAMES)

中轴文化
CENTRAL AXIS CULTURE

北京奥林匹克公园设计方案 Beijing Olympic Green

■ 规划设计构思

规划目的：
建设 21 世纪世界城市示范区
建设首都大型公共文化娱乐活动区
创造具有纪念性意义的平民化空间
加强规划用地的综合辐射带动能力

规划思想：
完成北京中轴线空间系统的转换——由城市空间系统转换为自然空间系统。
强调纪念性空间轴线：土城历史空间序列＋人工化绿轴＋入口广场＋高台广场系列＋水面绿地＋森林公园。
突出空间的"市民性"和多元化，发展地下空间。
建设生态城市：太阳能系统、降低热岛效应、可控自然通风系统、城市污水处理系统、建筑集中与绿化集中。
完善城市整体发展战略,强化中关村与 CBD 之间的空间节点。
远近结合，综合效益最大化。

■ Planning and Design Concept

Planning order:
Building a Model District for the 21th century.
Building up a larger public cultural and enterainment center for the capital city.
To create a monumental place for the populace.
Integrate the site with its surrounding areas together.

Planning Ideology:
Transforming the grand central axis of Beijing from artificial based pattern to natural based pattern.

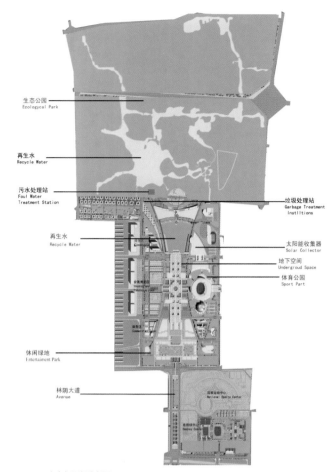

生态与环保分析图
ECOLOGICAL SEYSTEM & ENVIRONMENT DESIGN

Beijing Olympic Green 2008

参赛作品 Works

北方交通大学建筑系 BA 研究室（中国）
ARCHITECTURAL BA STUDIO, NORTHERR JIAOTONG UNIVERSITY(CHINA)
东南大学建筑系（中国）
ARCHITECTURE DEPARTMENT, SOUTHEAST UNIVERSITY(CHINA)
U ＆ A 设计集团（澳大利亚）
U ＆ A DESIGN GROUP(AUSTRALIA)
中厦建筑国际有限公司（中国）
ZHONGXIA ARCHITECTURE INTERNATIONAL CO,.LTD(CHINA)

Enhancing the spatial monumentality of the axis through a serial characteristic places from the Yuan earth-wall heritage to the entrance square, and from the rised central square to the forestry park.

Emphasizing the populace features of the space, developing underground space for public use.

Co-city development by using of solar energy, water-saving and reuse system, controlled natural ventilation and other techniques.

Integral strategy with CBD and Zhongguancun Scientific and Technological Zone.

Beneficial to both today and the future.

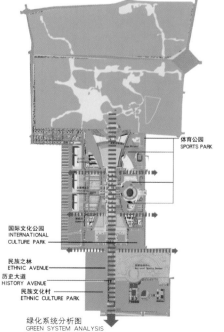

绿化系统分析图
GREEN SYSTEM ANALYSIS

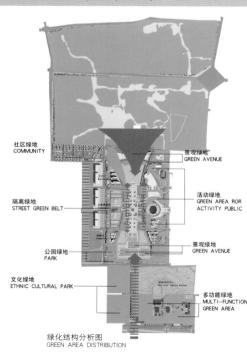

绿化结构分析图
GREEN AREA DISTRIBUTION

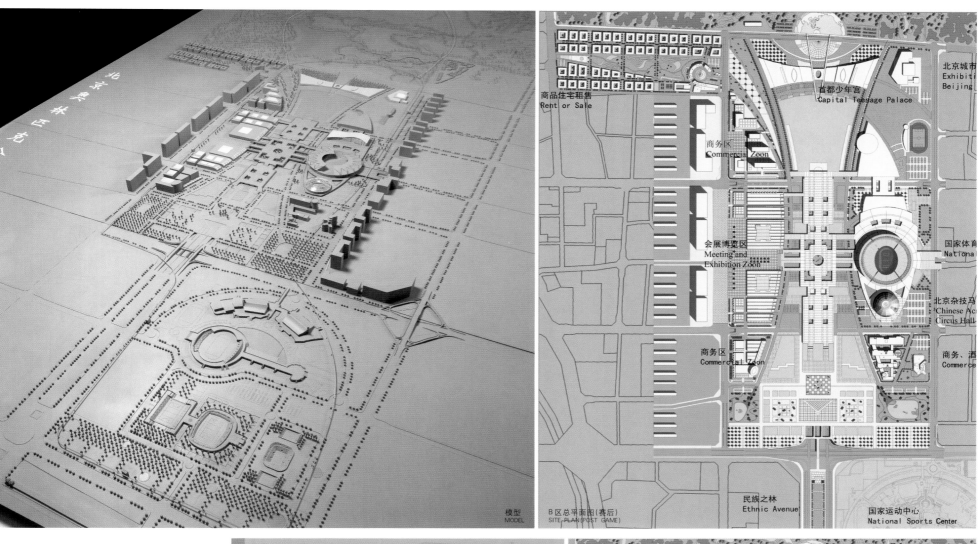
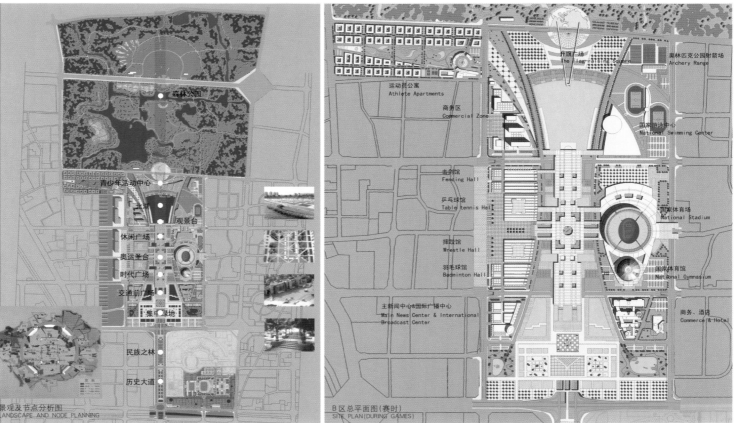

北京奥林匹克公园设计方案
Beijing Olympic Green

参赛作品 Works

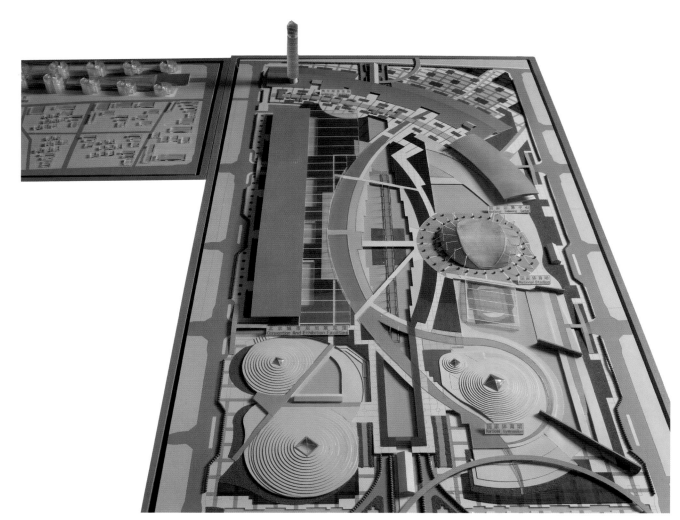

模型
MODEL

Beijing Olympic Green 2008

株式会社综合计画研究所（日本）
SYSTEM PLANNING (JAPAN)

■ 规划设计构思

规划将弧形的景观轴立体地叠加在北京的城市直线轴上，既延续了北京市固有的城市轴线，又避免了宽大的直线轴将地块硬性地分隔。奥运会设施分置在立体的景观环境轴上。奥运会中，提供最便捷的自然流畅的宽敞动线；奥运会后，顺着宽敞的动线，可最经济、迅速地改造成自然景观环境空间……体现绿色奥运、环保奥运、经济奥运的主题，体现城市与自然共存的主题。

将奥林匹克公园与既存区域、森林区域用大弧形景观线连接、融合为一体，体现奥林匹克所追求的"和的融合"。

圆弧朝天，在城市的脉络中勾画出如同"玉如意"似的象征吉祥、美满的一团瑞气，以动态景观轴的构成来回应城市和历史的中轴，中华民族的精神在北京——中国的心脏聚积、升腾。

将北京市的城市轴、历史轴、文化轴汇集至景观轴上。

积极采用立体景观绿化，以都市的绿洲为基本方针的规划。

象征胜利的V字形选手村建筑群，与动线、景观相协调，生机勃勃。

利用地铁、新交通、客车等各种交通工具、运送手段，在奥运会其间，为专业人员提供最短距离、便捷合理的动线，为普通游客提供最短时间内即可到达疏散目的地的合理动线

地面、架空人工地面为公园环境景观，地下构成设施、动线轴，将景观、设施利用地面、地下的立体布置，形成自然的、诱导性的整体布局。

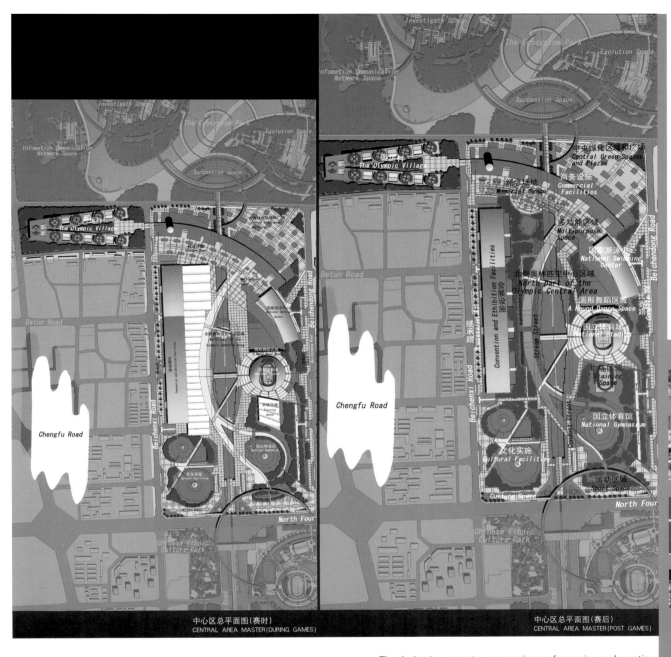

中心区总平面图(赛时)
CENTRAL AREA MASTER (DURING GAMES)

中心区总平面图(赛后)
CENTRAL AREA MASTER (POST GAMES)

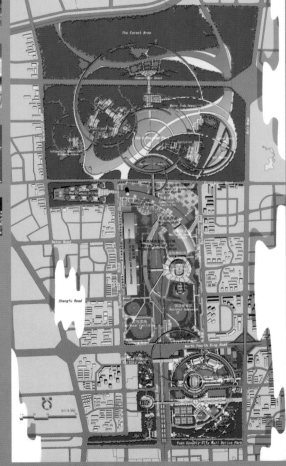

景观规划平面图(赛后)
MASTER LAYOUT AND LANDSCAPE PLAN (POST GAMES)

■ Planning and Design Concept

Taking a bird's-eye view of the Olympic venue, the planned site is clearly visible. By holding the Olympics, we have achieved our goal on the city axis, and the goal for the Forest Area is achieved by turning it into a green zone. We will have achieved our goal for the first half of the 21st century. Also, the plans allow for the continued development of the city of Beijing in the future, and they stress not the lines but the "plane", like Tienanmen which forms the "framework of the city".

In the present plans, importance is given to the post-Olympic stage and consideration is given to speedy conversion after the Olympics, keeping costs to a minimum, and making allowance for the environmental ecology from the construction stage. All the systems will be verified from the planning stage to ensure smooth progress to maturity and growing together after the Olympics.

The existing area and The Forest Area are fused into a large continuous arc representing the "fusion of peace", like the Olympic symbol.

The arc ascends upward, stressing future development by the idea of a "mace" or "circle".

The axes of history and culture, which each have their own site, are combined and replaced by the landscape axis.

The design incorporates aggressive roof greening and creation of an urban oasis.

The Olympic Village is V-shaped, representing the victory of the competitors, and the design is dynamic in terms of both architecture and landscape.

The traffic lines are kept as short as possible during the Olympics with a new transport system, buses and paths that keep competitors completely separate from visitors.

Parks are on ground level and facilities and traffic lines underground, and the design is such that visitors are led naturally from one to another.

北京奥林匹克公园设计方案
Beijing Olympic Green

■ 规划设计构思
规划中的奥运公园将是一首交响乐，从中轴线一幕幕拉开。

中轴交响乐
规划展示了从天坛到天安门广场，从故宫到元大都遗址，从中轴线到奥林匹克公园以及从过去到现在、再向未来的发展序列。

地球村——空间地标
运用了几何规划系统，以万分之一地球半径（637.11m）的模式半径，确定了公园中心区、奥运场馆区、森林公园与城市间的联系，并以这个模式半径布局了有效的路网系统与分区运行管理组织系统。

规划轴线
规划中采用了轴线理论。中央轴线（即中轴线）作为历史人文轴线；在田园轴线上通过水平和竖向景观与青少年科普教育中心相配合；绿色轴线则无论是在欣赏角度上还是在步行系统上更明显地将奥运公园与其南侧的现状奥体中心再次连接。从东西向穿过奥运公园中心点的城市轴线，将奥运公园东西两侧的发展中地块有序地组织与联系起来。

■ Planning and Design Concept
The Olympic Park is planned from central axis area to the side area as when you go for a symphony the stage curtain is pulled open from center stage.

Central axis symphony
Central axis will carry and act on the continuity and direction of the human culture, history and images of Beijing.

模型 MODEL

柏涛建筑设计有限公司（澳大利亚）
PTW ARCHITECTS (AUSTRALIA)

Beijing Olympic Green 2008

The plan of Olympic Park displays the sequence of development stages from Heaven Temple to Tian An Men Square, from the Forbidden City to Yuan Dynasty City wall Relics Park, from Central Axis to the Olympic Park, and from the past to the present and towards the future.

Global village-Spatial ground elevation
In the design of this master plan, a geometrical planning system, one ten thousandth of the earth's radius (637.11 meters) is used as the radius to define the center of the park and the connection of Olympic venues, forest park and the city. By using this radius model, we distribute the effective road and traffic system and division organization system.

Planning axis using axis theory in the plan.
The central axis is the axis of history and human culture.

The field axis coordiates the horizontal and vertical scenery with the Youth Technology Education Center.

The green axis clearly shows the connection between the Olympic Park and the southern existing Olympic Sport Centres both from the view point, and the pedestrian system.

The city axis crosses the Olympic Park from the east to the west, and allows the future developments to be more organized, and the connection of both sides of the Olympic Park to become more structured.

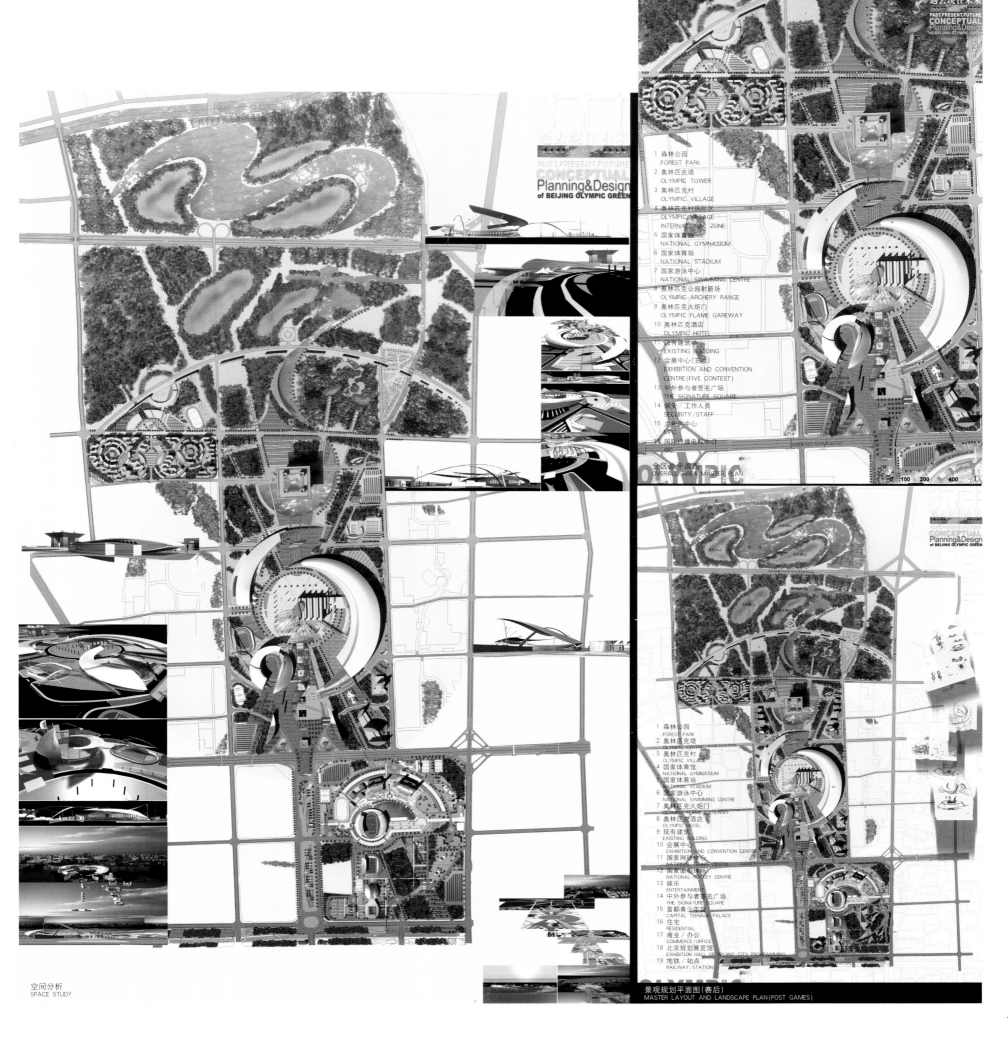

北京奥林匹克公园设计方案 Beijing Olympic Green

参赛作品 Works

■ 规划设计构思

本规划方案在构思过程中,巧妙地把体育和科技、环保、教育、艺术、文化以及商业运作结合在一起,突破了传统公园的设计模式,既富有新意而又切实可行,为"举办一届历史上最出色的奥运会"奠定了坚实的基础,也对"绿色奥运、科技奥运、人文奥运"的原则进行了很好的阐释。

未来的人类将不再满足于生活在由单调的线条所围成的框架结构中。在整个奥林匹克公园的景观和环境设计方面,我们突出了其园林化和人性化,并体现了可持续发展的原则。

本方案还把文化和艺术的因素带入到公园的建设规划中。

在奥运会相关活动的组织规划方面,本方案体现了设计者承办大型活动的能力和丰富经验。奥运期间,我们将在公园中着力渲染一种各国人民大会聚的节日的欢庆气氛。

世界各国的人们广泛合作的精神在奥运会的前期资金积累和奥运设施的赛后利用中得到了充分的体现。

■ Planning and Design Concept

Create the world's premiere venue for staging large-scale special events

Our plan expresses a rational use of space and management of the site for "Hosting the best-ever Olympics" in Beijing, and sets a solid foundation for staging other expositions, sporting and major entertainment events in the future.

Balance opportunities for economic growth with a sustainable, healthy environment.

Enhance Beijing's reputation as an attractive international tourist destination.

Establish a permanent cultural legacy in the Olympic Green.

Enhance Beijing's international leadership in health and environmental technology.

Create an marvelous landmark that defines Beijing's vision of the future.

Create an research destination to attract leading scientists from around the world.

Beijing Olympic Green 2008

WORLD QUEST 工程公司(美国)
WORLD QUEST ENGINEERING(USA)

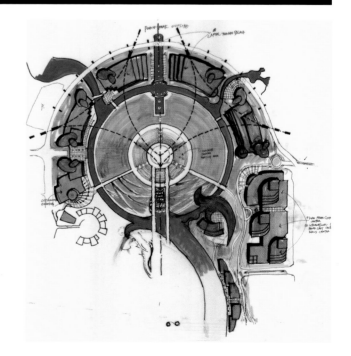

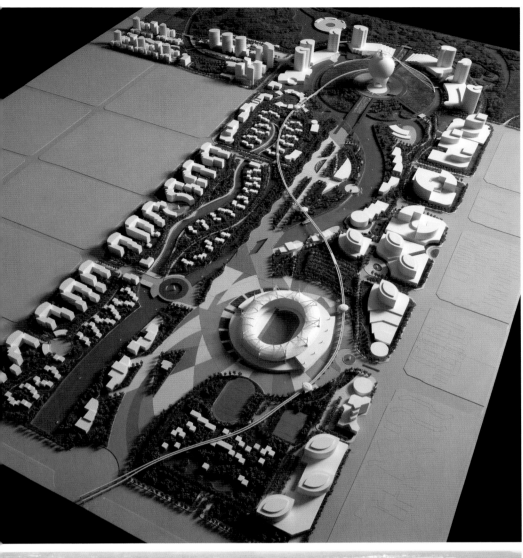

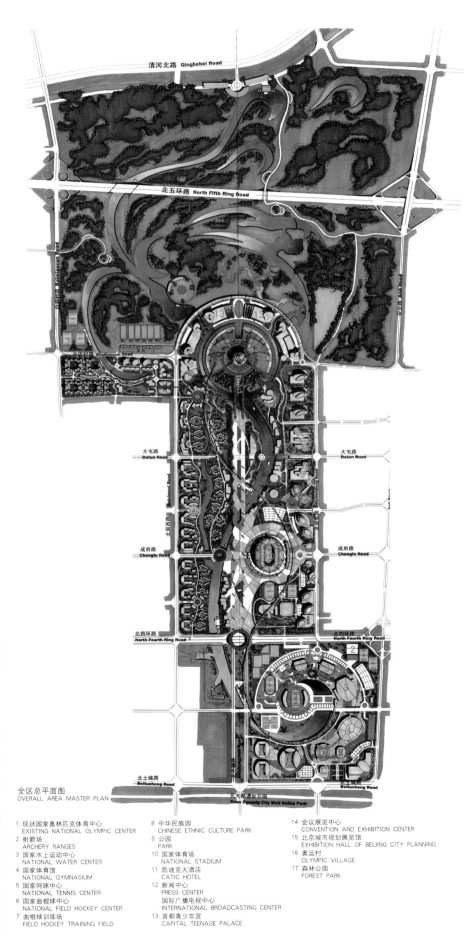

全区总平面图
OVERALL AREA MASTER PLAN

1 现状国家奥林匹克体育中心
  EXISTING NATIONAL OLYMPIC CENTER
2 射箭场
  ARCHERY RANGES
3 国家水上运动中心
  NATIONAL WATER CENTER
4 国家体育馆
  NATIONAL GYMNASIUM
5 国家网球中心
  NATIONAL TENNIS CENTER
6 国家曲棍球中心
  NATIONAL FIELD HOCKEY CENTER
7 曲棍球训练场
  FIELD HOCKEY TRAINING FIELD
8 中华民族园
  CHINESE ETHNIC CULTURE PARK
9 公园
  PARK
10 国家体育场
   NATIONAL STADIUM
11 凯迪克大酒店
   CATIC HOTEL
12 新闻中心
   PRESS CENTER
   国际广播电视中心
   INTERNATIONAL BROADCASTING CENTER
13 首都青少年宫
   CAPITAL TEENAGE PALACE
14 会议展览中心
   CONVENTION AND EXHIBITION CENTER
15 北京城市规划展览馆
   EXHIBITION HALL OF BEIJING CITY PLANNING
16 奥运村
   OLYMPIC VILLAGE
17 森林公园
   FOREST PARK

北京奥林匹克公园设计方案 Beijing Olympic Green

参赛作品 Works

■ 规划设计构思

北京奥林匹克公园位于北京古迹中心线的北端，紫禁城、景山、天安门、钟鼓楼等重要地标连成一线，承继自元朝流传下来的文化传统，具有重大历史、政治和经济意义，现选为2008年奥运会的基地，可视为北京市区生机重现的里程碑。因此，本规划方案既考虑到基地上的资源如何能在奥运会及以后都能有效地被运用，更着眼于北京市长远的发展策略。

可持续性－为暂时性的节目铺排长远性的策略。

短暂性的活动可使用流动的设备，以及一些可改装的永久性建筑物。

为了加强奥林匹克公园的社群性，本方案曾尝试利用户外空间的潜质。中心的斜道延伸至森林区的南端，以及在这里以一个大型平台结尾，充分利用森林区景观的优势。B区不同平台及广场的园林设计容纳不同的户外个人及团体活动，例如太极及象棋。区内也设置了一系列大大小小不同形状的水池。以水池砌成的不规划形状是回应森林区的自然形态。在北京的闹市中提供一个较清静舒适的空间。

■ Planning and Design Concept

Strategically located at the northern end of Beijing,s historical central axis, the site of Beijing Olympic Green aligns with the most important monuments, such as the Forbidden City, Prospect Hill, Tienanmen Square and the Bell and Drum Tower just to name a few, inheriting the culture and heritage of China dating back to Yuan Dynasty. This very site of the greatest historical, political and economical importance, dedicated to the summer Games in 2008, is the steppingstone for Beijing, urban rejuvenation that will flourish well beyond he games. Our proposed solution addresses the concern of the utilization of the facilities during and after the Olympics, as well as complementing Beijing, urban growth in the long term.

Sustainability - temporary program for temporary event.

The national stadium with a capacity of over one hundred thousand seats, which is used mainly for the opening/closing ceremony and track and field matches, will not find much utilization after the Olympic Games.

To further strengthen the sense of community in Olympic Green, the solution attempts to exploit the potential of the outdoor space. It takes advantage of its proximity to the forest area to the north of the site by extending the ramp beyond site, ending with a large piaza at the southern edge of the forest area with panoramic view. Landscape on site B has been accentuated to attract various individual and group activities, such as tai chi and Chinese chest games, in different plazas throughout the site. Water feature has been added to create a more soothing environment midst of the hectic Beijing.

 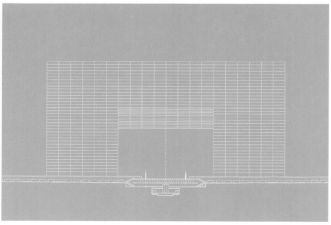

Beijing Olympic Green 2008

利安建筑设计及工程开发顾问（中国）有限公司（中国）
LEIGH & ORANGE DESIGN & PROJECT DEVELOPMENT CONSULTANTS (CHINA) LTD.(CHINA)

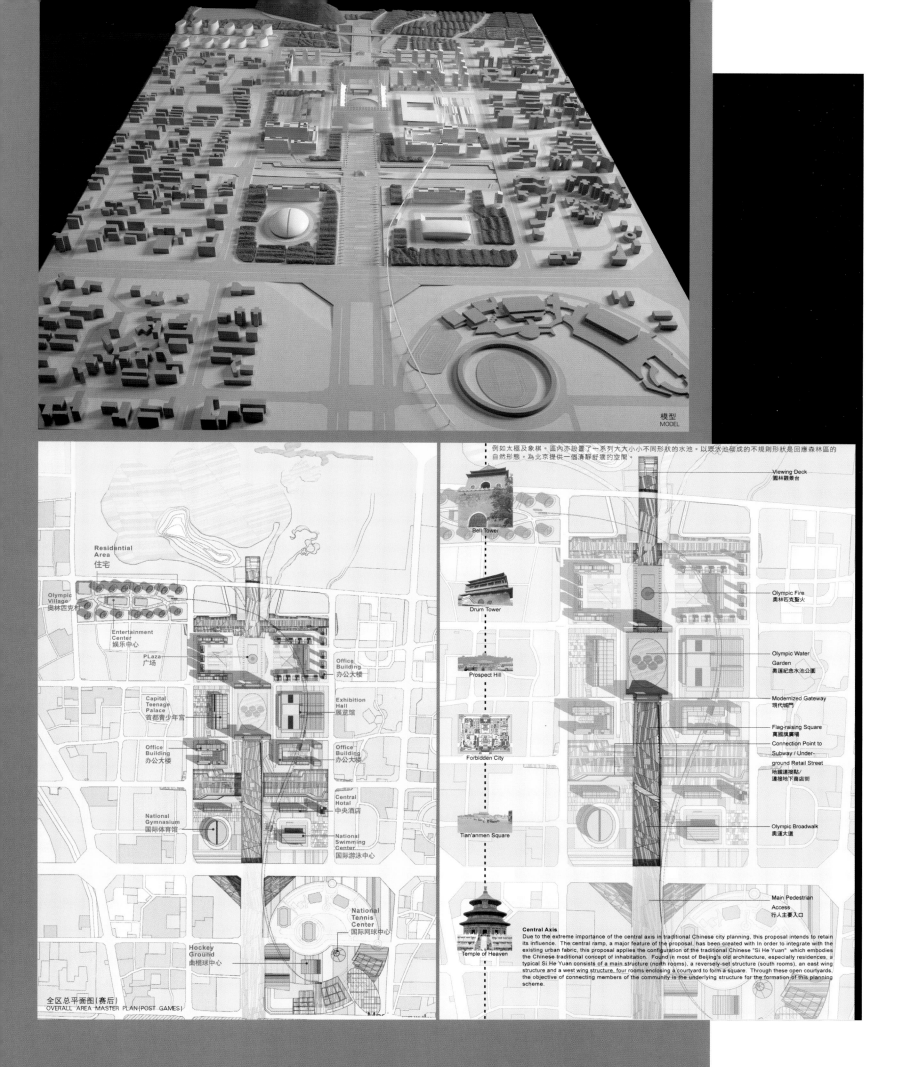

北京奥林匹克公园设计方案 Beijing Olympic Green

参赛作品 Works

■ 规划设计构思

新国际交流区的建设以"新奥运"、"新北京"为基本战略，去规划简洁易懂、崭新明快的城市空间。

将有效利用邻近首都北京机场极佳的立地条件和21世纪城市不可缺少的水和绿环境条件的优越，去规划吸引世界人们的高品质新样板城市。

在举办奥林匹克运动会这项体育运动的一大国际交流活动后，会促进体育运动、文化、游乐、交易等广泛国际交流活动据点地区的发展。

■ Planning and Design Concept

Construction of a New International Exchange District.

Beijing Olympic Green is planned as an original and distinctive urban space that embodies the key strategy elements, i.e., New Olympics and New Beijing.

Taking advantage of the convenient access to the international airport and the abundant natural environment which is essential for a city of the 21st century, BOG is planned as a new and high quality model city that attracts people from all over the world.

After the Olympic Games, a great international sports event, BOG will be developed into a center of international exchange activities centering on culture, recreation and trade, besides sports events.

地下规划
WORLD PLAZA UNDERGROUND PLAN

Beijing Olympic Green 2008

TAM地域环境研究院（日本）
TAM AREA ENVIRONMENTAL RESEARCH INSTITUTE(JAPAN)

模型
MODEL

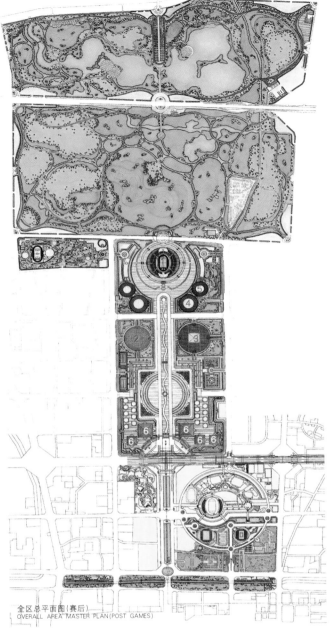

全区总平面图（赛后）
OVERALL AREA MASTER PLAN (POST GAMES)

1 国家体育场
  NATIONAL STADIUM
2 国家体育馆
  NATIONAL GYMNASIUM
3 国家游泳中心
  NATIONAL SWIMMING CENTER
4 首都青少年宫
  CAPITAL TEENAGE PALACE
5 北京城市规划展览馆
  EXHIBITION HALL OF BEIJING CITY PLANNING
6 北京会展博览设施
  BEIJING INTERNATIONAL EXHIBITION AND CONVENTION CENTER (BIEC)
7 商业服务设施
  COMMERCIAL SERVICES FACILITIES
8 升旗广场
  FLAG-RAISING

国家体育场
NATIONAL STADIUM

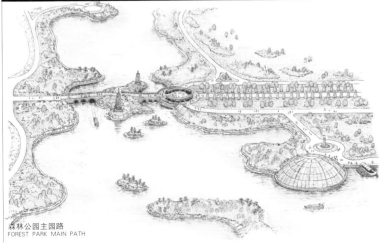

森林公园主园路
FOREST PARK MAIN PATH

175

北京奥林匹克公园设计方案 Beijing Olympic Green

模型 MODEL
北四环路 North Fourth Ring Road

■ 规划设计构思
　　以绿色奥运、科技奥运、人文奥运为宗旨，构造一个利于环境保护和可持续发展的生态工程，容科技、人文于绿色之中。在保持整体统一性的同时充分体现其民族性和时代性，形成北京新景观。

营造绿色的生态工程
　　在森林公园(C1区)人工建造100ha水面，取名为云湖。在五环路北堆筑龙山，成为城市中轴线的制高点。这一景观区极具生态意义。在四环路以北(B区)。以中轴线为中心，在东、西各场馆间规划建设面积为2360m×400m的中华广场。广场上有大草坪、乔木林等。使四环路以南的中轴线绿化带得以向北延续，与森林公园连为一体。除奥运村宿泊区外，在满足体育场馆功能空间的前提下，其他新建建筑均不采用高层建筑模式。围绕B1、B3区，按中水管网线开凿与森林公园云湖水相连的云水渠。在奥林匹克公园内增添水系，有利于生态平衡和小环境内干旱状态的改变。为保护环境，减少污染，用建筑垃圾和建筑挖掘土方堆龙山。在龙山和云湖区装置风力发电网。各体育场馆设置太阳能利用设备。

营造民族特色
　　中轴线是中华传统的龙脉，是本规划方案的重点。中华广场景观区有五座拱形汉白玉石桥、传统的牌楼、80m直径半球形透明建筑——世纪宫。主体建筑国家体育场坐落在高架坪台的北段，场前有高40m的地标性雕塑。长1600m的高架平台与主体建筑构成中轴线上的又一景观高潮。森林公园的中轴线上规划了景点云门和云湖，中心设置直径60m的透明建筑——云宫。

■ Planning and Design Concept
　　With the purpose of Green, Science and Technology, and Humane Olympics, to create an environment-friendly and sustainable ecological project, involving science and technology, humanism into

Beijing Olympic Green 2008

中国装饰（集团）公司（中国）
CHINA NATIONAL DECORATION (GROUP) CORPORATION(CHINA)

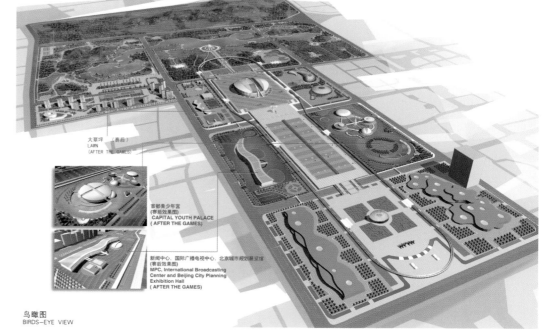
鸟瞰图 BIRDS-EYE VIEW

大草坪（赛后） LAWN (AFTER THE GAMES)
首都青少年宫（赛后效果图） CAPITAL YOUTH PALACE (AFTER THE GAMES)
新闻中心、国际广播电视中心、北京城市规划展览馆（赛后效果图） MPC, International Broadcasting Center and Beijing City Planning Exhibition Hall (AFTER THE GAMES)

参赛作品 Works

green. To preserve the general consistence reveal national and time characteristics.

Building a Green Ecological Project

In Zone C1 of the Forest Park is a man-made lake of 100 hectares named Cloud Lake. A hill, named The Dragon Hill, will be built on the north of Wuhuan Road, as the highest point of the Central Axis. This view is of ecological significance. On the north of Sihuan Road (Zone B) Zhonghua Square will be built on both sides of the Central Axis, partially covered with large grass fields and arbor trees. The high rising building mode will not be adopted except the accommodation and parking area of the Olympic Village. Around Zone B1 and B3, a canal named Yunqu will be built, connected with The Clould Lake. This facilitates the ecological balance. Wind power plants will be installed. Solar equipment will be set up in all the sports facilities.

Establishing National Characteristics

The Central Axis is of great significance in Chinese tradition. Accordingly, it plays a key role in the plan. There are five arch bridges

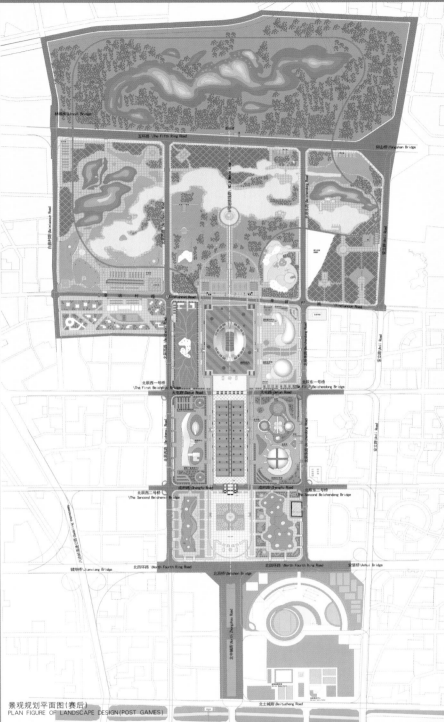

景观规划平面图(赛后)
PLAN FIGURE OF LANDSCAPE DESIGN (POST GAMES)

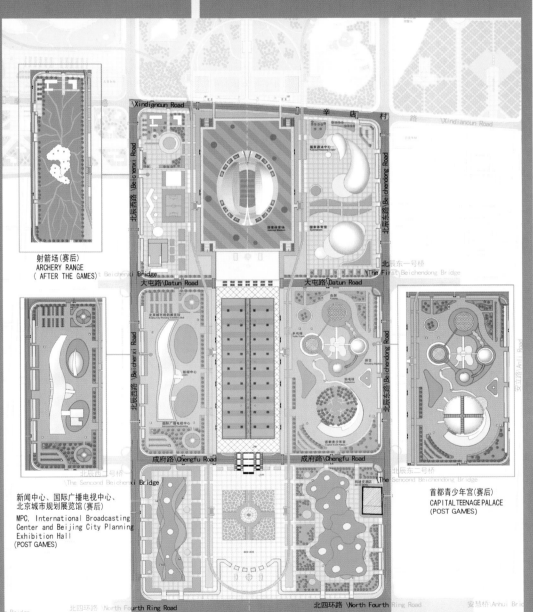

射箭场(赛后)
ARCHERY RANGE (AFTER THE GAMES)

新闻中心、国际广播电视中心、北京城市规划展览馆(赛后)
MPC, International Broadcasting Center and Beijing City Planning Exhibition Hall (POST GAMES)

首都青少年宫(赛后)
CAPITAL TEENAGE PALACE (POST GAMES)

全区总平面图(赛时)
OVERALL AREA MASTER PLAN (DURING)

made of white marble and traditional decorated archways in the Zhonghua Square. Besides, Century Palace, a transparent hemisphere with a diameter of 80 m, stands there as well. The major building, the National Stadium, is located on the northern section of the platform. In front of it lies a landscape sculpture. The 1600-meter-long platform and the major building stand as another scenery on the Central Axis.

The Cloud Gate and The Cloud Lake are located on the Central Axis of the Forest Park. In the middle there rests The Cloud Palace, a transparent building with a diameter of 60 m.

北京奥林匹克公园设计方案 Beijing Olympic Green

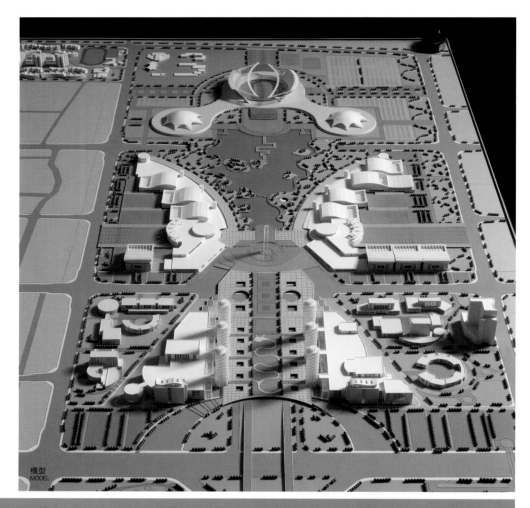
模型
MODEL

■ 规划设计构思
    延续和发展城市中轴线，保持城市文脉。
    突出"新北京、新奥运"的申办理念，体现"绿色奥运、科技奥运、人文奥运"的宗旨。
    充分考虑奥运会赛前与赛后的结合。
    注重生态环保，坚持可持续发展。

■ Planning and Design Concept
    Enhancement of the Central Axis reflects the culture heritage. Beijing Green locates to the west of the Central Axis, which has a significant impact on the construction and the development of the City.
    Outstand our faith and reflect our tenet.
    Beijing's faity for olympics in 2008 is "New Beijing, New Olympics", and the tenet is "Green Olympics, High Tech Olympics and People's olympics" The faith and the tenet are implemented all through the planning course.
    Integration of the pre-Olympics and the Post-Olympics.
    Environment protection and long-term development.

参赛作品 Works

Beijing Olympic Green 2008

中国武汉市建筑设计院（中国）
WUHAN ARCHITECTURAL DESIGN INSTITUTS CHINA(CHINA)

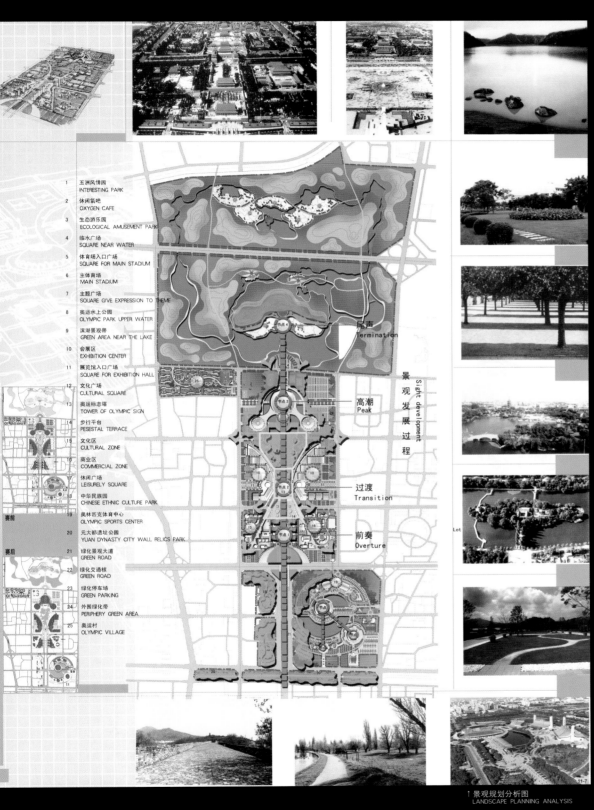

北京奥林匹克公园设计方案 Beijing Olympic Green

参赛作品 Works

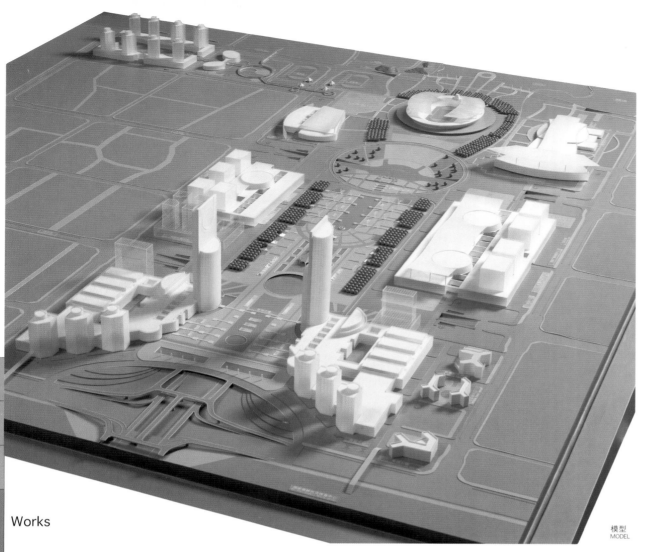

模型 MODEL

Beijing Olympic Green 2008

北方工业大学建筑学院（中国）
COLLEGE OF ARCHITECTURE OF NORTH CHINA UNIRERSITY OF TECHNOLOGY(CHINA)

北京中色北方建筑设计院（中国）
BEIJING ZHONGSE NORTH ARCHITECTURAL DESIGN INSTITUTE(CHINA)

■ 规划设计构思

明确的功能分区奠定了持续发展的基础

功能分区明确，各区均有发展空间，相互之间交融互动，奠定了赛事举行的空间基础，又最大限度地考虑了赛后城市中心的有机形成。

人车分流系统和基础设施通盘规划

在现状基础上慎重整合总体交通路网，通盘规划区域内完善的人车分流系统，整个B区及奥运森林公园区设轻轨内部交通车环线，总体公交站场设置合理，市政基础设施形成强大发展支撑。

本方案最大特点之一就是设计出整个B区与奥运森林山水区连成一个很大面积的人流安全步行区。赛时高峰20多万人流直接就近步行至地面城市干道（北辰西路，北辰东路等）两侧的公交站场、出租车站经过步行下沉广场到达地铁站及地下停车场疏散。出现临时紧急情况时，可将人流快捷引入奥林匹克森林区疏散。

丰富的人文景观环境和楔形绿地山水体系

研究了景观规划与空间设计、轴线设计及景观掩蔽设计、空间的开放与围合三个方面的问题。设计结合交通核心及下沉广场群，形成丰富的立体空间。最突出的特点是楔形绿地山水体系及生态环境设计。

■ Planning and Design Concept

The definite functional clusters for sustainable development

There is a definite and feasible function layout in this planning, which ensures ample space for long-term development in every district and provides opportunities for association or changes.

The integral planning of advanced system of separate roads for vehicle and pedestrian and the utilities of infrastructure.

This proposal shows the distinctive features in its transport and traffic planning such as the integral transport networks based on the present conditions, the advanced system of separate roads for the vehicle flow and pedestrian flow, a close system of light railway in the internal space from the Forest Area to Site B, the rationality in arranging the public Traffic Station and the solid support coming from the infrastructure of utilities.

One of the biggest features is that the safe zone for the pedestrian flow, which runs through all the Site B and the Forest Area, is created on both sides of the central axis. And only a few light railway and cycle routes are permitted in the internal area. These distinctive details show the spirits of Green Olympic and people's Olympic a lot.

The system of plentiful humanistic landscape and wedge-shaped green land and natural landscape

In this section we focus on the following three parts: landscape planning and space design; some considerations on axis, straight backdrop and veiling; the opening and enclosing in space design. The integration of traffic cores and sunken spaces brings about the plentiful spatial levels, and the most distinct feature lies in the wedge-shaped natural landscape and ecological environment planning.

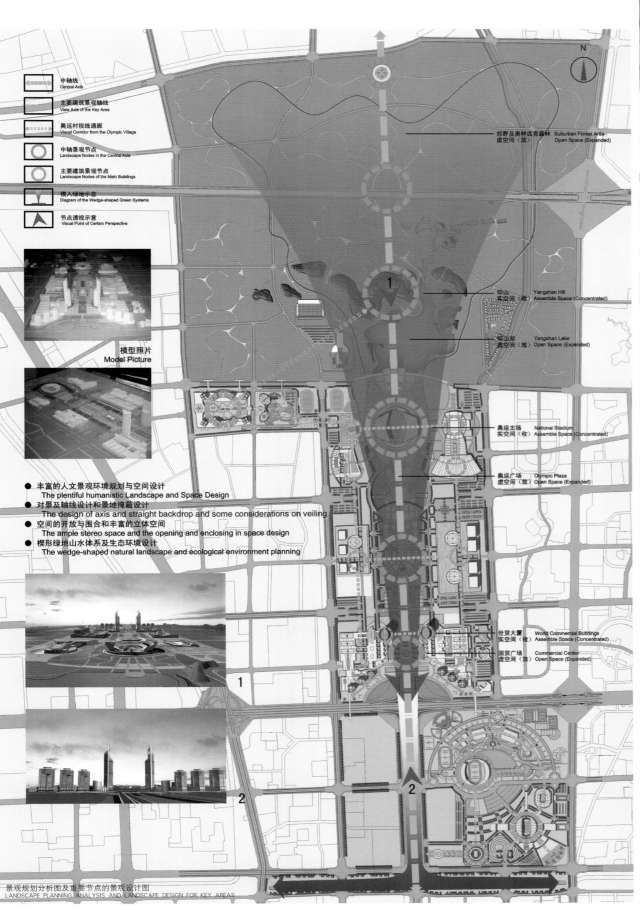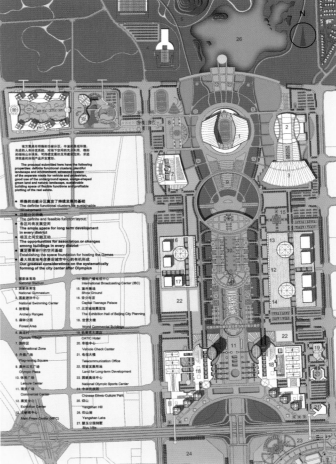

■ 规划设计构思

强调综合性多功能建筑群体的协调,简明的功能分区。

注重人文环境建设,突出中轴线文化,将国家体育场、会展博览设施、中心广场和高240m的仁和塔鳞次栉比地、有序地、有机地融入中轴线序列中,并从城市空间、城市景观、建筑风貌、建筑结构艺术、建筑材料和整体色彩上表现出对五千年中华悠久文化传统的继承,同时突出体现21世纪"新北京、新奥运、新风貌"的城市文化意义。

注重生态环境建设,契合"绿色奥运"的理念,在奥林匹克体育公园内建立丰富多彩的自然水系景观空间,形成优美的水景环境。

注重高科技在奥运场馆建设中的有机运用,用高科技来体现奥运精神,场馆建筑群追求功能、环保节能和城市建筑艺术的完美结合,以"科技奥运"促进"绿色奥运",凸现"科技奥运"。

不同功能的建筑群体以中轴线的空间序列变化及文化历史传承为联系纽带,将各个功能区融于一体,体现出奥林匹克公园建筑群的整体严谨性。

飘逸的运动轴将运动员村、奥林匹克中心区、现有的奥林匹克体育设施有机的连成一体,形成富于动感的奥林匹克运动轴。

■ Planning and Design Concept

The scheme places a stress on relational harmony and functional division among comprehensive and multi-functional architectures.

The scheme places a stress on construction of humanity and cultural setting and the axial culture. To reach the aim, the National Sports Center, the Exhibition Center, the Central Square and the 240m high Renhe Tower will be orderly arranged into the axial sequence. The orderly arrangement will indicate the continuation of 5000 year long history of Chinese tradition through the urban space, urban landscape, architecture style, structural art, building materials and color and at the same time reflect the urban cultural significance of the 21st century "new Beijing, new Olympics and new outlooks".

The scheme will give special priority to ecological environmental buildings to realize the "green Olympics" thesis. Inside the Park, rich and

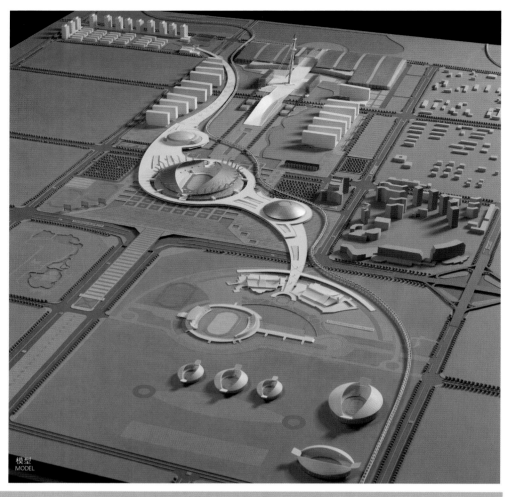

Beijing Olympic Green 2008

北京工业大学建筑勘察设计院(中国)
BEIJING POLYTECHNIC UNIVERSITY INSTITUTE OF ARCHITECTURAL EXPLORATION & DESIGN(CHINA)

colorful natural water landscapes will be built to create the elegant and beautiful water landscape environment.

The scheme puts emphasis on application of hi-tech into the constructions and constructions of functionality. Environmental protection, energy saving and urban architecture arts so as to realize the aim of promoting "green Olympics" through "sci-tech Olympics" and reflecting.

The scheme is supposed to locate different functional architecture groups along the axle through the spatial and timing sequence and the historic and cultural relationship to reach the effect of wholeness of the architecture groups in the Park, while the buildings of the same function remain as one independent group.

The silky-belted overhead transport conveyor connects the Olympic Village, the Olympic Center and stadiums together, forming a dynamic Olympic axel, which adjoins the axle of Beijing.

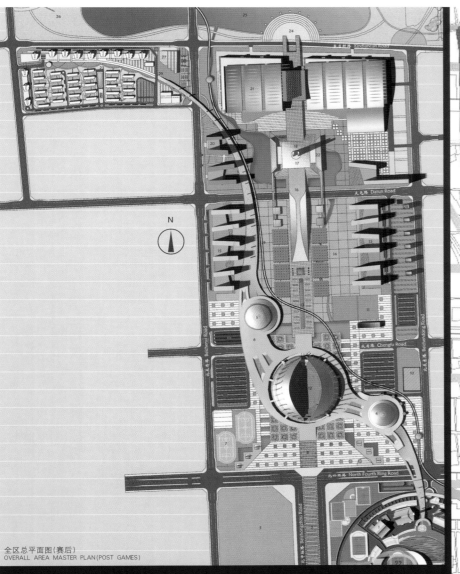

全区总平面图（赛后）
OVERALL AREA MASTER PLAN (POST GAMES)

1 国家奥体中心 NATIONAL OLYMPIC SPORTS CENTER
2 奥林匹克大道 THE ROAD OF THE OLYMPIC
3 中华民族园 CHINESE ETHNIC CULTRUE PARK
4 中国杂技马戏馆 CHINESE ACROBATIC AND CIRCUS HALL
5 戏水乐园 THE PADDLING PARADISE
6 运动轴 THE SPORTS AXIS
7 训练场 TRAINNING FIELD
8 地和广场 DIHE PLAZA
9 停车场 PARKING
10 凯迪克酒店 CATIC HOTEL
11 室外水幕电影 THE OUTDOOR WATER SCREEN FILM
12 国家体育场 NATIONAL STADIUM
13 中心广场 THE CENTRAL PLAZA
14 绿肺 THE GREEN LUNGS
15 商业 BUSINESS SERVICE AREA
16 北京城市规划展览馆 THE EXHIBITION HALL OF BEIJING PLANNING
17 仁和广场 RENHE PLAZA
18 仁和塔 RENHE TOWER
19 轻轨浏览专线 THE TOURIST SPECIAL LINE OF LIGHT TRAJECTORY
20 国际中心 THE INTERNATIONAL CENTER
21 会展中心 CONVENTION AND EXHIBITION CENTER
22 首都青少年宫 CAPITAL TEENAGE PALACE
23 奥运村 THE OLYMPIC VILLAGE
24 五环广场 THE FIVE RINGS PLAZA
25 五环湖 THE FIVE RINGS LAKE
26 森林湖 FOREST PARK
27 商场 SHOPPING CENTER
28 商务办公开发 BUSINESS AND OFFICE AREA
29 奥运之门 THE GATE OLYMPIC

中轴线景观规划与节点设想图
LANDSCAPE PLANNING FOR THE CENTRAL AXIS AND LANDSCAPE DESIGN FOR KEY AREA

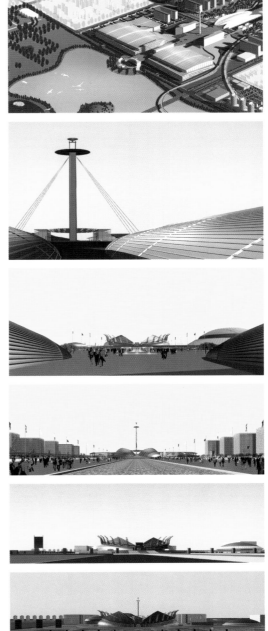

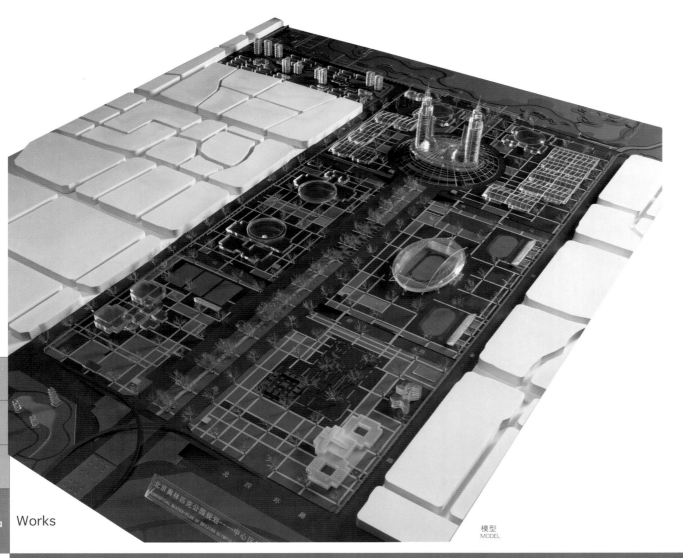

模型 MODEL

Beijing Olympic Green 2008

北京天下原色艺术设计有限责任公司(中国)
TIANXIA CORP. LTD.(CHINA)

■ 规划设计构思

本案通过讲述人与自然的故事阐述主题——共舞。

整体空间形态力求气势恢宏而又不失亲切舒缓,洗练有序而又不失丰富细腻。在平面上,通过中轴线与北京古城相互对应;在立面上,与天空形成相互对应。整体空间构成,由以建筑和建筑广场为主体的品字形空间,与以植物和山水为主体的品字形空间交错相套而成。

中轴线在此先"收"后"放"。以"给我一个支点我可以撬动地球"(阿基米德)为立意而成的景观,作为城市轴线的高潮与终结。完美表达了"绿色奥运、科技奥运、人文奥运"。

"院"套"院"的景观结构布局,产生了整体和谐有序的边缘空间系统,非常适宜人与人的交往,也适宜"上下五千年"的文明显示(人的故事),以及"生命进化史"(自然的故事)的景观演绎。

建筑和建筑广场组团沿公园边界"分散集团式"布局,地下空间规划也依托而成。这有助于交通、市政等问题的系统解决,有助于利用最先进的科学技术对土地资源最有效地利用。也与"依形就势,精细利用土地"的古代城市规划思想一脉相承。

绿地系统分为"一轴"、"一界"、"一点"、"一壶"、"一林"、"一题"和"一村",注重"春新绿、夏浓荫、秋彩叶和冬常绿"的季相景观营造。

中轴路四环以北段在赛后的封闭,以及成府路和大屯路的下沉穿越,保证了赛时需求和赛后形成最少限制的步行区,使公园真正成为市民喜爱的公共活动空间。

■ Planning and Design Concept

The scheme expounds the subject-Joint Dancing through description of the story between man and nature.

The shape of the overall space seeks grandeur without the lack of intimacy and mildness, succinctness and orderliness without the lack of richness and minuteness. Correspond to Beijing the ancient city through the axial line on the surface. Set off the sky on the vertical side. The overall space is of inverted T-shaped space on the basis of buildings and squares and matches alternatively with the inverted T-shaped space on the basis of plants, mountains and rivers.

The axial line first "shrinks" and then "extends" here. As the high tide and finish of the city axial line, the landscape established on the idea "I can pry the globe on a fulcrum"(Archimedes) perfectly expresses "Green Olympics, High-Tech Olympics, and People's Olympics".

The landscape layout of "yard" in "yard" produces the marginal space system of comprehensive harmony and is most proper for communication between people, the exhibition of civilization for "the history of five thousand years"(story of man), and the landscape show of "evolution history"(story of nature).

The building and square clusters are districted in "scattered groups" along the garden boundary, on which basis the planning of underground basis comes into being, which helps to solve traffic and municipal problems systematically and realize most effective use of land resources by most advanced technology, and conforms to the ancient city planning "take advantage of the layout to make fine use of land".

The green land system consists of "one axle", "one boundary", "one point", "one kettle", "one forestry", "one subject", and "one village", and emphasizes on building up seasonal landscapes of "fresh green in spring, thick shadow in summer, colored leaves in autumn, and evergreen in winter".

The after-match enclosure of the section north of the fourth roundabout along the axial road and the descending cross-over of Chengfu Road and Datun Road guarantee the demand of matches and formation of walking area with least limitation after the matches, so that the garden can really be the space of public activities loved by the city peole.

北京奥林匹克公园设计方案 Beijing Olympic Green

参赛作品 Works

 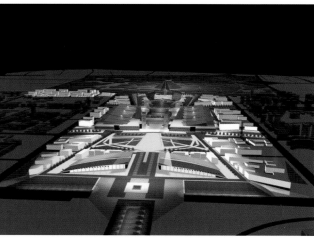

■ 规划设计构思

我们规划设计的起点是创造一个新型的城市规划布局,充分体现进步、灵活和可持续发展的特点。基于此,我们的总体规划主要考虑以下内容:

与当代的北京城市设计相结合,以北京市的历史文脉作借鉴。

探讨消极和积极的娱乐因素,创造一个灵活的多功能空间以确保商业活动的长期持续发展。

为北京市创造一个充满活力的公共活动场所,其吸引力将不受任何体育、文化和娱乐活动的影响。

在广场和建筑设计上采用先进的环境设计原则。

通过对居民生活环境的改善,并结合社区的日常使用来提高社区的可持续性发展。

强化奥林匹克运动的理想。

将北京和中国展现给全世界。

■ Planning and Design Concept

Our starting point in the design process was to create a new type of urbanlayout,one which reflected progress,sustainability and flexibility. In followingthese themes,our key considerations forthe Master Plan included:

To weave and link into the existing urban design fabric of Beijing, drawing inspiration from the rich historical context of the city.

A multi-functional,flexible spaceto ensure longer term commercial sustainability,whilst ensuring planningopportunities for both passive and active recreation.

Creating an active destination point within the City of Beijing, which was an attractive place to be irrespective of any sporting, cultural or entertainment events occurring within the precinct.

To create opportunities to utilize current advancements in Environmental Design principles in the use of the site and the built structures it contained.

Beijing Olympic Green 2008

HOK 体育建筑设计公司（澳大利亚）
HOK SPORT+VENUE+EVENT(AUSTRALIA)

A place that would enhance community sustainability by allowing improvements to civil life and integration with daily community use.

Reinforcement of the Olympic Movement ideals.

Showcase Beijing and China to the world.

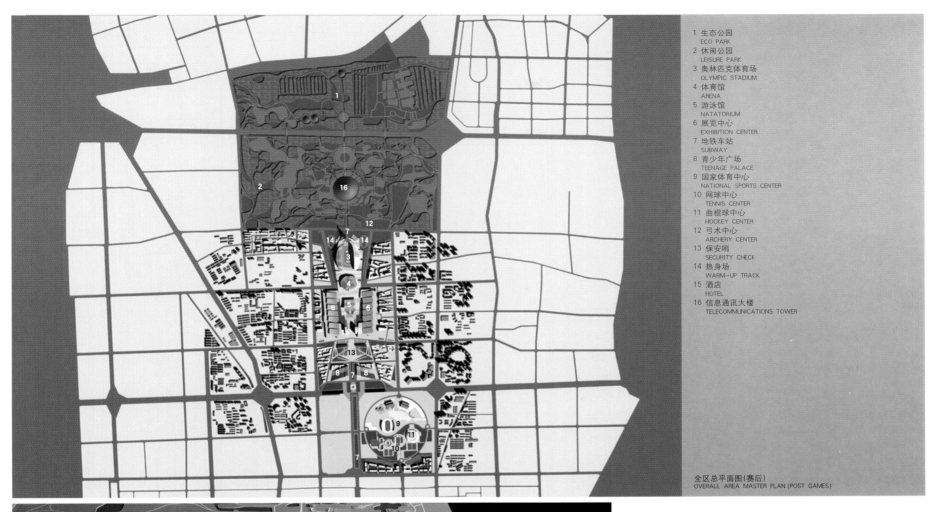

1	生态公园 ECO PARK
2	休闲公园 LEISURE PARK
3	奥林匹克体育场 OLYMPIC STADIUM
4	体育馆 ARENA
5	游泳馆 NATATORIUM
6	展览中心 EXHIBITION CENTER
7	地铁车站 SUBWAY
8	青少年广场 TEENAGE PALACE
9	国家体育中心 NATIONAL SPORTS CENTER
10	网球中心 TENNIS CENTER
11	曲棍球中心 HOCKEY CENTER
12	弓术中心 ARCHERY CENTER
13	保安哨 SECURITY CHECK
14	热身场 WARM-UP TRACK
15	酒店 HOTEL
16	信息通讯大楼 TELECOMMUNICATIONS TOWER

全区总平面图(赛后)
OVERALL AREA MASTER PLAN (POST GAMES)

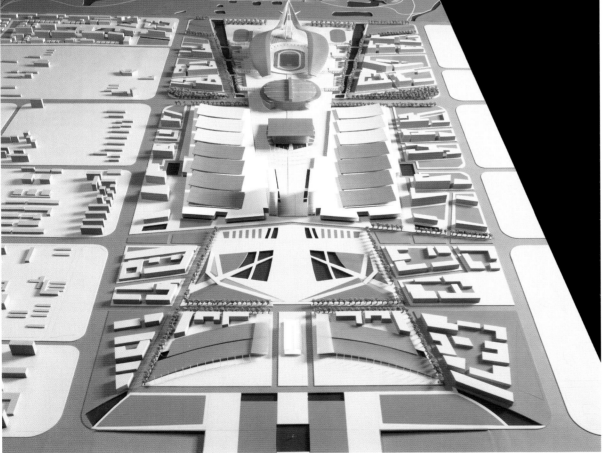

模型
MODEL

北京奥林匹克公园设计方案 Beijing Olympic Green

■ 规划设计构思

元大都北城墙(F区)是一条东西向为主的城垣遗址,将提供一处居民和旅游者凭吊古迹、休憩交往的公共活动场所。对古代遗迹,保护原有风貌;对周边现状,改善环境质量。

体育文化公园(A区)是一个以体育为主题的极具亲和力的文化公园。

中华民族园(D区)基本保持现状,它集中反映了中国各民族丰富多彩的生活风貌,但其风格有别于我们的总设计思路,故通过绿化等手段进行弱化处理。

E区是在绿化带的中央即中轴线上设置高10m左右的白色体育浮雕柱29根,均匀排列,代表2008年的第29届奥运会。

中心区(B-1区)的设计集中体现了人文、科技、自然的完美结合。

运动员村(B-2区),为各国运动员及未来社区居民提供舒适的休息环境,轻松的交流氛围。

森林公园(C区)作为中轴线的升华,采用古典园林与自然相结合的手法,山水交融,动静分明,形成一个"世外桃源"。

■ Planning and Design Concept

The Yuan Dynasty City Wall Relics Park (Site F) of Dadu of Yuan dynasty is an east-west direction city wall ruins. Our planning idea is to provide a relics for the citizens to visit and ponder on the past and a public activity place for rest and association so as to let the citizens quietly exchange their modern life with the imposing style and feathers of relics on both bands of the moat.

Sports and Culture Park (Site A) is to create a cultural park with sports as the chief topic exceedingly possessing affinity to make the people produce enthusiasm to participate in sports and realize the interest of sports, while being affected by the sports culture, and they are completely relaxed amongst the space of park with the pleasant shade made by green trees and the murmuring of running water.

Beijing Olympic Green 2008

中国寰球化学工程公司(中国)
CHINA HUANQIU CHEMICAL ENGINEERING CORP(CHINA)

Chinese Ethnic Culture Park (Site D) retains its existing state on the whole, it concentrated-reflects the rich and colorful living style and feathers of different nationalities, but their level differs from our design idea, and so we hope to proceed a weakening treatment by measures greening and the like.

Site E is the realized part road section of North Central Axis Road, both sides of the road centerline have already built greening belt with lawn as the principal established, and our design is to set up 29 white sports relief columns with height of about 10m in the center of greening belt, i.e on the central axis and they are uniformly arranged to represent the 29th Olympic Games in 2008.

The Central Area (Site B-1) concentrated-embodies the perfect combination of human culture, science and technology and nature, the huge dimensions of buildings and square are weakened as far as possible in the design to decrease the depression on human feeling, and emphasis is put on the combination of large area stereo-greening and natural landscape with the site function of buildings so that this not only makes the entire district have grand momentum, but also gives relaxed and comfortable feelings to those experiencing personally.

Olympic Village (Site B-2) is to be used as high-grade apartment community after the game. Therefore the purpose of design is aimed at providing a comfortable rest environment and a relaxed exchange atmosphere for athletes of various countries and the future in the community.

Forest Area (Site C) as a sublimation of the central axis, the skill of combination of classic gardening wit nature is used by harmonizing hill with water and distinction of activity from inertia to from a "fictitious land of peace' such that people are enchanted in undulated landform and enriched vegetation and feel variation and rhythm of the nature.

A 运动员村 OLYMPIC VILLAGE
B 会展办公楼 ASSISTANT BUILDING FOR D
C 会展办公楼 ASSISTANT BUILDING FOR D
D 会展博览中心 CONVENTION AND EXHIBITION FACILITIES
E 露天剧场 OPEN-AIR THEATRE
F 国家游泳中心 NATIONAL SWIMMING CENTER
G 国家体育场 NATIONAL STADIUM
H 场馆区商服中心 COMMERCIAL CENTER FOR SPORTS FACILITIES
I 训练场区 TRAINING AREA
J 国家体育馆 NATIONAL GYMNASIUM
K 公园广场 SQUARE OF OLYMPIC GREEN
L 首都青少年宫 CAPITAL TEENAGE PALACE
M 城市规划展馆 EXHIBITION HALL OF BEIJING CITY PLANNING
N 文化公园 CULTURAL PARK
O 奥运商区 OLYMPIC COMMERCIAL DISTRICT
P 凯迪克酒店 CATIC HOTEL

中心区总平面图（赛后）
CENTRAL AREA PLAN (POST GAMES)

别墅开发区 VILLA DEVELOPMENT AREA
开放自然公园 OPEN GREEN PARK
高档公寓社区 APARTMENT COMMUNITY
奥运公园 OLYMPIC PARK
文化公园 CULTURAL PARK
商务开发区 COMMERCIAL DEVELOPMENT AREA
体育主题公园 PHYSICAL PARK
古今交流公园 COMMUNICATION PARK

功能空间分析
MASTER FUNCTION ARRANGEMENT AND SPACE ANALYSIS

鸟瞰图
BIRDS-EYE VIEW

北京奥林匹克公园设计方案 Beijing Olympic Green

参赛作品 Works

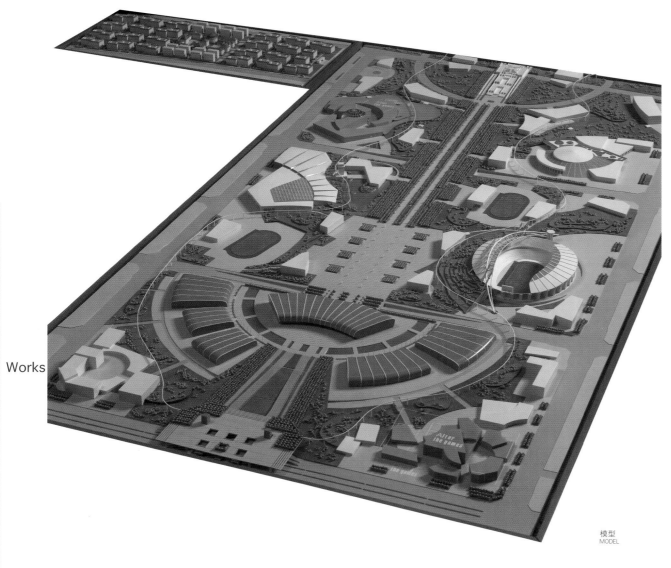
模型
MODEL

■ 规划设计构思

中轴景观规划设计主题：来自远古、回归自然。

与传统中轴相似的建筑形式（稳重安详）、空间节奏（平缓舒展），延续传统中轴线的城市文化精神。

元大都遗址公园尽量保留原址面貌。

从封闭性（中规中矩的空间秩序，四面围合的广场，大段隔阻的传统中轴）走向开放性（回归自然的北中轴）的城市空间形式，对北京有着极典型的象征意义。

森林公园景观设计构思：自然环保、原始生态。

以地方野生植物为主，仿自然生态森林配植。尽可能地保留现有自然水系。

五环大道下沉。

森林公园没有任何体育、娱乐项目设施，只有低影响度的旅游内容。

B区设计构思城市山水、山水城市。

以中国独特的山水画意境营造北京特有的城市中轴线经及城市山水风格的奥林匹克公园。

致力于园林化、山水化、自然化、绿地面积的最大化，创建生态型山水城市的样板地块。

■ Planning and Design Concept

The theme of central axis in From and To Nature.

From Past is displayed by similar space rhythm of the layout in the Forbidden City, traditional culture features, methods of implide meaning and symbolizaition, all of which extend urbanization spirit in traditional central axis.

In Yuan Dynasty City Wall Park, original site and plants remain with redecorated bank and added bridge balustrades.

The urban structure and space from close to open is typically significant to Beijing city.

The design ideology of the Forest Area is Environment Protection and Original Reservation.

The forest, most of which is local wild plants and the priority task.

The Fifth Ring Road is suggested to go underground

No sports and recreational facility is in the Forest Area

Design ideology of District B: Urban Oasis and Oasis Urban

Olympic Green and central axis are also highlighted by green mountains and blue waters of Chinese characteristics.

The boundless and large-scale plants, elegant gardens, green mountains and blue waters create an ideal living place in the capital.

Beijing Olympic Green 2008

深圳市建筑设计研究总院（中国）
SHENZHEN GENERAL INSTITUTE OF ARCHITECTURAL DESIGN ANG RESEARCH(CHINA)

深圳市景观设计装饰工程有限公司（中国）
SHENZHEN JINGGUAN LANDSCAPE DECORATION AND DESIGN LTD.(CHINA)

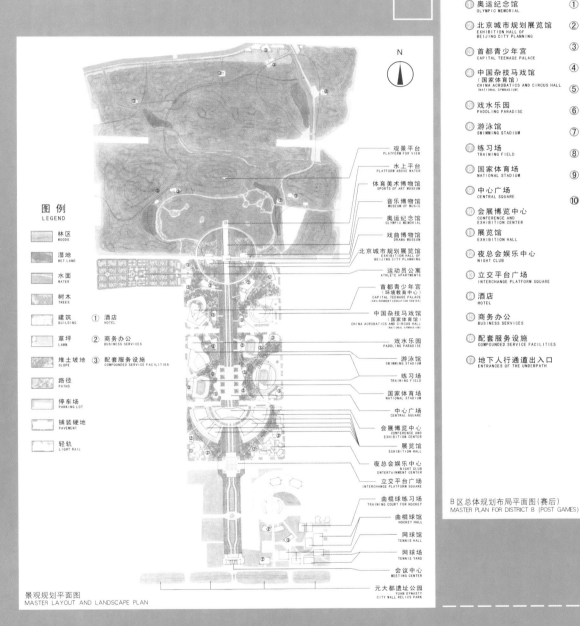

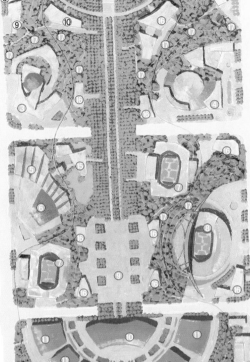

景观规划平面图
MASTER LAYOUT AND LANDSCAPE PLAN

B区总体规划布局平面图（赛后）
MASTER PLAN FOR DISTRICT B (POST GAMES)

北京奥林匹克公园设计方案
Beijing Olympic Green

Beijing Olympic Green 2008

DP建筑设计有限责任公司（新加坡）
DP ARCHITECTS PTE LTD(SINGAPORE)

参赛作品 Works

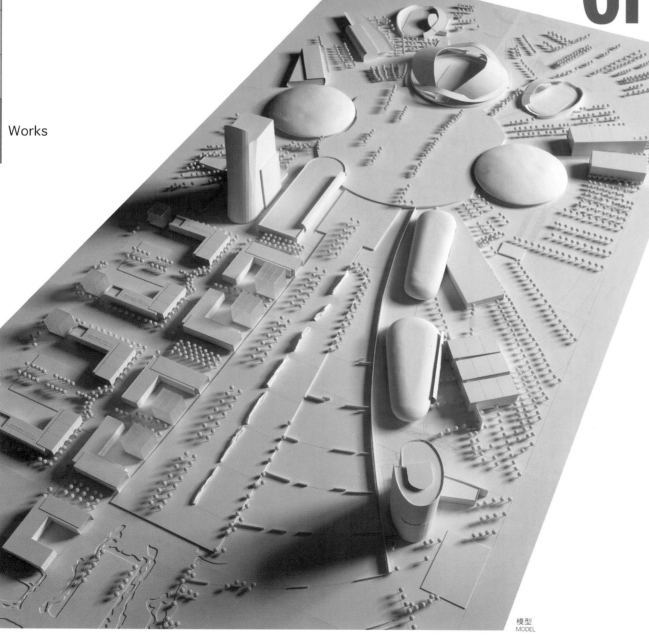
模型 MODEL

■ 规划设计构思

总体上，规划思路在于以"光"字的形体作为整体布局的策略。以"光"的字体构造而言，笔划由一个明确的中心向外伸展。"光"字的上部，固定了国家体育场及两所辅助体育场的位置，而国家游泳中心及国家体育馆则立于两旁，形成字中的横划。"光"字的下部则形成两道向外延伸的曲轴线，固定着其余设施与园林设计的布局。左旁的建筑立面随着曲轴线弯曲，而右旁曲轴线则奠定了两所会议中心的位置。

此外，"光"字的中央，也成为另一道轴心线的终点。轴心线从台阶广场为首，再由中部的南北行人大道延续，最后以中央广场画上句点，形成了一条鲜明的直线，更突出了中央广场的重要性。沿着南北行人大道的两旁则是商业基础设施以及有临时帐篷的露天广场，为行人大道注入了无限的生命力。

■ Planning and Design Concept

The concept of "radiance" goes along with the idea of the ever-lasting Olympic Flame. The intention of our scheme is to terminate the central axis at the Olympic Green, where the termination point is represented by the Chinese ideogram "π,"(radiance).These ideogram's strokes are developing outward from an obvious central point. The upper strokes of "light" represent the locations of the State Gymnasium and the supporting facilities. The horizontal stroke of "light" represents the location of the State Swimming Center. The bottom strokes of "light" are two curved axes setting the rest of the facilities and the layout of the gardens. The left storke of the sweeping axis denotes the facades along the concourse, while the right sweeping axis delines the location of the two conferece centers and the ramp.

Besides, the central axis of "light" is also the end of another axis. This axis is begun with the steps rising up to the sidewalks and is ended with the Central Plaza. This layout is a definite line to stress on the importance of the Central Plaza. Beside the sidewalks, there are commercial infrastructures and open courtyard with tents set up temporarily. These buildings give infinite vigor to the sidewalks.

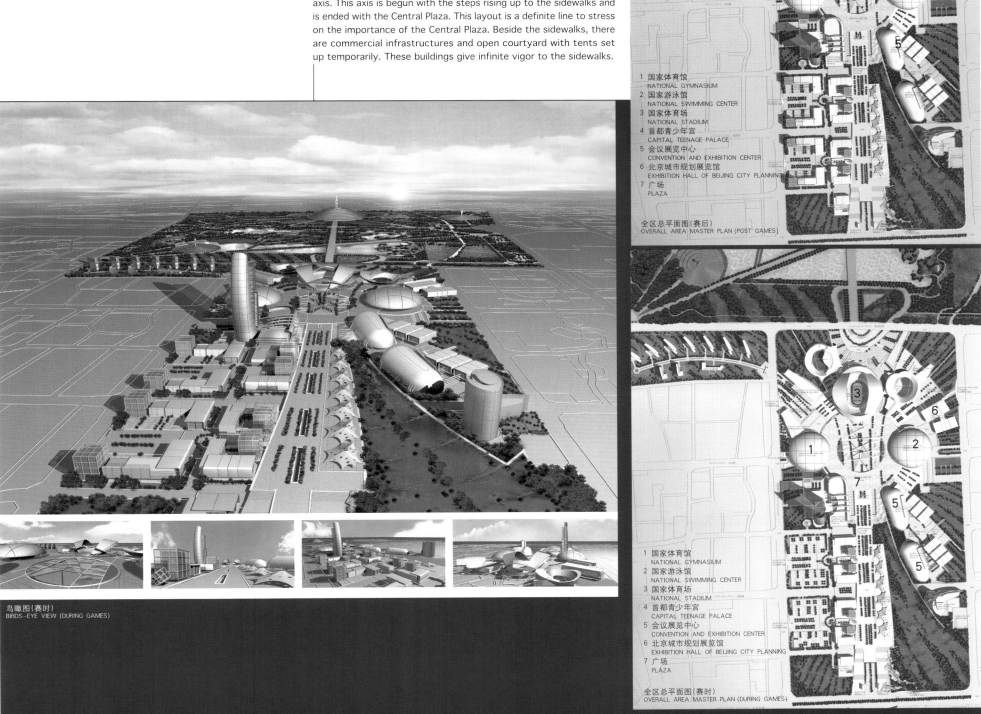

北京奥林匹克公园设计方案 Beijing Olympic Green

参赛作品 Works

模型 MODEL

■ 规划设计构思

本次规划征集书点明了"绿色、科技、人文"三点，做为北京奥运的主题，体现国际城市风范，并以国际级文化古老都市特色为主要精神，结合21世纪流行风潮之现代化科技及绿色环保，作为设计北京新国际舞台之目标。

■ Planning and Design Concept

In this proposal it has been stated clearly that "green, high-tech and humanities" are the main themes of Beijing Olympics, showing the elegance and beauty of an international city like Beijing during the Olympics period. By catching the spirit of Beijing as an international cultural and historic city, combining the prevailing practice of modern technology and green environmental protection of the 21st century, Olympic Green is designed to make Beijing an international stage.

Beijing Olympic Green 2008

中兴工程顾问股份有限公司（中国台湾）
SINOTECH ENGINEERING CONSULTANTS, LTD. (TAIWAN CHINA)

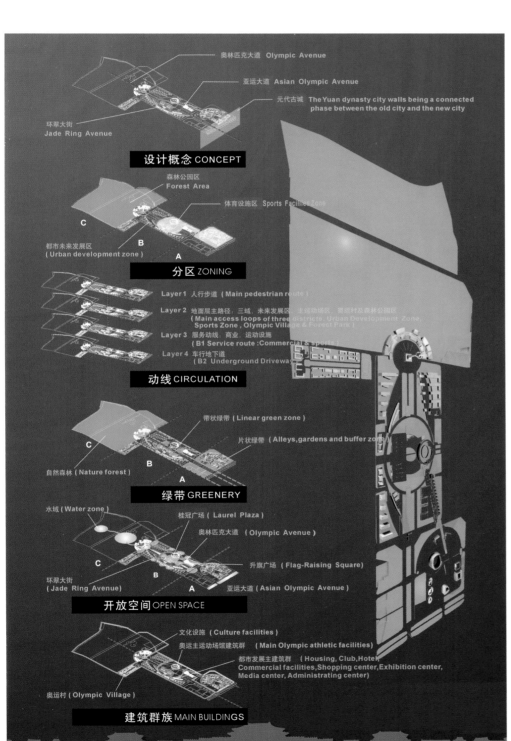

重点范围总体规划布局构想
THE CONCEPT OF MASTER LAYOUT

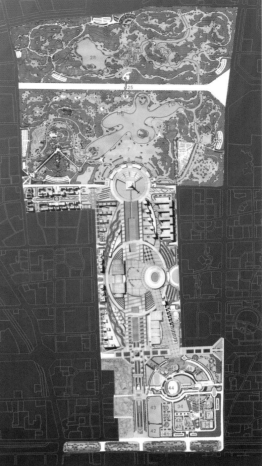

城市景观规划平面图
COMPREHENSIVE LAYOUT PLAN

重点范围总体规划布局(赛后)
MASTER LAYOUT PLAN OF THE KEY AREA (POST GAMES)

北京奥林匹克公园设计方案 Beijing Olympic Green

参赛作品 Works

■ 规划设计构思

本设计的主题是"河"与"岸",这出自对于城市结构的关注和对华夏文明的理解,同时也倾注了对于自然环境的渴望和对不息生命的追求。这条运动着的城市之河孕育着生命,积淀着历史,召唤着未来。

它是城市空间的延展,它提炼了生命与自然的寓意,展示着文明古城片断和奥运精神的内涵,融合了人工的建筑街区与自然的生态环境,表述着现代建筑理念和科学技术的语言。

规划中,蓝色水域托起一系列大小不一的主题花园绿岛,岛上漂浮着点点白帆,流水绿茵帆影始终伴随着漫步者的行踪,变换着的画面将风格各异的片断剪接成主题鲜明建筑意境。主要的功能性建筑以岸的形式由高渐低的围合着中央之"河",规整的建筑形式与自由的河岸线形成鲜明的对比,寓意人工美与自然美的共生。

■ Planning and Design Concept

The theme of our design is river and bank. It comes from our awareness of Beijing's structure and comprehension of China's culture. It also represents humanity craving for a natural environment and life eternal flow. Running waters create life, record history in layers of sediment and greets the future.

It is an Extension of Urbanization. It gleans the virtues of life and nature; It displays a profile of this ancient civilized city and has an essence of Olympic sprit; It harmoniously blends man-made architecture and natural ecology; It expresses modern architectural concepts as well as technologically advanced features.

The blue water body embraces a chain of differently sized green Garden Islands and white sails flutter freely over the islands. Green environment, flowing water, and the shadows of sails accompany strolling pedestrians. The various backdrops are distinctively divided into artistic concepts, creating a vivid theme.

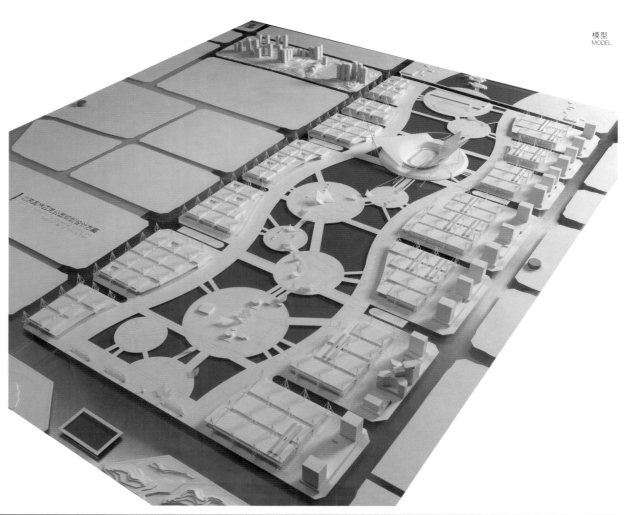

模型 MODEL

Beijing Olympic Green 2008

哈尔滨工业大学建筑设计研究院(中国)
ARCHITECTURAL DESIGN AND RESERARCH INSTITUTE OF HIT(CHINA)
莫斯科第四国立设计院(俄罗斯)
GUE MNIIP "MOSPROECT-4"(RUSSIA)

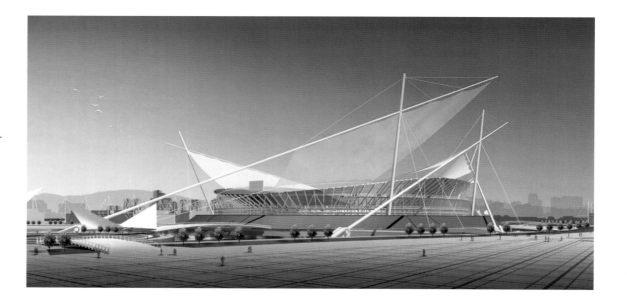

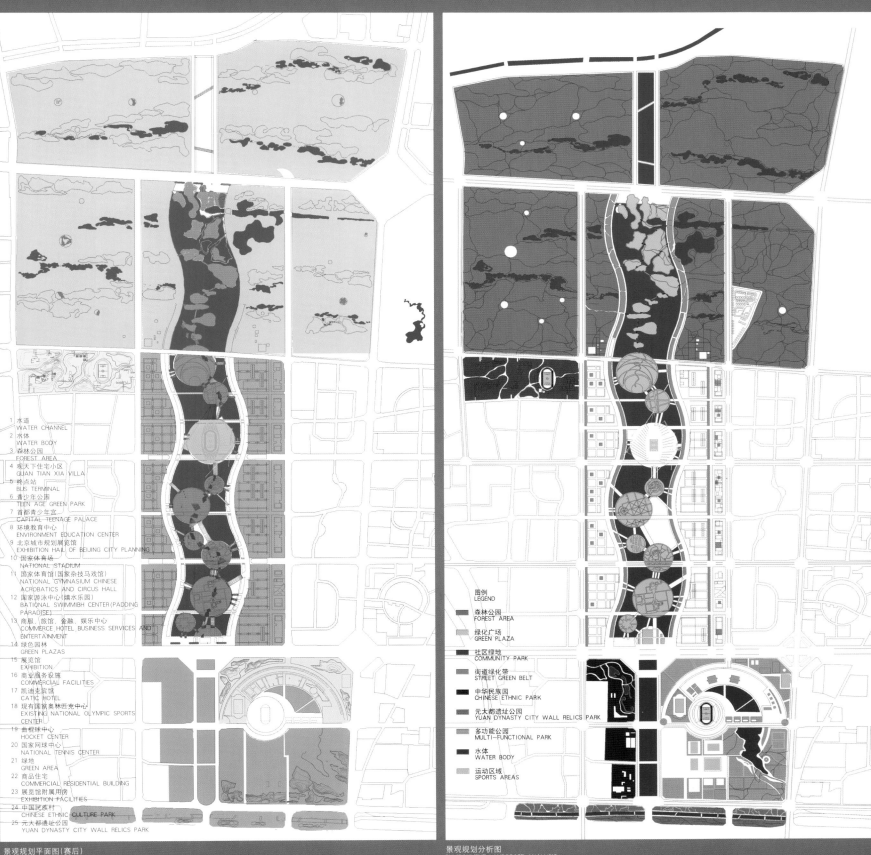

景观规划平面图(赛后)
LANDSCAPE PLAN (POST GAMES)

景观规划分析图
ENVIRONMENT LANDSCAPE ANALYSIS

北京奥林匹克公园设计方案 Beijing Olympic Green

参赛作品 Works

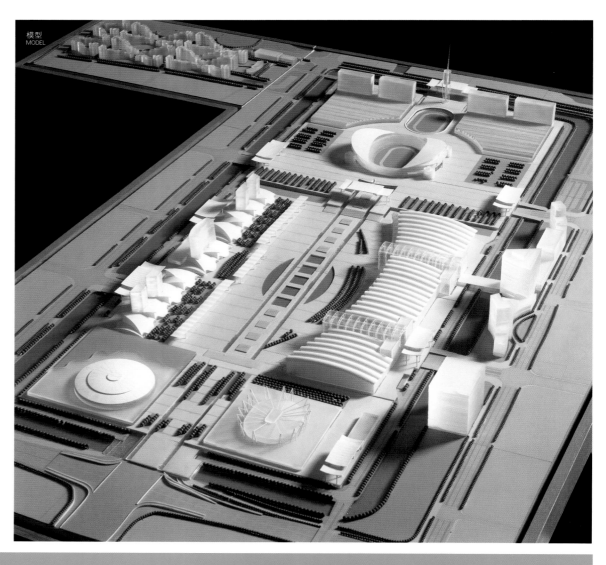

■ 规划设计构思

设计目标不仅仅局限于为北京此次世界级体育盛事规划基础设施、引入先进技术以及提供充足的绿地，我们更致力于将这项宏伟项目的适应性延伸至奥运会之后。它将基于使用者的美学与感受需要，平衡效用，吸引人文与商业，并为北京市营造一种场所感与归属感。

设计采用了垂直分离与线性分离，从而可以减少步行与车行交通流量之间的冲突，同时也便于对大量人流的管理组织。

设计着重强调北京的城市中轴线，并将其延续至奥林匹克公园内，规划一条奥林匹克大道，穿过公园主入口（北京大门），两侧以形态不同的标志性建筑来界定，营造一种激动人心的奥运体验。

■ Planning and Design Concept

Our Design seeks to provide a world-class Sporting Event with the adaptability to ensure that its legacy continues into the Post-Olympic period as an important addition to the infrastructure, technlolgy and green spaces of Beijing.It balances the need for efficiency with the aesthetic and emotional needs for these places to attract people/business and develop a sense of "place"and "belonging"for the people of Beijing.

We have created a design that has utilized both vertical and linear segregation to reduce traffic/pedestrian conflicts and assisted with crowd management:

Our Design focuses on strengthening the central Beijing axis and providing an exciting "Olympic experience"through the creation of a long Olympic Promenade broken by magnificent gateways and landmark buildings of varying heights and sizes.

Beijing Olympic Green 2008

同济大学建筑设计研究院（中国）
ARCHITECTURAL DESIGN INSTITUTE OF TONGJI UNIVERSITY(CHINA)
迪克森·罗斯希尔 PTY 有限公司（澳大利亚）
DICKSON ROTHSCHILD PTY LTD (AUSTRALIA)

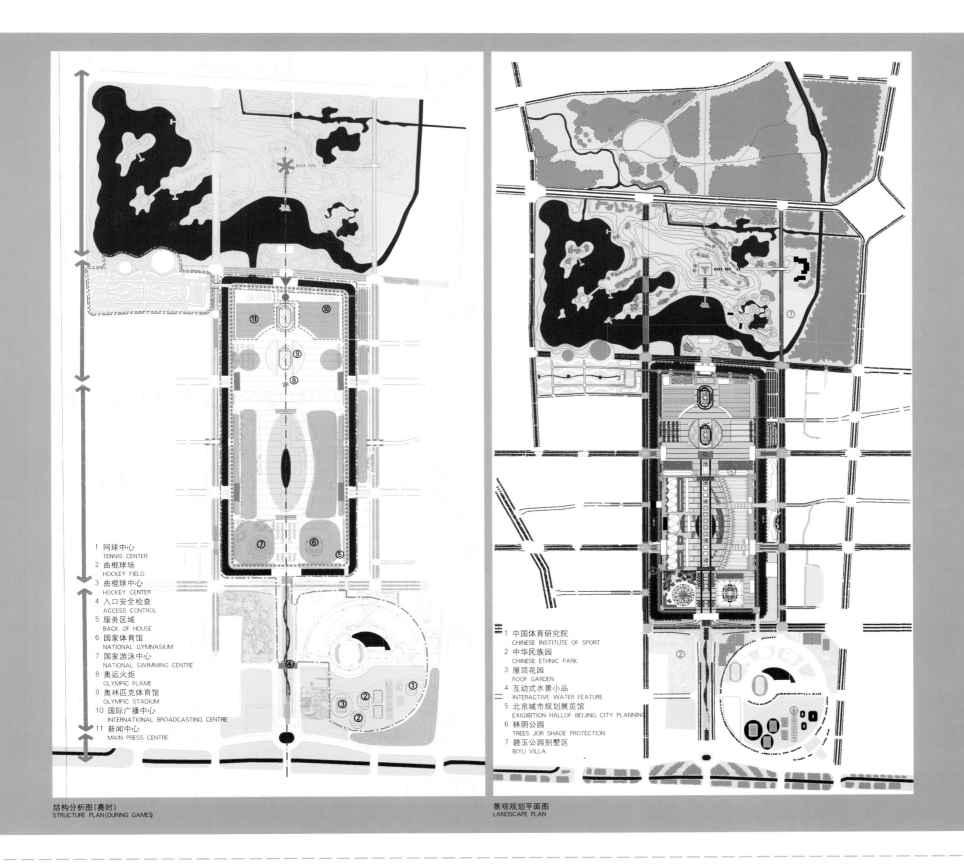

结构分析图(赛时)
STRUCTURE PLAN (DURING GAMES)

1 网球中心 TENNIS CENTER
2 曲棍球场 HOCKEY FIELD
3 曲棍球中心 HOCKEY CENTER
4 入口安全检查 ACCESS CONTROL
5 服务区域 BACK OF HOUSE
6 国家体育馆 NATIONAL GYMNASIUM
7 国家游泳中心 NATIONAL SWIMMING CENTRE
8 奥运火炬 OLYMPIC FLAME
9 奥林匹克体育馆 OLYMPIC STADIUM
10 国际广播中心 INTERNATIONAL BROADCASTING CENTRE
11 新闻中心 MAIN PRESS CENTRE

景观规划平面图
LANDSCAPE PLAN

1 中国体育研究院 CHINESE INSTITUTE OF SPORT
2 中华民族园 CHINESE ETHNIC PARK
3 屋顶花园 ROOF GARDEN
4 互动式水景小品 INTERACTIVE WATER FEATURE
5 北京城市规划展览馆 EXIGIBITION HALL OF BEIJING CITY PLANNING
6 林阴公园 TREES JOR SHADE PROTECTION
7 碧玉公园别墅区 BIYU VILLA

Beijing Olympic Green 2008

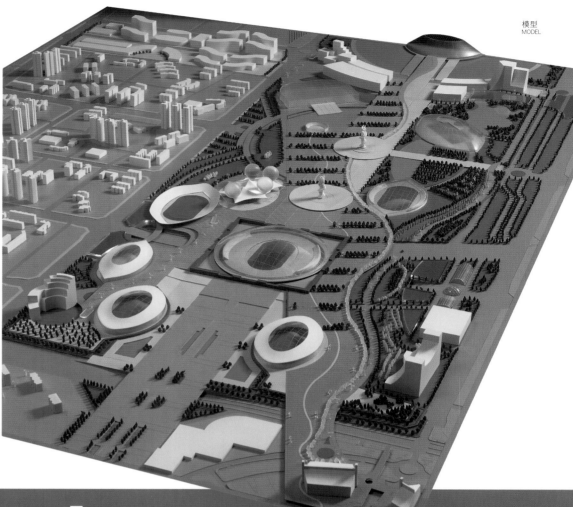
模型 MODEL

■ 规划设计构思

总体规划将通过以下措施实现"绿色奥运、科技奥运、人文奥运"这一宗旨：

总体布局和景观规划确保建设具国际先进水平的设施，为成功举办第29届奥运会创造条件；

21世纪充满活力的城市中心设计满足市民娱乐的需要，把运动、文化、展览、娱乐和旅游等设施结合在一起；

21世纪多功能城市中心发展规划设计了公共空间和大规模绿色开放空间，有利于城市的长期发展；

优美城市景观设计确保提供高质量的环境，体现中国传统文化以及现代生活观念。

设计试图给北京市民以及世界游人提供一个机会，让他们游览奥林匹克公园时犹如在时间中穿行，不经意地观赏各处景点。运用北京"地区精神"和深刻"故事主题"，使奥运公园既有功能齐全的物质设施，又有深厚的精神内涵。设计把奥运公园里的体验和沿中轴线穿越时空的概念很好地结合在一起。

设计用了三段时间为标志——过去、现在和未来，三个区内南区代表了"过去"，中部的奥林匹克场馆代表"现在"，北部的森林区意味着"将来"。

■ Planning and Design Concept

The concept of the overall Landscape Master plan is to achieve the set objectives of the design brief particularly in terms of the "Green Olympics, High Tech Olympics and People's Olympics" through the following:

Master layout and landscape planning to provide world - class facilities for successfully holding the 29th Olympic Games;

Design of a vibrant 21st Century city centre for the enjoyment of the public and incorporating sporting, cultural, exhibition, leisure and tourism facilities;

景点规划设计公司（澳大利亚）
PLACE PLANNING DESIGN(AUSTRALIA)
植被生态建筑行（澳大利亚）
PLANTE & ASSOCIATES ARCHITECTS(AUSTRALIA)
国际建设有限公司（澳大利亚）
JIANSHENG INTERNATIONAL PTY LTD.(AUSTRALIA)

Development of a multi-functional 21st Century city centre with public spaces and spacious green open spaces benefiting the long - term development of the city; and Creatinng an image of an elegant city with a high quality environment representative of Chinese culture and history, as well as Chinese contemporary lifestyles.

Our design attempts to involve the people of Beijing, and indeed the rest of the world, in a travel through time as a natural visitation event for all who venture into The Olympic Green. We have embraced the "Spirit of Place" of Beijing and included a powerful storyline or "narrative" to provide the project with fully functional physical components, as well as intellectual depth and purpose. The experience of moving through the site has been integrated with the concept of moving through time along the central axis.

Our design philosophy is to embrace the three fundamental time landmarks - "the past", "the present", and "the future". This has been achieved by suggestive and interpretive landscape planning by dividing the site into 3 general areas where both historical references are provided as clues or innuences within the design process rather than literal translations. The three precincts are the southern precinct representing "the past", the central Olympic facilities representing "the present", and the northern forest precinct representing "the future".

景观设计风格图片(赛时)
OLYMPIC LANDSCAPE CHARACTER IMAGES (DURING GAMES)

景观设计平面图(赛时)
LANDSCAPE MASTER PLAN (DURING GAMES)

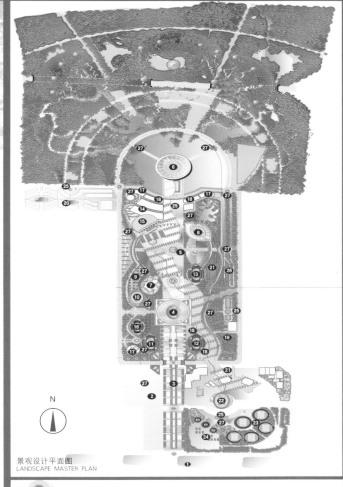

景观设计平面图
LANDSCAPE MASTER PLAN

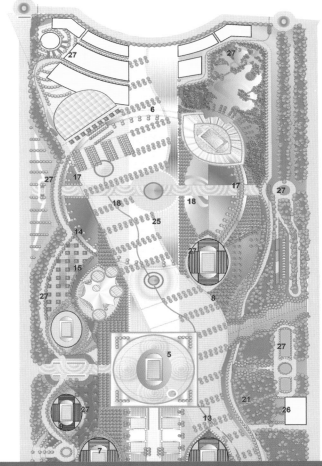

1 元大都遗址公园 YUAN DYNASTY CITY WALL RELICS PARK
2 中华民族园 CHINESE ETNIC CULTURAL PARK
3 庆典大道 CELEBRATION PARADE
4 国家体育场 NATIONAL STADIUM
5 庆典广场 CELEBRATION PLAZA
6 音乐学院 CONSERVATORY
7 国家体操馆 NATIONAL GYMNASIUM
8 国家游泳中心 NATIONAL SWIMMING CENTRE
9 箭木 ARCHERY
10 临时热身跑道 TEMPORARY WARM-UP TRACK
11 会议中心 CONVENTION CENTRE
12 展览中心 EXHIBITION CENTRE
13 多功能体育场 MULTI-USE STADIUM
14 主新闻中心 MAIN PRESS CENTRE
15 首都青少年宫 CAPITAL TEENAGE PALACE
16 北京规划展览厅 BEIJING PLANNING EXHIBITION HALL
17 旅馆 HOTELS
18 餐馆 RESTAURANTS
19 零售 RETAIL
20 奥运村 OLYMPIC VILLAGE
21 国际饮食中心 INTERNATIONAL FOOD COURT
22 国家奥运体育中心 NATIONAL OLYMPIC SPORTS CENTRE
23 国家曲棍球中心 NATIONAL HOCKEY CENTRE
24 国家网球中心 NATIONAL TENNIS CENTRE
25 地铁站 METRO STATION
26 公共汽车站 BUS STATION
27 高架单轨铁站 MONORAIL STATION

北京奥林匹克公园设计方案 Beijing Olympic Green

■ 规划设计构思

着眼于建设一个以林地为主的绿色环境，摒弃对建筑形象的追求。

建立北京大中轴：南起永定河畔石佛寺，北至军都山黄花镇长城，中间与城市中轴线相接，山川与都市同存，展现出新北京作为国际大都市的恢宏气势。

国家体育场置于奥运山内，成为神圣的地标。其他体育场馆隐于丘陵之下与山林之中。

将简单实用的生态技术运用到景观环境建设与养护管理之中，并使其成为未来城市发展的模式。

建造高水准的绿色通道，鼓励绿色交通工具的使用，缓解北京交通长期面临的困境。

建立完整的立体交通系统，保持绿地的完整性。

■ Planning and Design Concept

Emphasize on the construction of a green environment based mainly on woodland, discarding the pursuit towards construction image.

Setup of Great Central Axis: Beginning from the Shifo Temple by the riverside of Yongding River in the south, to the Huanghuazhen Great Wall at Jundu Mountain in the north. The center links the city central axis, the city co-exists with mountains & rivers, displaying the magnificent vigor of the New Beijing as an international metropolis.

The National Stadium is located inside the Olympic Hill, becoming a sacred landmark; other stadiums will be hidden under the shades of green hills and mountainous forests.

Simple and practical ecological technology is applied into the landscaping environmental construction and maintenance control, making it the mode for future urban development.

Construction of high-level greenway, encouraging the use of green traffic means to relieve Beijing of the traffic dilemma that has been around for a long time.

Setup of complete vertical traffic system to maintain the completeness of green land.

Beijing Olympic Green 2008

参赛作品 | Works

北京中国风景园林规划设计研究中心（中国）
CHINA GARDENING DESIGN & RESEARCH CENTER(CHINA)

中央美术学院（中国）
CENTRAL ACADEMY OF FINE ARTS(CHINA)

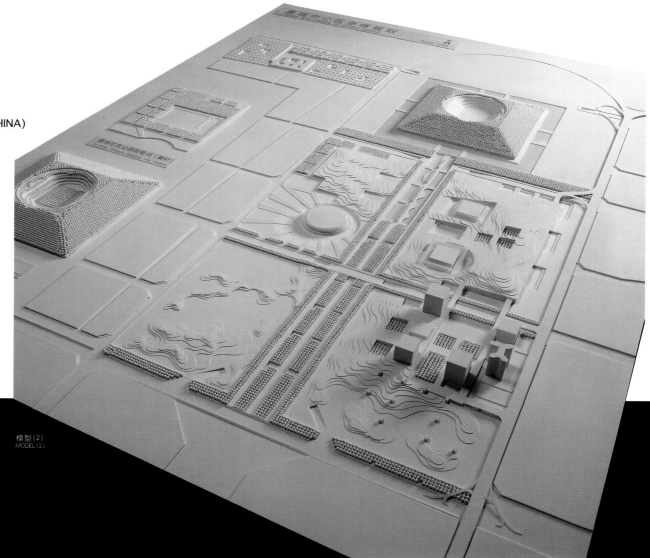

模型(2)
MODEL (2)

国家体育馆 NATIONAL GYMNASIUM
奥运山 OLYMPIC MOUNTAIN
北京奥林匹克公园景观规划(赛后) Landscape Planning of Beijing Olympic Green (After the Game)
模型(1) MODEL(1)

北京奥林匹克公园设计方案 Beijing Olympic Green

核工业部第四研究设计院深圳设计院（中国）
THE FOURTH DESIGN INSTITUTE OF THE MINISTRY OF NUCLEAR INDUSTRY(CHINA)
伍凸设计师事务所（深圳有限公司）（中国）
WUTU ARCHITECTS ASSOCIATES (SHENZHEN LIMITED)(CHINA)
加盟单位：
SPONSOR:
深圳宗灏建筑师事务所（中国）
SHENZHEN ZONGHAO DESIGN ARCHITED'S OFFICE(CHINA)
北方设计研究院（中国）
NORINDAR INTERNATIONAL (CHINA)
深圳市裕华建筑设计有限公司（中国）
SHENZHEN YUHUA ARCHITECTURAL DESIGN CO.,LTD(CHINA)
蒂奥特型建筑科技（深圳）有限公司（中国）
DIAO SPECIAL BUILDING TECHNOLOGY(SHENZHEN)CO,. LTD(CHINA)

参赛作品 Works

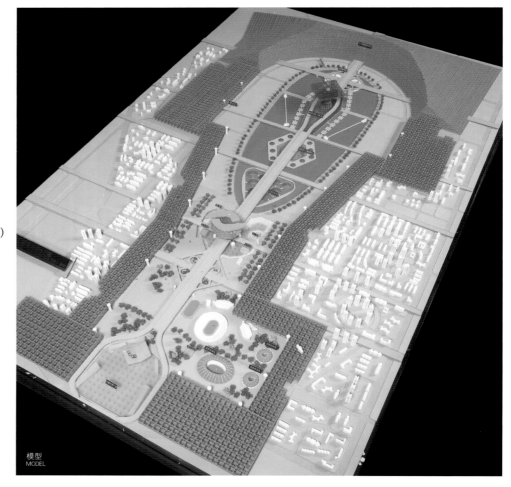

模型
MODEL

Beijing Olympic Green 2008

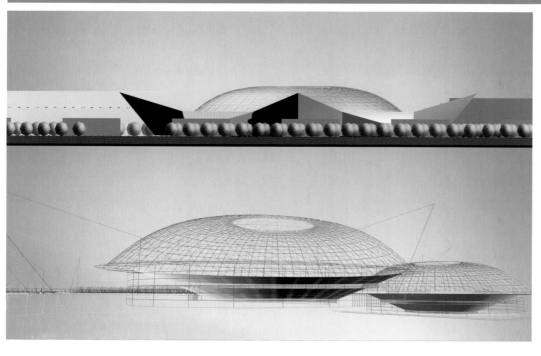

■ 规划构思说明
　　以绿色奥运、科技奥运、人文奥运为内涵，以人为自由意图力作对称几何图形的形态美布置，以及一系列世界前瞻的新概念建筑群，所形成巨大而持久的全球旅游凝聚（相似或大于艾菲尔铁塔、悉尼歌剧院的全球旅游凝聚和辐射所形成的经济桥梁）作经济杠杆，实现可持续发展长远目标。

■ Planning and Design Concept
　　Surrounding the aim of "Green Olympics, Hi-Tech Olympics and Human Olympics",the proposals target to realize sustainable development through design of a permanent world tourist hit with natural symmetric beauty of geometric design and a series of world leading new concept building groups (similar to or hotter than the Eiffel and Sydney Opera hit).

←太空舱绿覆盖看台体育馆
SUPER CAMP STADIUM WITH STANDS COVERDE BY GREEN VEGETATION

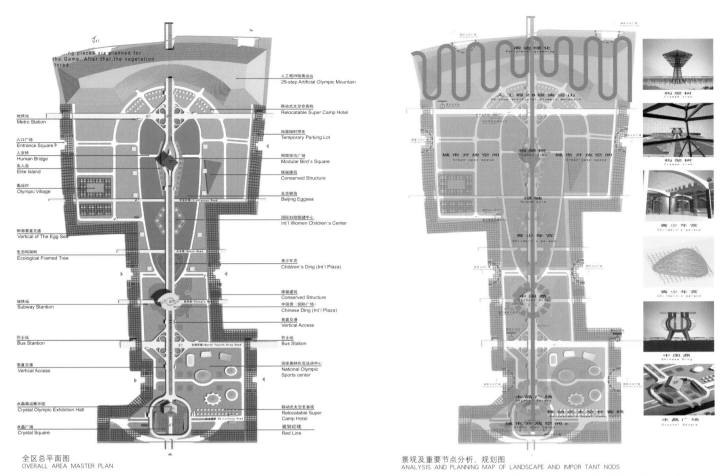

全区总平面图
OVERALL AREA MASTER PLAN

景观及重要节点分析，规划图
ANALYSIS AND PLANNING MAP OF LANDSCAPE AND IMPORTANT NODS

鸟瞰图
BIRDS-EYE VIEW

■ 规划设计构思

我们的建议是：

建筑设施及环境规划应遵循中国传统设计为主要原理。

建筑设施及环境规划应符合北京的规模及空间特点。

规划结果及实施步骤应当灵活可变。我们应当确定行动计划、建立相应的体系，而不是将一幅美学构图简单机械地加以实施。

■ Planning and Design Concept

Our proposals are based on those conclusions:

Buildings and environments should follow the deep logic of the Chinese planning tradition.

Buildings and environments should conform to the scale and spatial characteristics of Beijing.

Product and process need to be flexible and adaptable. We need to set strategies and establish systems, rather than impose an aesthetic composition.

Beijing Olympic Green 2008

何弢设计国际有限公司（中国）
TAOHO DESIGN INTERNATIONAL LTD.(CHINA)

鸟瞰图
BIRDS-EYE VIEW

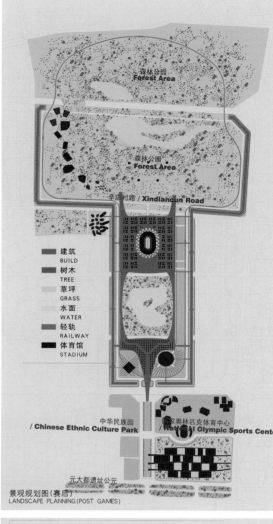

景观规划图（赛后）
LANDSCAPE PLANNING (POST GAMES)

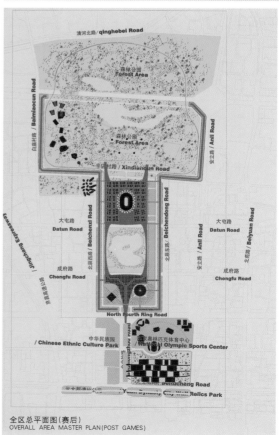

全区总平面图（赛后）
OVERALL AREA MASTER PLAN (POST GAMES)

■ 规划设计构思

奥林匹克运动会为一个国家提供了向世界展示它的文化、技术和成就的机会。"绿色奥林匹克"的概念反映了中国社会秩序正常、协调和与大地相连的传统价值。概念的关键元素包括与下面中国文化元素相应的类似物：围墙、天坛、故宫。

通过技术的使用，这些元素直接与奥林匹克的信念——"人类与自然和谐共处"结合在一起。

"北京绿色奥林匹克"的绿色区域场所将使中国具有能力展示领先的优势技术。这将包括建筑、环境、通讯和运输技术。

在"绿色奥林匹克"内环境方面的主动性集中在使用最少的材料和减少能量消耗，从而减少污染物。这包括有源系统和无源系统两个方面。"绿色奥林匹克"的概念与奥林匹克运动会所有的必备条件对应。

■ Planning and Design Concept

The Olympic Games provide a one off opportunity to showcase a nation's culture, technology and achievements to the rest of the world. The concept for the Olympic Green reflects the traditional values of Chinese society of order, harmony and the connection to the land. Key elements of the concept include analogies to the following cultural elements of China: Walls, Temple of Heaven, Forbidden City.

These elements correspond directly to Olympic ideals of man living in harmony with nature through the use of technology.

The Beijing Olympic Green's green field site will give the capability of China to showcase leading edge technologies. This will include building, environmental, communication and transportation technologies.

Environmental initiatives within the Olympic Green will focus on minimizing material and energy consumption and reduction of pollutants. This will include both active and passive systems.

Beijing Olympic Green 2008

考克斯集团（澳大利亚）
THE COX GROUP (AUSTRALIA)

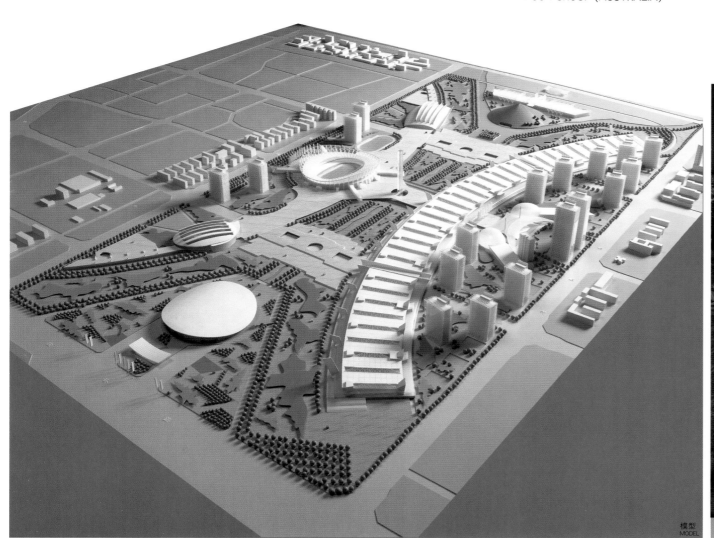

模型
MODEL

鸟瞰图
BIRDS-EYE VIEW

Contents 目录

五棵松文化体育中心 BEIJING WUKESONG CULTURAL AND SPORTS CENTER

■一等奖 First Prize
☐ 空缺 VACANT

■二等奖(2名) Second Prize (Two)
☐ SASAKI设计有限责任公司（美国） 210
　SASAKI ASSOCIATES,INC (USA)

☐ BURCHKHARDT建筑设计集团（瑞士） 216
　BURCHKHARDT+PARTNER AG (SWITZERLAND)

优秀奖(3名) Honorary Mention (Three)
☐ 总后勤部建筑设计研究院（中国） 220
　ARCHITECTURAL DESIGN INSTITUTE OF THE ARMY LOGISTICS (CHINA)

☐ 北京市城建建筑设计研究院有限责任公司（中国） 224
　BEIJING URBAN DEVELOPMENT AND ARCHITECTURAL DESIGN INSTITUTE LTD. (CHINA)
　ABCP 设计公司（加拿大）
　ABCP DESIGN INC., (CANADA)

☐ 中国建筑设计研究院（中国） 228
　CHINA ARCHITECTURAL DESIGN AND RESEARCH INSTITUTE. (CHINA)

■参赛作品(29名) Works (Twenty Nine)
☐ 总装备部工程设计研究总院（中国） 232
　ARMY GENERAL EQUIPMENT ENGINEERING DESIGN INSTITUTE (CHINA)
　哈尔滨工业大学天作建筑研究所（中国）
　TIANZUO ARCHITECTURAL INSTITUTE OF HARBIN INDUSTRIAL UNIVERSITY (CHINA)

☐ 中心建筑设计公司（法国） 235
　CR ARCHITECTURE (FRANCE)
　汤姆建筑艺术设计公司（法国）
　ATS TOM SHEEHAN (FRANCE)
　建筑艺术、城区规划与地理概况研究所（法国）
　ACD GIRARDET & ASSOCIATES (FRANCE)

☐ 株式会社久米设计（日本） 238
　KUME SEKKEI (JAPAN)

☐ 龙安建筑规划设计顾问公司（美国） 241
　J.A.O. DESIGN INTERNATIONAL / ARCHITECTS & PLANNERS LIMITED (USA)
　北京未明众志城市规划与建筑设计研究所（中国）
　BEIJING WEIMINGZHONGZHI URBAN PLANNING & ARCHITECTURAL DESIGN INSTITUTE (CHINA)

☐ GMP建筑师事务所（德国） 244
　VON GERKAN, MARG U. PARTNER ARCHITECTS (GERMANY)

☐ 派森思基础设施技术公司（美国） 247
　PARSONS INFRASTRUCTURE TECHNOLOGY GROUP INC. (USA)
　佩迪建筑师有限公司（英国）
　PEDDLE THORPE ARCHITECTS (ENGLAND)
　北京中建建筑设计院（中国）
　CSCEC-BUILDING DESIGN INSTITUTE, BEIJING. (CHINA)

☐ 万家工程有限公司（香港） 250
　ARTORIA ROAD AND TUNNEL LIGHTING LTD. (H0NGKONG)

☐ 广西建筑综合设计研究院（中国） 252
　GUANGXI ARCHITECTURAL COMPREHENSIVE DESIGN & RESEARCH INSTITUTE (CHINA)

☐ 北京建筑工程学院建筑设计研究院（中国） 254
　ARCHITECTURAL DESIGN AND RESEARCH INSTITUTE OF BEIJING
　INSTITUTE OF ARCHITECTURAL AND ENGINEERING (CHINA)

☐ 中元国际工程设计研究院（中国） 256
　IPPR ENGINEERING INTERNATIONAL (CHINA)

☐ 中国建筑科学研究院建筑设计院（中国） 258
　THE INSTITUTE OF BUILDING DESIGN OF CHINA ACADEMY OF BUILDING RESEARCH (CHINA)
　巴马丹拿国际公司（香港）
　PALMER & TURNER INTERNATIONAL INC. (HONGKONG)
　百和纽特公司（澳大利亚）
　BLIGH VOLLER NIELD PTY.LTD. (AUSTRALIA)

□ 北京清华城市规划设计研究院(中国) 260
　URBAN PLANNING AND DESIGN INSTITUTE OF TSINGHUA UNIVERSITY (CHINA)

□ 北京工业大学建筑勘察设计院(中国) 262
　BEIJING POLYTECHNIC UNIVERSITY INSTITUTE OF ARCHITECTURAL EXPLORATION & DESIGN (CHINA)

□ 哈尔滨工业大学建筑设计研究院(中国) 264
　ARCHITECTURAL DESIGN AND RESERARCH INSTITUTE OF HIT (CHINA)
　莫斯科第四国立设计院(俄罗斯)
　GUE MNIIP "MOSPROECT-4" (RUSSIA)

□ 北方工业大学建筑学院(中国) 266
　COLLEGE OF ARCHITECTURE OF NORTH CHINA UNIRERSITY OF TECHNOLOGY (CHINA)
　北京中色北方建筑设计院(中国)
　BEIJING ZHONGSE NORTH ARCHITEET URAL DESIGN INSTITUTE (CHINA)

□ DP建筑设计有限责任公司(新加坡) 268
　DP ARCHITECTS PTE LTD. (SINGAPORE)

□ 总参工程兵第四设计研究院(中国) 270
　FOURTH DESIGN INSTITUTE, CORPS OF ENGINEER, GS (CHINA)

□ 北京首钢设计院(中国) 272
　BEIJING SHOUGANG DESIGN INSTITUTE (CHINA)

□ 浙江省建筑设计研究院(中国) 274
　ZHEJIANG PROVINCE ARCHITECTURAL DESIGN AND RESEARCH INSTITUTE (CHINA)

□ 考克斯集团(澳大利亚) 276
　COX GROUP (AUSTRALIA)

□ 大地建筑事务所(国际)(中国) 278
　GREAT EARTH ARCHITECTS & ENGINEERS, INTERNATIONAL (CHINA)
　异空建筑师事务所(韩国)
　BEYOND SPACE ARCHITECTS (KOREA)
　沈祖海建筑师事务所(中国台湾)
　HAIGO SHEN & PARTNERS, ARCHITECTS AND ENGINEERS (TAIWAN CHINA)
　核工业第二研究设计院(中国)
　BEIJING INSTITUTE OF NUCLEAR ENGINEERING (CHINA)

□ 北京大学城市规划设计研究中心(中国) 280
　THE CENTER FOR URBAN PLANNING AND DESIGN (CUPD), PEKING UNIVERSITY (CHINA)

□ 北京中标勘察设计咨询公司(中国) 282
　BEIJING ZHONGBIAO RECONNAISSANCE DESIGN AND CONSULTATION COMPANY (CHINA)

□ HOK体育建筑设计公司(澳大利亚) 284
　HOK SPORT+VENUE+EVENT (AUSTRALIA)

□ 北京市建筑设计研究院(中国) 286
　BEIJING ARCHITECTURAL DESIGN AND RESEARCH INSTITUTE (CHINA)
　交通部规划研究院(中国)
　PLANNING INSTITUTE OF MINISTRY OF COMMUNICATIONS (CHINA)

□ 核工业第四研究设计院深圳设计院(中国) 288
　THE FOURTH DESIGN INSTITUTE OF THE MINISTRY OF NUCLEAR INDUSTRY (CHINA)
　伍凸设计师事务所(深圳)有限公司(中国)
　WUTU ARCHITECTS ASSOCIATES (SHENZHEN LIMITED) (CHINA)

□ 上海同济城市规划设计研究院(中国) 289
　URBAN PLANNING & DESIGN INSTITUTE SHANGHAI TONGJI UNIVERSITY (CHINA)
　高柏伙伴规划园林和建筑顾问公司(荷兰)
　KUIPERCOMPAGNONS, OFFICE FOR URBAN PLANNING, ARCHITECTURE AND LANDSCAPE CONSULTANCY (HOLAND)

□ 北京凯帝克建筑设计有限公司(中国) 290
　BEIJING CADTIC ARCHITECTURE DESIGN CO,. LTD (CHINA)

□ ANTHONY BECHU建筑设计公司(法国) 291
　AGENCE D'ARCHITECTURE ANTHONY BECHU (FRANCE)

五棵松文化体育中心 Beijing wukesong Cultural and Sports Center

二等奖 Second Prize

Sasaki公司始建于马萨诸塞州首府波士顿。其创始人为Hideo Sasaki先生。最初的专业人员仅限于同Sasaki先生一起在哈佛大学设计研究院学习的景观建筑师和规划师。之后，随着业务的不断扩展，吸引了一大批建筑师、土木工程师、城市设计师以及环境科学家，为Sasaki公司带来了新的生机。

Hideo Sasaki established the firm in 1953 in Boston Watertown, Massachusetts. The original professional staff included landscape architects and planners who had studied with Mr. Sasaki at the Harvard Graduate School of Design. As the practice flourished, architects, civil engineers, urban designers and environmental scientists brought additional breadth to the firm's capabilities.

■ 规划设计构思

设计潜力——基地和潜力

规划必须体现奥林匹克的运动、文化和环境的宗旨，建筑必须有感召力。

体现复兴路作为城市最为重要的一条大道与场所的价值。

在竞技场、体育场附近建造市民空间，为石景山地区提供一个区域中心。

绿色建筑、基地整体性的策略。

对气候、朝向、阳光、风向的考虑。

在地铁附近布置综合使用建筑、绿色策略。

在奥林匹克后期的有效利用计划。优化奥林匹克宾馆、场馆景观。

标识远期城市形态的象征性。

重要设计概念——城市公园

这一设计概念通过使奥运设施形成一个引人入胜的公园环境而响应"绿色奥运"的主题。这个公园将成为这一地区奥运活动的中心，同时也是奥运后期的活动中心。总体规划意识到了复兴路对于北京的重要性而把主要开放空间朝向南边。

公园的主要特征是一个人工湖，它将来访者吸引到新的公共空间的中心。开挖人工湖所产生的土壤用来沿四环路形成两个山丘，篮球馆和棒球场便结合这两个山丘而设计。这两个山丘不仅对四环路上的噪声形成了隔离带，同时在某种程度上阻挡了春天从西北吹来的风沙。

总体规划利用了在四环路和复兴路交汇处的地铁站，将场馆尽量靠近交通站点。位于场地西南角的地铁站为场馆和其他使用功能提供了便利的交通。视觉和实体的联系提供了贯穿场地的连续道路。

体育场馆的建筑形式自然而成，他们融入山丘之中，减少了视觉阻碍。建筑的大小体量模仿自然地形而形成山丘和起伏地形。建筑的屋顶为山丘坡形的延续以减少对地形的影响。

■ Planning and Design Concept

The Opportunity

The plan must appeal to the Olympic ideals of Sport, Culture and Environment; architecture must be evocative.

Capture the value of Fu Xing Road: the city's most important avenue and "address".

Make civic space adjacent to arena/stadium; provide a district center for Shijingshan.

Adopt strategies for green buildings and site integration
Respond to climate/orientation/sun/wind.

Wukesong 2008

SASAKI设计有限责任公司（美国）
SASAKI ASSOCIATES, INC.(USA)

区域交通分析图
DISTRICT TRAFFIC DIAGRAM

图例 KEY — 快速干道 EXPRESSWAY — 主要干道 ARTERIAL ROAD — 次级干道 COLLECTOR ROAD ◆ 立体交叉路口 GRADE SEPERATE INTERCHANGE ✛ 平面交叉路口 AT GRADE INTERCHANGE ⊢ 立体分隔路口 GRADE SEPERATION--NO INTERCHANGE ⊤ 地铁线路及站点 SUBWAY AND STATION

鸟瞰图(赛时)
BIRDS-EYE VIEW (DURING GAMES)

鸟瞰图(赛后)
BIRDS-EYE VIEW (POST GAMES)

Strategically locate uses nearer the transit station; green strategy.

In post-Olympic mode, plan to reuse buildings and spaces wisely. Optimize Olympic Hotel/Arena landscape.

Identify long-term symbolic value of urban form.

Design Concept The City Park

This concept responds to the "Green Olympics" theme by creating a spectacular park setting for the Olympic facilities. This park will become the focus for the Olympic events in this part of the city as well as for the district in the post-Olympic period. The master plan recognizes the importance of Fu Xing Road as a significant civic address in Beijing by orienting the major open spaces towards the south.

The main feature of the park is an artificial lake which draws visitors to the heart of the new public space. The earth that is excavated to form the lake is used to create two hills, along the Fourth Ring Road, into which the baseball stadium and basketball arena are integrated. These hills not only buffer noise from the busy roads but also shelter the park, to a degree, from the spring dust storms that come from the northwest.

The master plan takes advantage of the transit station at the junction of Fu Xing Road and Fourth Ring road, locating the arena in close proximity. The transit station, which is located on the southwest corner of the site, provides easy access to the venues and other uses. Connections, visual and physical, provide continuous paths throughout the site.

The architecture of the sports venues is naturalistic; built into and merging with the landfill, it minimizes visual bulk. The size of the facilities relates more to the nature of landforms, of cliffs and hills and undulating land. The architecture intervenes lightly on the land, as the shell roofs continue the hill form.

模型
MODEL

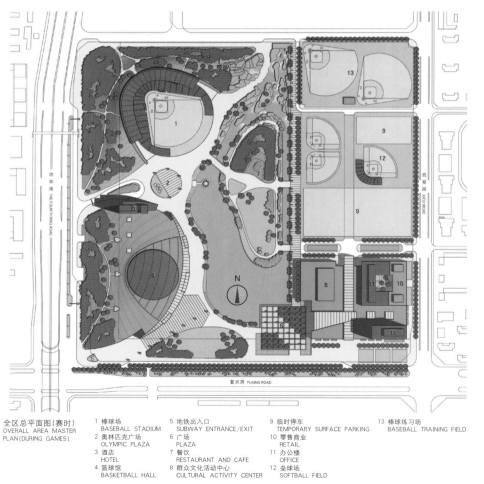

全区总平面图（赛时）
OVERALL AREA MASTER PLAN (DURING GAMES)

1 棒球场 BASEBALL STADIUM
2 奥林匹克广场 OLYMPIC PLAZA
3 酒店 HOTEL
4 篮球馆 BASKETBALL HALL
5 地铁出入口 SUBWAY ENTRANCE/EXIT
6 广场 PLAZA
7 餐饮 RESTAURANT AND CAFE
8 群众文化活动中心 CULTURAL ACTIVITY CENTER
9 临时停车 TEMPORARY SURFACE PARKING
10 零售商业 RETAIL
11 办公楼 OFFICE
12 垒球场 SOFTBALL FIELD
13 棒球练习场 BASEBALL TRAINING FIELD

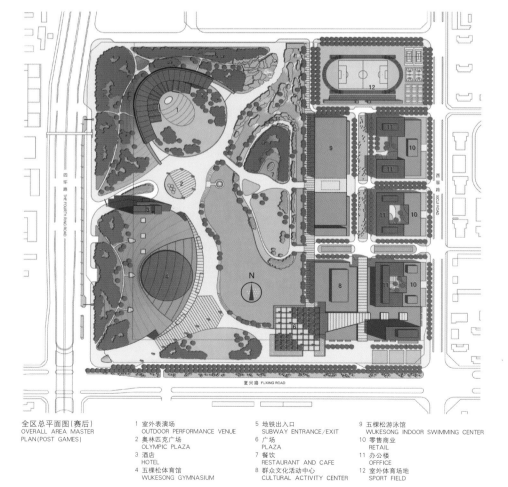

全区总平面图（赛后）
OVERALL AREA MASTER PLAN (POST GAMES)

1 室外表演场 OUTDOOR PERFORMANCE VENUE
2 奥林匹克广场 OLYMPIC PLAZA
3 酒店 HOTEL
4 五棵松体育馆 WUKESONG GYMNASIUM
5 地铁出入口 SUBWAY ENTRANCE/EXIT
6 广场 PLAZA
7 餐饮 RESTAURANT AND CAFE
8 群众文化活动中心 CULTURAL ACTIVITY CENTER
9 五棵松体育游泳馆 WUKESONG INDOOR SWIMMING CENTER
10 零售商业 RETAIL
11 办公楼 OFFFICE
12 室外体育场地 SPORT FIELD

沿四环路立面
FORTH RING ROAD ELEVATION

沿复兴路立面
FUXING ROAD ELEVATION

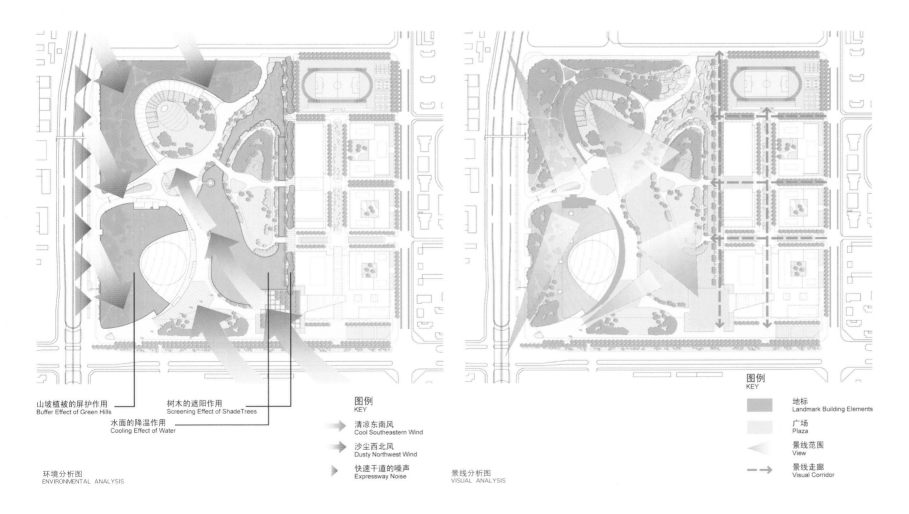

环境分析图
ENVIRONMENTAL ANALYSIS

山坡植被的屏护作用
Buffer Effect of Green Hills

树木的遮阳作用
Screening Effect of ShadeTrees

水面的降温作用
Cooling Effect of Water

图例
KEY

清凉东南风
Cool Southeastern Wind

沙尘西北风
Dusty Northwest Wind

快速干道的噪声
Expressway Noise

景线分析图
VISUAL ANALYSIS

图例
KEY

地标
Landmark Building Elements

广场
Plaza

景线范围
View

景线走廊
Visual Corridor

大厅室内透视图
ENTRY HALL INTERIOR PERSPECTIVE

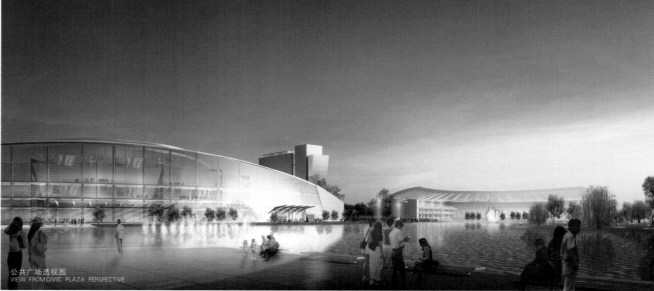
公共广场透视图
VIEW FROM CIVIC PLAZA PERSPECTIVE

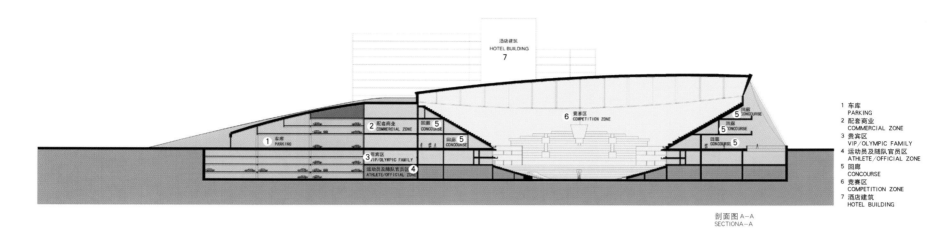

剖面图 A-A
SECTION A-A

1 车库
  PARKING
2 配套商业
  COMMERCIAL ZONE
3 贵宾区
  VIP/OLYMPIC FAMILY
4 运动员及随队官员区
  ATHLETE/OFFICIAL ZONE
5 回廊
  CONCOURSE
6 竞赛区
  COMPETITION ZONE
7 酒店建筑
  HOTEL BUILDING

剖面图 B-B
SECTION B-B

1 酒店客房
  HOTEL ROOM
2 贵宾区
  VIP/OLYMPIC FAMILY
3 竞赛区管理区
  COMPETITION MANAGEMENT ZONE
4 配套商业
  COMMERCIAL ZONE
5 竞赛区
  COMPETITION ZONE
6 运动员及随队官员区
  ATHLETE/OFFICIAL ZONE
7 与现有地铁连接
  CONNECTING TO EXISTING SUBWAY STATION

东立面图
EAST ELEVATION

215

BURCHKHARDT+PARTNER AG 成立于1951年，创始人是Martin H Burckhardt。公司包括建筑师和规划设计师在内有180名员工，是瑞士第二大建筑设计企业。

BURCHKHARDT+PARTNER AG was founded in 1951 By Martin H.Burckhardt.The firm is comprised of architects and planners and, with a staff of 180,is the second-largest architectural firm in Switzerland.

■ 规划设计构思

规划概念——环形山（公园）体系

规划设计中的环形山格栅体系是为了最大限度地灵活开发场地，统一场地，并为创造开放式感受而设计的图形。环形山的某些空间将被预留，以保证未来的需要。娱乐文化用地和商业用地将被整体考虑，两者的开发比例可依据未来的发展动态调整。

整个用地将被划分成两个不同标高：高处用于建设交通流线网，在上面可以观看各项体育活动及演出；低处建设五棵松篮球馆、临时棒球和垒球场、未来公共体育设施（五棵松室内游泳馆）等公共文化和体育中心或公园。

交通流线规划——同心环形路分流车辆和人员

就交通流量而言，在规划用地中采用同心环形路建设是有效的措施。在环形路之内，有必要采取措施阻止车流量过大，其中包括禁止私人汽车在环形路上行驶等措施。

市政设施规划——互换功能

高标高用地将建设永久性基础设施网；低标高用地内则考虑适应社会发展的需要，保持可动态调整的可能性。

景观规划——为居住在附近居民建设有实用价值的菜园。

景观构想的宗旨综合考虑健康社会的基本构成因素：运动设施和健康饮食。

这个场地不仅建设各种体育设施，而且还为人们提供水果和蔬菜；绿地不仅具有观赏价值，而且还有实用价值。所有的树木和植物不是果树就是蔬菜。某些地块专门用于种植从国外引进的特殊有趣品种，并标明果树和蔬菜的名称和原产地，成为人们学习相关植物知识的基地。

五棵松篮球馆概念设计——建设城市居民的聚集场所

五棵松多功能体育馆是环形山格栅体系总规划中的一部分。外部空间规划成为露天运动场和具有魅力的多媒体景观聚会场所。

奥运会期间，这个地方成为观众喜爱的开放式"体育馆"，巨型室外屏幕成为人们观赏比赛的聚集地。

主要建筑设计概念——多功能体育馆的功能和结构概念旨在提高不同时间内的使用价值。

这个综合建筑物在功能上被分为两部分：商业和娱乐辅助设施及多功能体育场。递升排列的圆形看台的设计旨在避免观看比赛时妨碍他人观看，结构设计采用悬浮双曲线体，主结构：周长环形结构钢梁支撑建筑物负荷。次要结构：一个由12个双曲线体组成的系统悬挂在主结构上，支撑体育馆承载的所有负荷。

■ Planning and Design Concept

The Crater Grid system is a diagram designed to maximise the site,option for flexible development, unifying the site, and creating an open horizontal perception. Some of the crater spaces will be left unassigned of program, over the years, functions will be assigned to insure an accurate cover of future necessities. The recreational and cultural activities will be programmed to enhance the site, leisure domain while commercial activities will be assigned when necessary to respond to society, demands and maintain a mixed-use dynamic development, respecting a permanent and unchangeable minimum of 30% of Green space.

Topography: The ground is levelled selectively to produce upper and lower horizontal planes of activity ranging from playing and sports fields to wild parks or avenues. The upper level surfaces are used for the main direct transportation network, which allows an overview of the lower surfaces of activities. The lower level surfaces form an amphitheatre bowl to accommodate large-scale permanent or temporary performance venues, as for example the Wukesong Basketball Hall, the temporary Baseball and Softball fields, future development of public sports facilities (Wukesong Indoor Swimming Complex), the Public Cultural and Sports Centre or parks.

Transport and Traffic Planning—Concentricrings to Organise Vehicle and Pedestrian Traffic

The fundamental measure, which has been applied on the site in terms of traffic and transport, is to place several concentric rings, with measures taken within them to dissuade traffic, including the prohibition of private cars in the innermost circle. Only special vans, ambulances, and security vehicles are allowed in after the first outer slope. The elevated grid of avenues and paths allow a fluid pedestrian flow and an easy orientation to visitors.

Utility Planning—Exchangeable Functions

The upper level surface defines a permanent infrastructure network, allowing the lower surfaces (interior of crater-bowl containers) to change continuously, adapting the site to new societies and unpredictable necessities. This allows a constant renovation of the site use.

Landscape Concept—Vegetable Garden

Olympics for the people:vegetables and fruit for the neighbourhood and visitors.

The landscape concept puts together basic elements for a healthy society:sports facilities and healthy eating.

The site not only offers a variety of sport facilities but also offers fruit and vegetables to visitors and users. The greenery is not decorative but useful. All trees and plants are either fruit trees or vegetable plants. Some of the craters are dedicated to special interesting varieties imported from forging countries.

The Wukesong Multipurpose Sport

The Wukesong Multipurpose Sport Hall is integrated in the crater grid master-plan of the site. The exterior open space becomes open-air seat stands and an inviting gathering area for the multimedia spectacle.

During the Games it becomes an open-air "stadium" for All those interested in Olympic events. It is an open welcoming space for people following events on the outdoor giant screens.

The functional and structural conception of the Multipurpose Sports Hall is based on the idea of having a mixed use inter-linked complex in order to enhance its use during different times of the day or year.

The complex is divided fundamentally in two spaces:The ancillary business and entertainment facilities and The Multipurpose Stadium.The curved rows facilitate better viewing.

Main structure: A perimeter ring of frame steel girders support the loads of the building.

Secondary structure: A system of 12 hyperboloids hang from the main structure and support all loads which are placed over the stadium auxiliary structure: beam steel grid structures are used as reinforcement.

五棵松文化体育中心 Beijing wukesong Cultural and Sports Center

二等奖 Second Prize

Wukesong 2008

BURCHKHARDT 建筑设计集团（瑞士）
BURCHKHARDT+PARTNER AG(SWITZERLAND)

鸟瞰图（赛后）
BIRDS-EYE VIEW (POST GAMES)

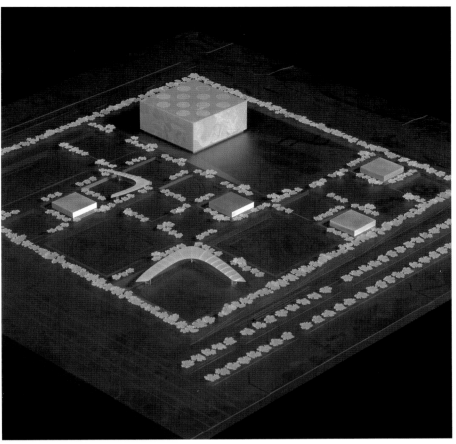
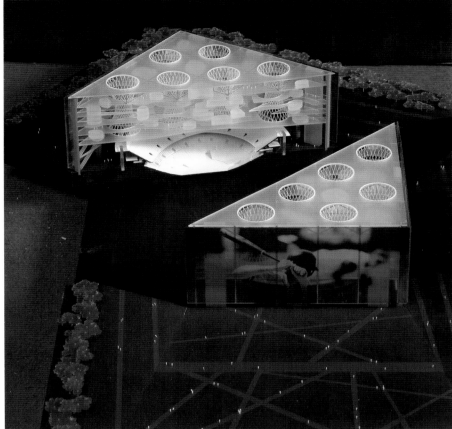

沿复兴路立面图(赛时)
ELEVATION ALONG FUXING ROAD (DURING GAMES)

沿复兴路立面图(赛后)
ELEVATION ALONG FUXING ROAD (POST GAMES)

沿西四环路立面图(赛时)
ELEVATION ALONG THE WEST FOURTH RING ROAD (DURING GAMES)

沿西四环路立面图(赛后)
ELEVATION ALONG THE WEST FOURTH RING ROAD (POST GAMES)

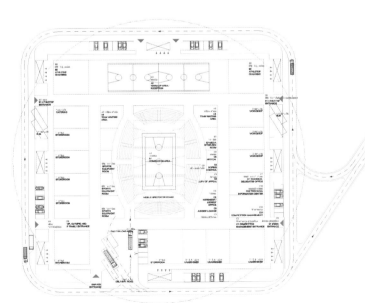

馆内服务街 −23.56
INTERIOR SERVICE STREET −23.56

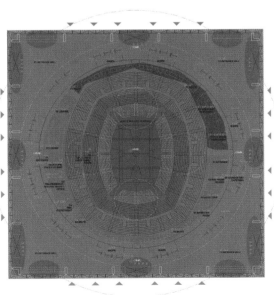

体育馆出入层 −13.00
SPORT HALL ENTRANCE/EXIT −13.00

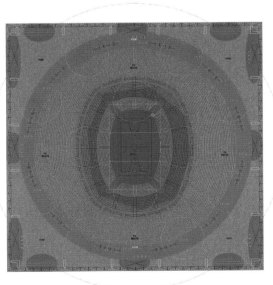

上层看台 +0.00
UPPER STAND LEVEL +0.00

绿化系统规划图(赛时)
MASTERPLAN AND GREEN PLAN (DURING GAMES)

绿化系统规划图(赛后)
MASTERPLAN AND GREEN PLAN (POST GAMES)

1 篮球馆 BASKETBALL HALL
2 足球场 FOOTBALL FIELD
3 足球训练场 FOOTBALL TRAINING CENTER
4 垒球训练场 SOFTBALL TRAINING CENTER
5 棒球训练场 BASEBALL TRAINING CENTER
6 商务、商业娱乐设施 ANCILLARY BUSIBESS AND ENTERTAINMENT FACILITIES
7 文化体育中心 CULTURAL SPORTS CENTER
8 水上活动中心 SWIMMING COMPLEX

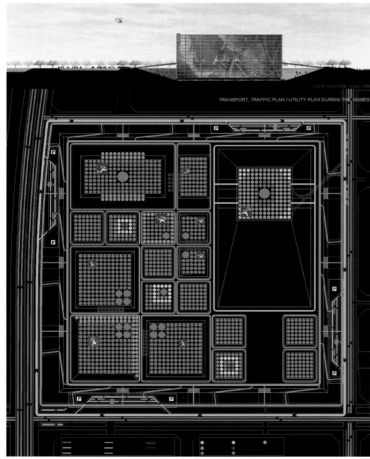

交通系统规划 市政设施规划图(赛时)
TRANSPORT, TRAFFIC PLAN/UTILITY PLAN (DURING GAMES)

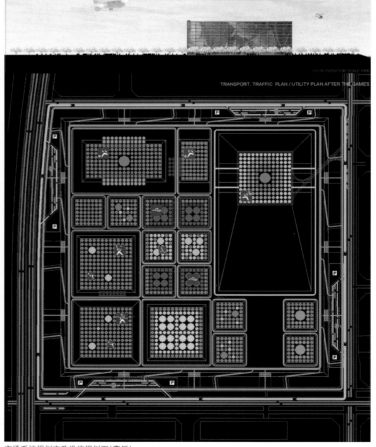

交通系统规划市政设施规划图(赛后)
TRANSPORT, TRAFFIC PLAN/UTILITY PLAN (POST GAMES)

Wukesong 2008

五棵松篮球馆概念建筑设计
CONCEPTUAL ARCHITECTURAL DESIGN FOR THE WUKESONG BASKETBALL HALL

多功能体育馆比赛大厅
THE MULTIPURPOSE SPORTS HALL

五棵松篮球馆概念建筑设计
CONCEPTUAL ARCHITECTURAL DESIGN
FOR THE WUKESONG BASKETBALL HALL

北京奥林匹克公园设计方案 Beijing Olympic Green

优秀奖 Honorary Mention

鸟瞰图（赛后）
BIRDS-EYE VIEW (POST GAMES)

Wukesong 2008

总后勤部建筑设计研究院（中国）
ARCHITECTURAL DESIGN INSTITUTE OF THE ARMY LOGISTICS(CHINA)

总后勤部建筑设计研究院成立于1950年11月，以大型民用建筑和军队专用建筑为主的勘察、设计、科研单位。全院现有310人，其中各类专业技术人员240人。

The PLA General Logistics Department Architectural Design & Research Institute was established in November 1950, is a surveying, design and academic research unit mainly dealt with large-scale civil architecture and military building. The Institute has 310 employees, among them 240 are professional technicians in various fields.

■ 规划设计构思

充分满足举办奥运会的要求，突出"绿色奥运、科技奥运、人文奥运"的主办宗旨。贯彻生态设计理念，把保护和维建筑周边环境生态系统的平衡，作为规划新区与其周围自然环境共生的设计基础。设计结合地形、气候等情况，充分考虑规划建筑建成后对周围地区光环境、热环境、风环境、水环境等的影响，通过对建筑的布局、分区、形态、结构、材料、设备等进行合理设计，并利用各种自然条件，采用生态手段，在节省能源和环保的前提下，创造良好的室内外环境，努力实现可持续发展，提高规划地块整体环境质量。

■ Planning and Design Concept

Adequately meet the needs of holding the Olympic Games by emphasizing the goal to host a Green Olympics, a Hi-tech Olympics, and the People's Olympics, carry through the concept of eco-design, to keep and maintain the balance of the eco system around the architectural complex, and take it as the designing principal of the symbiosis of the newly planning area and the natural surroundings. In view of the topography and climate, fully consider the influence to the environmental factor of lighting, thermal, ventilation and water system by the architecture built according to the planning. Rationally design the architectural layout, form, structure, material and equipment, utilizing multiple natural resources while performing the ecological strategy. Create the favorable exterior environment, as well as the interior, in the principle of energy saving and environmental protection. Endeavor to realize sustainable development, increasing the whole environmental quality within the planning plot.

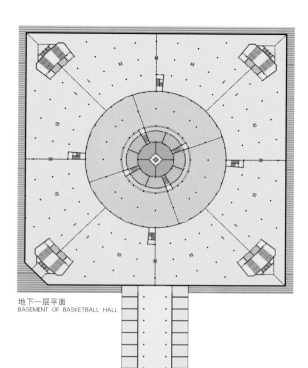
地下一层平面
BASEMENT OF BASKETBALL HALL

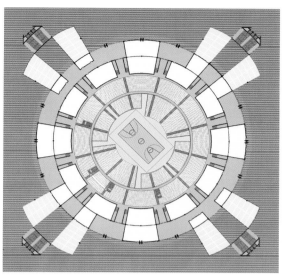
二层平面
SECOND FOLLR OF BASKETBALL HALL

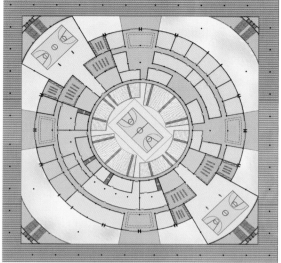
首层平面
FIRST FLOOR OF BASKETBALL HALL

篮球馆剖面
SECTION OF BASKETBALL HALL

篮球馆立面
ELEVATION OF BASKETBALL HALL

e视点 篮球馆室外透视图
PERSPECTIVE OF BASKETBALL HALL FROM E VIEW POINT

鸟瞰图(赛后)
BIRDS-EYE VIEW (POST GAMES)

Wukesong 2008

优秀奖 Honorary Mention

北京市城建建筑设计研究院有限责任公司(中国)
BEIJING URBAN DERELOPMENT AND ARCHITECTURAL DESIGN INSTITUTE LTD(CHINA)
ABCP 设计公司(加拿大)
ABCP DESIGN INC.,(CANADA)

北京市城建建筑设计研究院成立于1958年，现有技术人员150多名，是一家综合性设计、研究、咨询机构，可承担城镇和住宅区的总体规划，大型公共建筑及高层住宅设计，工业厂房设计，公路、桥梁、车站设计，工程预算编制等任务。

The history of BUADRI can be traced back to 1958, which is separated from Beijing Urban Engineering Design & Research Institute. Now there are 150 professionals in the Institute. The institute has been examined interior design and the design of intelligent building system.The Institute is able to undertake general planning of towns and residential areas; design of large-scale public buildings and towers; design of residential areas,industrial buildings,roads & bridges;and the calculation of engineering budgets.

■规划设计构思

本项目作为一项新的设施应该反映新千年的一些价值理念，并包含可持续性场地规划和设计方面的知识。同时新的设施也应该能够满足北京市民对体育和文化活动的不断增长的需求。同时在奥运会盛会之后，这些设施也应继续为民众服务并使他们为之骄傲和自豪。

绿色奥林匹克这一点在我们的项目中将从两个方面得以体现：最优化绿地使其面积超过北京市所要求的面积百分比；在规划与设计整体环境，外部公共用地和楼房的过程中综合考虑可持续性设计的问题。

高科技奥林匹克这个将从我们在可持续性设计中采用的方法中得以体现。我们认为新的千年给我们带来一种新的意识：技术不再是为了征服自然界，而是为了在人类的创造和自然环境之间建立一个积极和受人们认同的平衡。在这个项目中我们将采用先进的视觉通信技术。

奥运会之后，本项目将对中国人民和来自世界各地的游客开放，并给人们带来快乐。拱形建筑面向西四环路，作为本工程项目与西四环路的屏障，也作为本项目的西大门。这个建筑将使项目内的园林与广场免受四环路上强烈噪声的影响。这个建筑最终将包括使其能够综合发挥多种功能，创造出一个动感十足的大型公共长廊，其中将容纳商店、餐馆、办公室、宾馆以及与宏伟壮观的篮球馆相关联的体育文化设施。

波形广场将用一大片林地把波形广场与地铁站连接起来。在这一片区域的西南角还将有一个沉降式小广场把波形广场与地铁站连接起来。这一都市空间将不受西北风的侵袭，大片林地将在暖热季节里提供阴凉和舒适。广场波状的起伏地势将人们的视线引导到篮

模型(赛时)
MODEL (DURING GAMES)

篮球馆室内
BASKETBALL HALL INDOOR

篮球馆透视图1
BASKETBALL HALL RENDERING 1

篮球馆透视图2
BASKETBALL HALL RENDERING 2

篮球馆透视图3
BASKETBALL HALL RENDERING 3

球馆入口处广场和两个露天球类体育场。

碧波园是与本项目东北南三面居住区建筑相联系的大片绿地园林,与西北部的远山相呼应。碧波园与周边建筑物的波状起伏和谐一致,构成整体的景观层次,将一个独特的景观呈现在人们眼前。奥运会期间的临时训练场地在会后也将改建成景观园林,使整个工程区域的用地比例更加优化。

篮球馆像一尊灯塔居于整个工程的中央,其巍然耸立的身形和外表面的木质覆层宣示着篮球馆在整个工程的建筑群落中独特的价值。从任何方面观赏,篮球馆都将突出景观的立体透视效果。作为工程区域的主要体育设施,一个造型简洁、气势恢弘的建筑矗立在一片渐近渐升的风景区中。馆内的看台区和内场区可用来举办从体育比赛到文艺表演等多种大型活动。

■ Planning and Design Concept

The new facilities should reflect the new Millenium values and knowledge with regard to sustainable site planning and design as well as support the growing needs of Beijing citizens for sports and cultural activities.

Green Olympics will be featured in our project as a main aspect. Optimization of green areas will exceed the percentages required by the City of Beijing. Sustainable design issues will be integrated in the planning and design of the landscape, exterior public spaces and buildings.

High-tech Olympics will be embodied in the approach to sustainable design. We believe the New Millenium brings a new awareness and that technology is no longer used to tame Nature but rather to create a positive and respectful balance between human creation and the natural environment.

Advanced visual communication equipment, comfort and energy control technologies will be used in this project.

A large number of non-paying individuals can have free access to the site without compromising the security and controls at the events.

The legacy of the Games will be one of open-mindedness and joy for the Chinese people and for visitors from all over the world.

The Arched Building will act as a barrier and facade to the Fourth west Ring road, and represent the West gate of the Wukesong project for the thousands of people passing by each day. The building will protect the garden and Plaza from the intense noise of the highway. It will ultimately contain the complem-entary functions described in the competition documents. Its urban scale and large public spaces will allow a successful mix of the planned uses, creating a lively and dynamic public gallery of shops, restaurants. Offices, hotel resort, sports and cultural facilities in relation with the overwhelming presence of the Basketball Hall.

The Undulated plaza will connect the project to Fuxing Lu through a pattern of plantations and pools. It will also connect to the subway station through a sunken plaza located on the southeast corner of the land. This urban space will be protected from western winds and the plantations will provide shade and comfort during the warm season. Its ascending waves bring the eye and visitors to the Basketball hall entrance plaza level and to the two ball stadiums.

The Rippled Gardens will constitute a large body of greenery in

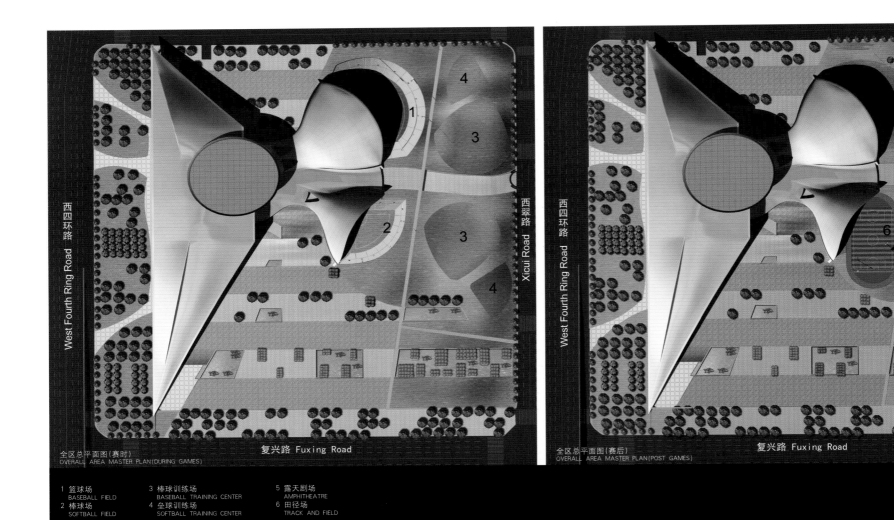

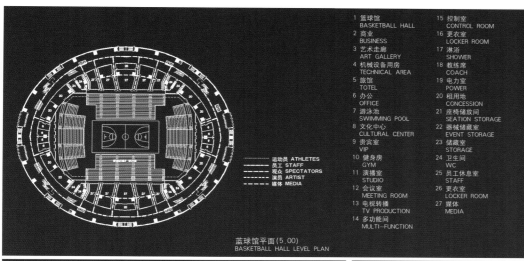

1	篮球馆 BASKETBALL HALL	15	控制室 CONTROL ROOM
2	商业 BUSINESS	16	更衣室 LOCKER ROOM
3	艺术走廊 ART GALLERY	17	淋浴 SHOWER
4	机械设备用房 TECHNICAL AREA	18	教练席 COACH
5	旅馆 TOTEL	19	电力室 POWER
6	办公 OFFICE	20	租用地 CONCESSION
7	游泳池 SWIMMING POOL	21	座椅储放间 SEATION STORAGE
8	文化中心 CULTURAL CENTER	22	器械储藏室 EVENT STORAGE
9	贵宾室 VIP	23	储藏室 STORAGE
10	健身房 GYM	24	卫生间 WC
11	演播室 STUDIO	25	员工休息室 STAFF
12	会议室 MEETING ROOM	26	更衣室 LOCKER ROOM
13	电视转播 TV PRODUCTION	27	媒体 MEDIA
14	多功能间 MULTI-FUNCTION		

篮球馆平面 (5.00)
BASKETBALL HALL LEVEL PLAN

relation to the surrounding residential buildings (existing and future) on the east, north and south side of the lot. It echoes the shapes of the northwest mountains. The garden following the shape of the curves defined by the buildings creates a whole scenery of layers guiding the eye through the picturesque composition. The temporary training fields that will be replaced by landscaped gardens after the Olympics will optimize its proportions .

The Basketball Hall is nestled in the middle of the composition as a beacon. Its singular shape and warm wood cladding are an affirmation of its value in the global composition. It will punctuate perspective views form all directions, a simple, majestic volume set up an ascending landscape as the main sports venue on the site. Its seating configuration and inner volume can be adapted to many types of events ranging from sports to performance arts.

篮球馆平面 (-2.00)
BASKETBALL HALL LEVEL PLAN

篮球馆
BASKETBALL HALL

北京奥林匹克公园设计方案 Beijing Olympic Green

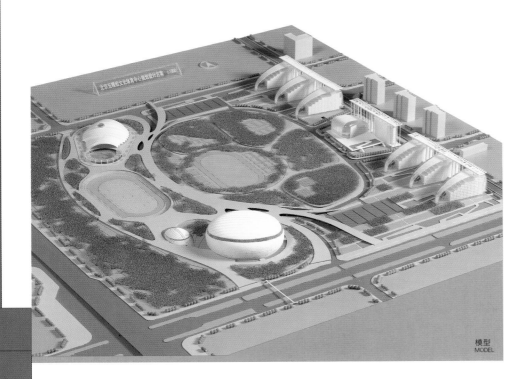

优 秀 奖 Honorary Mention

中国建筑设计研究院（中国）
CHINA ARCHITECTURAL DESIGN AND RESEARCH INSTITUTE.(CHINA)

Wukesong 2008

中国建筑设计研究院是以原建设部建筑设计院、中国建筑技术研究院为母体，吸纳中国市政工程华北设计研究院、建设部城市建设研究院为所属单位，于2000年4月组建的国家大型科技型企业。现有员工2000余人。其中工程院院士2人，全国工程勘察设计大师6人，技术人员占全院总人数的90%以上。

The China Architectural & Design Research Group is a large state-owned enterprise with scientific and technological abilities. It was established on the base of the former MOC Architectural Design & Research Institute and the former China Architectural Technology Research Institute, and admitting the China Municipal Engineering Huabei Design and Research Institute and the MOC Urbane Construction Research Institute as its subordinate unites. At present, it has more than 2000 employees with 2 members of the China Academy of Engineering and 6 national masters of engineering, surveying and design. More than 90% of the total employees of the Group are technicians.

■ 规划设计构思
体育公园、健身之岛
北京五棵松文化体育中心既是第29届奥运会的一处重要比赛场地，同时也是北京西部地区市民进行体育文化活动的重要场所。因此，设计伊始我们就极力将其规划为一个真正为市民喜爱的体育公园，一个绿色的公众健身之岛。在用地中央我们开辟出一块12.7ha的绿地，将各种运动项目布置其中。在这里不仅有气氛热烈的集中的大球比赛场，也有分布在林间绿地中较为安静的小球和器械练习场，人们可以远离城市的喧嚣，来到这里自由选择，尽情放松，融入自然。

同时，绿色的健身之岛在地段内南北方向开敞，东西方向视线亦可贯通，具有很强的向心性和很好的景观效果，势必成为四周居民区、医疗机构以及西四环路、复兴门外大街等紧张繁忙的交通线交会处一个充满活力的乐园。

奥林匹克大平台
根据体育场馆一般通过二层平台组织和疏散观众的特点，本方案在用地西侧一线主要布置的体育场馆被几个平台通过引桥和坡道连接起来，形成一个宽阔通畅的奥林匹克大平台。平台下面是机动车道，这一方式有效地将步行系统和车行系统进行了分离。

这个大平台不仅把几个场馆串联在一起，而且平台下的停车库、设备用房和其他服务设施用房部分采用了覆土建筑的模式，与西侧绿化隔离带的绿色坡地相连，营造出地段西侧连绵近700m的起伏的绿色景观，使建筑、交通、景观融为一体。

■ Planning and Design Concept
Sport Park
Wukesong cultural & sports center is not only an important competition venue during the 29th Olympic Games but also a gathering place that serves people that live in the west. At the very beginning, we aim at it becoming a sport park that will be visited and loved by its people. A green island with an area about 12.7 ha, is set to hold various cultural and sports events, including large ball fields in the center of the site, and other small ball fields spread among the greens. The faith is that we leave the choice to the people, they choose to exercise in the sun and on the lawns.

The green island is designed free of views through south to north, while it is perceived interrupted by buildings that are on the east. A sense of cohesion was created by the center-focused space, the convenient accessibility to the city, two main thoroughfares and the potential of commercial benefits from the nearby inhabitation.

Olympic Platform
According to stadium design feature of up-leveled entrance and ramp exits, we placed main competition arenas and stadiums along the West Side of the site, which are connected with passage bridges and gently ramps. A horizontal Olympic Platform is formed and it gives a sweeping feeling of circulating effectively when vehicle and pedestrian are separated respectively and safely.

Under the platform, there are driveways, indoor parking, and other auxiliary rooms for sport facilities. Based on the different lighting requirements, we determined to use earthed-up building type to go harmoniously with the green sloping lawns that extend over 700m, which is a prominent feature in topography in the west site. A peaceful image conformed by architecture, ramps and landscape is just what we try to immerge.

透视图
RENDERINGS

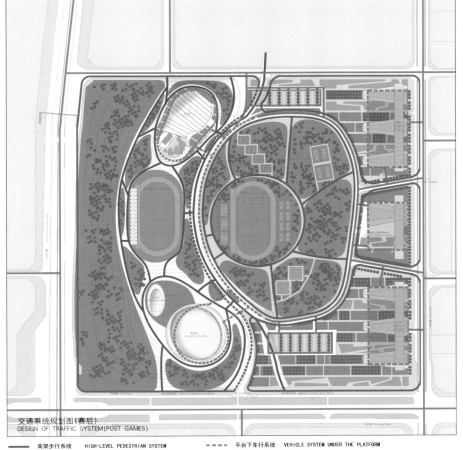

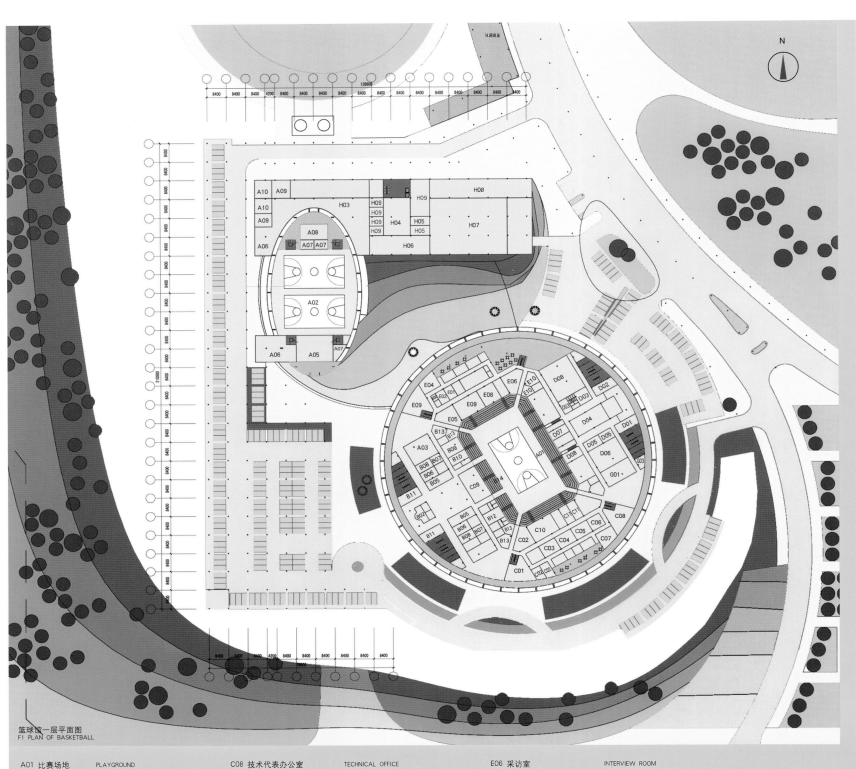

篮球馆一层平面图
F1 PLAN OF BASKETBALL

A01	比赛场地	PLAYGROUND	C08	技术代表办公室	TECHNICAL OFFICE	
A02	热身馆	WARMING-UP ROOMS	C09	国际篮联秘书处	RETINUES ROOMS	
A03	团队等候	RECEPTION ROOM	C10	咖啡厅	COFFEE SHOP	
A04	接待厅	RECEPTION HALL	C11	空调机房	A-C ROOM	
A05	休息厅	LOUNGE HALL	C12	配电室	TRANSFORMER ROOM	
A06	卫生间	TOILET	C13	水泵房	PUMP ROOM	
A07	小卖部	GROCERY STORE	C14	消防控制室	FIRE PROTECTION SYSTEM CENTER	
B01	运动员入口	ATHLETES ENTRANCE	D01	贵宾入口	VIP ENTRANCE	
B02	检录大厅	REGISTRATION	D02	贵宾休息室	VIP LOUNGE	
B03	运动员更衣室	ATHLETES CHANGE ROOM	D03	奥林匹克大家庭休息室	OLYMPIC MEMBERS LOUNGE	
B04	随队官员休息	RETINUES LOUNGE	D04	接见大厅	MEETING HALL	
B05	运动员席	ATHLETES SEATS	D05	国际篮联主席办公室	VIP OFFICE 1	
B06	接待大厅	RECEPTION HALL	D06	国际篮联秘书长办公室	VIP OFFICE 2	
			D07	服务用房	SERVICES	
C01	竞赛管理入口	ADMINISTRATION ENTRANCE	D08	卫生间	TOILET	
C02	裁判员休息室	REFEREES LOUNGE				
C03	卫生间	TOILET	E01	新闻记者入口	PRESS MEDIA ENTRANCE	
C04	兴奋剂检查	DRUG TESTING CENTER	E02	新闻中心	PRESS MEDIA SERVICES	
C05	医疗按摩	INFINMARY	E03	休息厅	INTERVIEW ROOM	
C06	仲裁室	ARBITRATION ROOM	E04	新闻工作用房	PRESS MEDIA ROOMS	
C07	技术信息中心	TECHNICAL CENTER	E05	电信服务	PRESS MEDIA SERVICES	
E06	采访室	INTERVIEW ROOM				
E07	咖啡厅	COFFEE SHOP				
E08	卫生间	TOILET				
F01	观众席(活动座位)	MOVABLE SEAT				
G01	俱乐部成员入口	CLUB MEMBER ENTRANCE				
G02	休息室	LOUNGE ROOM				
H01	景观环廊	LANDSCAPE GALLERY				
H02	健身服务	FITNESS SERVICES				
H03	餐饮	FASTFOOD				
H04	办公	OFFICE				
H05	休息厅	LOUNGE HALL				
H06	咖啡厅	COFFEE SHOP				
P01	运动员停车区	ATHLETES PARKING				
P02	内部人员停车区	STAFF PARKING				
P03	俱乐部停车区	CLUB PARKING				
P04	贵宾停车区	VIP PARKING				
P05	新闻转播停车区	BROADCAST VANS				

沿西四环路立面（赛时）
FACADE ALONG WEST FORTH RING ROAD (DURING GAMES)

沿复兴路立面（赛时）
FACADE ALONG FUXING ROAD (DURING GAMES)

沿西四环路立面（赛后）
FACADE ALONG WEST FORTH RING ROAD (POST GAMES)

沿复兴路立面（赛后）
FACADE ALONG FUXING ROAD (POST GAMES)

Olympic Platform

篮球馆西立面
BASKETBALL HALL WEST ELEVATION

南立面
SOUTH ELEVATION

篮球馆 A-A 剖面
BASKETBALL HALL A-A SECTION

篮球馆 B-B 剖面
BASKETBALL HALL B-B SECTION

五棵松文化体育中心 Beijing wukesong Cultural and Sports Center

参赛作品 Works

■ 规划设计思想

　　布局——注重赛后的整体城市形态。一个贯通场地东西的超体量的综合体统领全局。各主要出入通路均与周边道路形成对位关系，从视觉和空间上与城市紧密联系在一起。

　　广场——面向城市的公共开放空间。在使用性质和空间形态上各不相同的南、北两个绿化广场具有共同的特征：面向城市空间全面开放。

　　纽带——自地坪升向空中的连续整体层面。

　　明珠——外表纯净而内涵丰富的主体赛馆，即简洁纯净的篮球馆，通体晶莹剔透表达出充满自信的、超越地域和时空的全新形象特征，成为城市中新的景观标志。

　　绿色掩映下的空间流动和交通组织场区内部，广场、绿化、建筑相互穿插，形成互相映视的流动的空间效果，从而使身处其间的人们有一种绿色无所不在的感受。

■ Planning and Design Concept

　　Layout-A mega-building complex stretching from east to west of the site stands out as a leading figure either during the events or after the events. Major roads correspond to the surrounding roads, which not only guarantees smooth traffic but also connects all parts of the site with the city visually and spatially.

　　Plaza-Although they differ in usage and form, the southern and the northern green plazas are similar in that both are open to the city.

　　Building Complex-With more attention to its post-Olympic utilization, the building complex consists of Indoor Swimming Complex, Public Cultural and Sports Center and Commercial Services Facilities so that it has a sustainable commercial life. Up from a green buffer in the west, a unique roof shaped irregularly flys toward the south-east sky. The roof not only connects the basketball hall with other parts of the building complex but also the two plazas. In addition, the roof visually connects the ground and the sky.

　　Main Stadium The bowl-like Basketball Hall is transparent, with inner structure faintly visible. Embraced by the building complex, the huge dome characterizes the architecture and becomes a new landmark in the city.

　　Space & Traffic Planning-Plants can be seen everywhere in the plazas. Together with the two urban public green spaces located to the west and the south of the site, the whole place is abundant in greenery. With respect to traffic, user-friendly traffic will be provided.

Wukesong 2008

总装备部工程设计研究总院（中国）
ARMY GENERAL EQUIPMENT ENGINEERING DESIGN INSTITUTE(CHINA)

哈尔滨工业大学天作建筑研究所（中国）
TIANZUO ARCHITECTURAL INSTITUTE OF HARBIN INDUSTRIAL UNIVERSITY(CHINA)

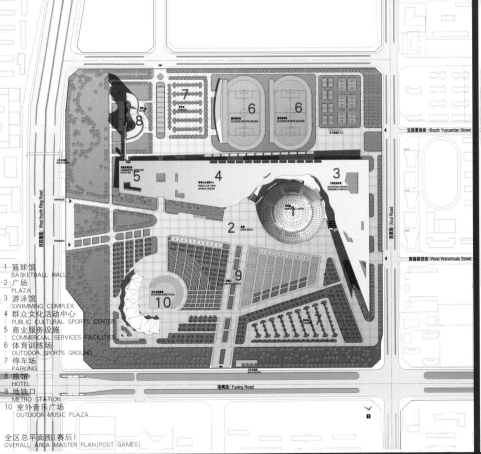

1 篮球馆 BASKETBALL HALL
2 广场 PLAZA
3 游泳馆 SWIMMING COMPLEX
4 群众文化活动中心 PUBLIC CULTURAL SPORTS CENTER
5 商业服务设施 COMMERCIAL SERVICES FACILITIES
6 体育训练场 OUTDOOR SPORTS GROUND
7 停车场 PARKING
8 旅馆 HOTEL
9 地铁口 METRO STATION
10 室外音乐广场 OUTDOOR MUSIC PLAZA

全区总平面图(赛后)
OVERALL AREA MASTER PLAN (POST GAMES)

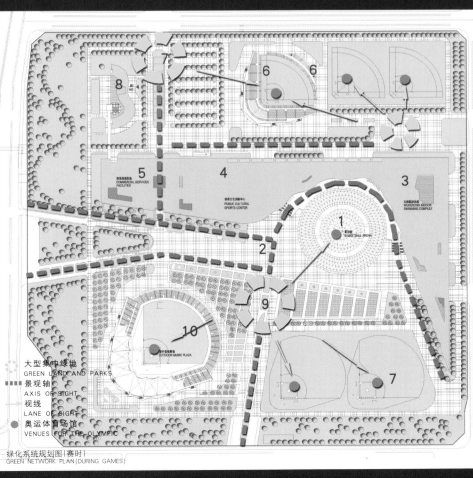

大型集中绿地 GREEN LAND AND PARKS
景观轴 AXIS OF SIGHT
视线 LANE OF SIGHT
奥运体育场馆 VENUES FOR THE OLYMPIC

绿化系统规划图(赛时)
GREEN NETWORK PLAN (DURING GAMES)

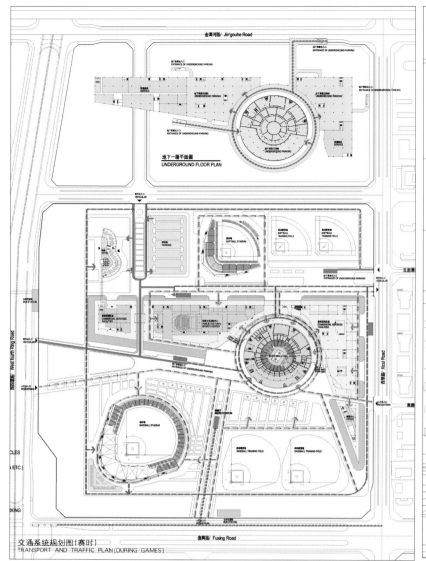

交通系统规划图（赛时）
TRANSPORT AND TRAFFIC PLAN (DURING GAMES)

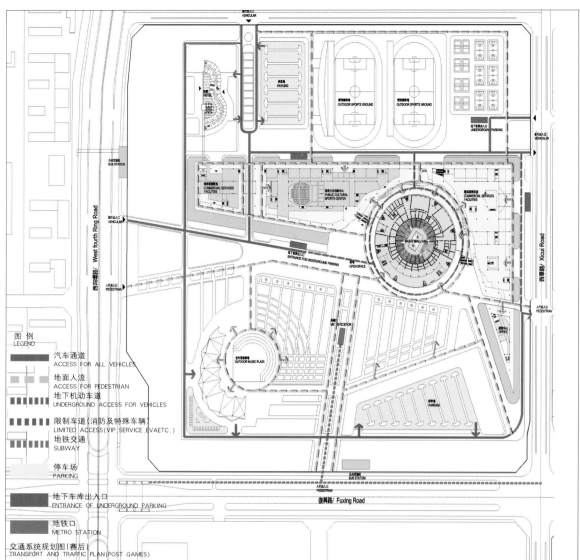

交通系统规划图（赛后）
TRANSPORT AND TRAFFIC PLAN (POST GAMES)

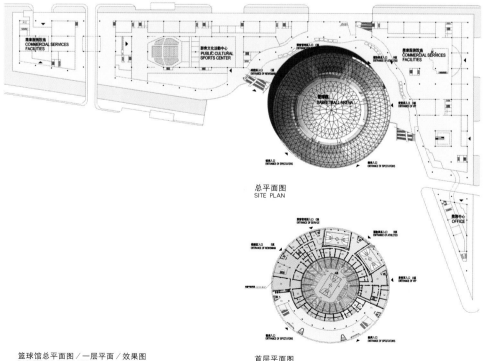

篮球馆总平面图／一层平面／效果图
BASKETBALL HALL SITE PLAN/FIRST FLOOR PLAN/PERSPECTIVE

首层平面图
FIRST FLOOR PLAN

模型
MODEL

■ 规划设计构思

我们的目标是设计出位于五棵松的独特的并有象征意义的城市发展方案，把自然与所有体育设施相结合，始终贯穿着整个方案，流水绿茵始终伴着漫步者的行踪。公园西边将带给我们一片宁静，使人脱离喧嚣和交通污染。但该项目不仅仅是在建造一座公园，它用新的方式把周围的环境与其联系在一起。

所有这些体育设施，不论是临时的还是永久性的，都根据地形的不同来确定它们的位置。在奥林匹克运动会结束后，临时的设施将被替换。在北京城的历史上，这片区域将因为奥运会这一盛事而留下足迹。

■ Planning and Design Concept

The goal of our project Is to create at the Wukesong site, echoing the central site, a unique and exemplary urban development scheme, where nature is to be abundantly present throughout the plan of major urban park in which are integrated all the sports facilities. Water and green vegetation will welcome the visitors. The artificially created surface features will allow for the creation of a zone of calm, cut-off from the noise and pollution of the traffic beltway along the west side of the site. But our project goes much further than the simple creation of a park in a developing neighbourhood; it offers a new way to integrate that neighbourhood into its environment.

All the sports installations, whether they be temporary or permanent, are located according to the notion of topographical variations that create the links between them. Once the Olympic Games are over and the temporary installations have been converted, the neighbourhood will still be marked by this unique event in the history of the city of Beijing, and this will be for the greatest benefit to the inhabitants their well being.

参赛作品　Works

Wukesong 2008

中心建筑设计公司（法国）
CR ARCHITECTURE (FRANCE)
汤姆建筑艺术设计公司（法国）
ATS TOM SHEEHAN (FRANCE)
建筑艺术、城区规划与地理概况研究所（法国）
ACD GIRARDET & ASSOCIATES (FRANCE)

五棵松文化体育中心　Beijing wukesong Cultural and Sports Center

篮球馆模型
MODEL OF BASKETBALL HALL

景观环境规划设计
LANDSCAPE

全区总平面图(赛时)
OVERALL AREA MASTER PLAN (DURING GAMES)

全区总平面图(赛后)
OVERALL AREA MASTER PLAN (POST GAMES)

颜色代码 SERVICE COLOR CODE

竞赛区 COMPETITION AREA
- 竞赛区 A1 COMPETITION AREA
- 热身区 A2 WARM-UP AREA
- 团队等候区 A3 TEAM WAITING AREA

运动员和随行官员区 ATHLETES AND OFFICIALS
- 入口 B1 ENTRANCE
- 运动员更衣室 B2 ATHLETES CHANGING ROOMS
- 运动员和随行官员休息区 B3 ATHLETES AND OFFICIALS LOUNGE
- 坐席 B4 SEATS OFFICIALS(196 PL)
- 运动员停车场 B5 PARKING ATHLETES(BUS)

竞赛管理区 COMPETITION MANAGEMENT
- 入口 C1 ENTRANCE
- 裁判室 C2 REFEREES/JUDGES OFFICE
- 裁判员休息室 C3 JUDGE LOUNGE
- 国际篮联秘书处 C4 FIBA SECRELARY
- 停车场 C5 PARKING
- 仲裁室 C6 JURY OF APPESL
- 技术代表办公室 C7 TECHNICAL DELEGATES OFFICE
- 医疗区 C8 MEDICAL CENTER
- 兴奋剂检查站 C9 DOPING CENTER
- 技术信息中心 C10 TECHNICAL INFORMATION CENTER
- 竞赛工作区 C11 COMPETITION MANAGEMENT

贵宾区 VIP/OLYMPIC AND FAMILY
- 入口 D1 ENTRANCE
- 国际篮联主席、秘书长办公室 D2 FIBA PRESIDENT/GENERAL SEC
- 奥林匹克休息室 D3 OLYMPIC FAMILY LOUNGE
- 坐席 D4 SEATS(675PL)
- 停车场 D5 PARKING(40PL)

媒体区 MEDIA
- 入口 E1 ENTRANCE
- 新闻中心 E2 SUB PRESS CENTER
- 休息室与咖啡厅 E3 LOUNGES AND CAFETERIA
- 坐席 E4 SEATS MEDIA
- 停车场 E5 PARKING MEDIA
- 混合区 E6 MIXED ZONE
- 采访室 E7 INTERVIEW ROOM

观众区 PUBLIC/SPECTATORS
- 入口 F1 ENTRANCE
- 洗手间 F2 RESTROOM
- 咖啡厅 F3 CAFE AND CONCESSION
- 坐席 F4 SEATS(17000 PL)
- 室外停车场 F5 PARKING EXTERIOR
- 问讯处 F6 INFORMATION BOOTHS
- 广场 F7 PUBLIC CONCOURSES

安全保卫区 SECURITY
- 安全保卫厅 G1 MAIN SECURITY ROOM
- 保卫中心 G2 DETENTION CENTER

后勤服务区 LOGISTICS
- 交通运行区 H1 DELIVERY AREA
- 维护保养室 H2 MAINTENANCE

商业区 COMMERCIAL SPACE
- 商店 J1 COMMERCE/SHOP
- 贮藏库 J2 STORAGE
- 交通运行区 J3 DELIVERY AREA

P 停车场 PARKING PUBLIC
I 问讯处 INFORMATION

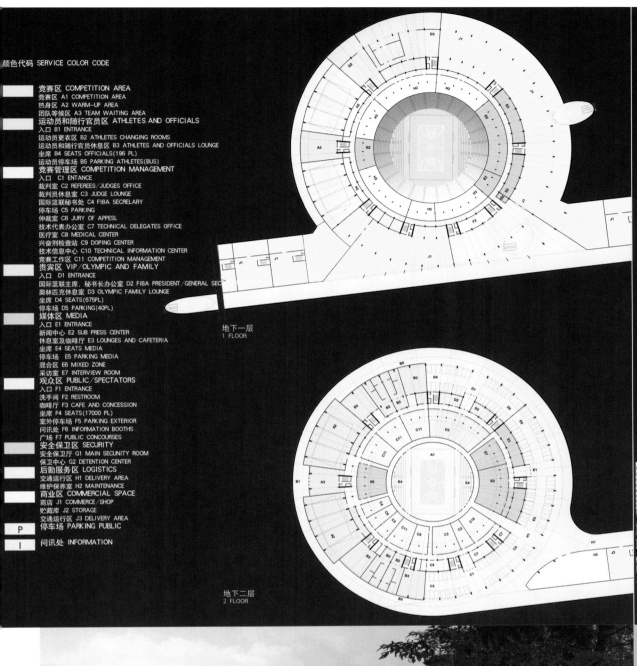

地下一层
1 FLOOR

地下二层
2 FLOOR

篮球馆
BASKETBALL HALL

BASKETBALL ARENA

五棵松文化体育中心 Beijing wukesong Cultural and Sports Center

参赛作品　Works

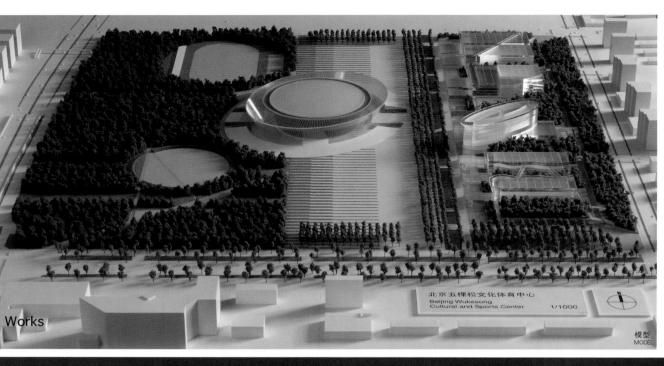

模型 MODEL

■ 规划设计构思

中国的"太极拳"以秩序井然的联贯动作，以求得内心的美，无限和永恒的宇宙之感。而我们本次的设计目标正是要创作出以"宇宙与中国"为思路，同时保持中国传统文化风格的、具有宇宙空间和谐之感的作品。

为此，设计强调传统文化、宇宙意识、奥运会宗旨的融合。

绿色奥运——创出规划合理的宽广绿地——GREEN MOTHER BOARD 提案

科技奥运——创出传统文化和宇宙意识融合的新空间——COSMOS CONSCIOUSNESS 提案

人文奥运——创出美、动、静的音调谐美的文化交流环境——PEOPLE'S OLYMPICS 提案

■ Planning and Design Concept

Practicing a string of Chinese Shadowboxing Acts, you can also enjoin a kind of feeling mixed with beauty, immensity and foreverness of the cosmos. This is just what we are planning to fulfill this time. We determine to create a piece of perfect works that is based on the concept of [COSMOS & CHINA], which has a traditional Chinese cultural character mixed with the cosmic consciousness harmoniously.

The traditional Chinese cultural and cosmic consciousness harmonize with the guiding principles of the Olympics.

Green Olympics: -Creating a large Profusive Green Area- Green Mother Board Proposal

High-tech Olympics: Creating a New Space composed by the traditional Chinese culture harmonizing with the cosmic consciousness- Scent Cosmos Proposal.

People's Olympics:-Creating a harmonious cultural communication space composed of beauty, sports-action, peacefulness - Community Stream Proposal

Wukesong 2008

株式会社久米设计（日本）
KUME SEKKEI (JAPAN)

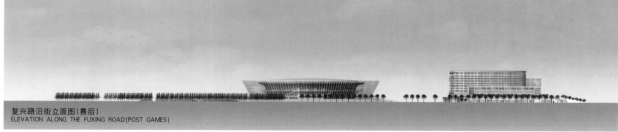

复兴路沿街立面图（赛后）
ELEVATION ALONG THE FUXING ROAD (POST GAMES)

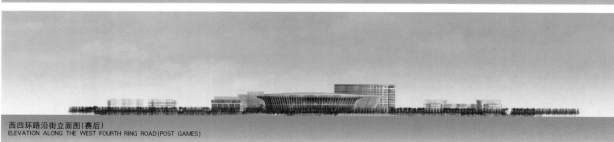

西四环路沿街立面图（赛后）
ELEVATION ALONG THE WEST FOURTH RING ROAD (POST GAMES)

全区总平面图(赛后)
OVERALL AREA MASTER PLAN (POST GAMES)

1 宇宙厅 COSMIC HALL
2 五棵松广场 WUKESONG PLAZA
3 绿色散步道 GREEN PROMENADE
4 采光长廊 LIGHT CORRIDOR
5 存车处 BICYCLE PIT
6 公交车站 BUSSTOP
7 奥林匹克广场 OLYMPIC PLAZA
8 树林 WOODS
9 商业娱乐设施 BUSNESS ENTERTAINMENT FACILITIES
10 酒店 HOTEL
11 群众文化活动中心 PUBLIC CULTURAL AND SPORTS CENTER
12 游泳馆 INDOOR SWIMMING COMPLEX
13 田径跑道 TRACK FIELD
14 室外体育场地 OUTDOOR SPORTS GROUND
15 顶层花园 ROOF GARDEN
16 底层公园 SUNKEN GARDEN

[COSMOS & CHINA]

1 循减式绿地 GREEN GRADATION
2 小溪 STREAM
3 五棵松广场 WUKESONG PLAZA
4 绿色散步道 GREEN PROMENADE
5 奥林匹克广场 OLYMPIC PLAZA
6 树林 WOODS
7 宇宙厅 COSMIC HALL
8 顶层花园 ROOF GARDEN
9 底层公园 SUNKEN GARDEN

绿化系统规划图(赛后)
GREEN NETWORK PLAN (POST GAMES)

全区总平面图(赛时)
OVERALL AREA MASTER PLAN (DURIN GAMES)

1 宇宙厅 COSMIC HALL
2 五棵松广场 WUKESONG PLAZA
3 绿色散步道 GREEN PROMENADE
4 采光长廊 LIGHT CORRIDOR
5 存车处 BICYCLE PIT
6 公交车站 BUSSTOP
7 奥林匹克广场 OLYMPIC PLAZA
8 树林 WOODS
9 商业娱乐设施 BUSNESS ENTERTAINMENT FACILITIES
10 酒店 HOTEL
11 服务亭 INFORMATION
12 垒球场 SOFTBALL FIELD
13 厕所 WC
14 热身场地 TRANING COURT
15 出租车专线 TAXI LINE
16 停车场 PARKING
17 棒球场 BASEBALL FIELD
18 顶层花园 ROOF GARDEN
19 底层公园 SUNKEN GARDEN

广场、树林、建筑相结合的新空间
A NEW SPACE COMBINATION OF SQUARE WOODS AND CONSTRUCTIONS

1. 底层公园将文化设施区和篮球馆连为一体（底层公园 SUNKEN GARDEN）
   A SUNDEN GARDEN CONNECTS CULTURAL BUSINESS ZONE AND BASKETBALL HALL
2. 文化设施区包含着水景和绿地（绿色散步道 GREEN PROMENADE）
   CULTURAL BUSINESS ZONE INCLUDES MANY SCENIC SPOTS OF WATER AND THE GREEN
3. 散步道人流往来，形成自然人流动线（人行道 PATH）
   PEOPLE WALK ALONG THE PROMENADE,COMPOSING A PEDESTRIAN FOLW

■规划设计构思

强调不同功能建筑之间的协调关系，功能分区简明。利用钢架支撑并中有明膜顶的开放式空间将篮球馆、棒球场和垒球场有机联系在一起，主要场馆东侧的带状市民体育文化休闲广场则将酒店商业娱乐区和主要场馆有机地联系在一起，位于东南角的五棵松奥运广场则与三个场馆遥相呼应再一次突显奥运精神，西侧的室外体育场地与市民休闲公园在山水绿地之中融于一体，在整个中心的西侧再次形成充满活力的、开放的活动中心。整个中心突出"以人为本"的精神，与追求"人文奥运"的思想相一致。

注重高科技在奥运场馆建设中的有机运用，用高科技体现奥运，场馆建筑群追求城市建筑艺术、功能和环保节能的完美结合，将"科技奥运"和"人文奥运"有机结合起来。

注意生态建设，追求"绿色奥运"，在整个文化体育中心内建立丰富多彩的自然水系景观空间，利用挖掘水面之土方堆积而成的自然斜坡，有效阻隔外界环境的污染，形成优美的自然山水环境，让人们在与自身体能挑战、发挥"更高、更快、更强"的奥林匹克精神的同时，融于自然的和谐之中，并且成为新北京、新形象的生态绿色中心。

用以支撑、牵引三个主要场馆采光屋顶的五根银灰色泽的圆柱，高耸于场馆屋顶的索网结构之上，文化体育中心内的五个焦点，丰富了文化体育中心建筑群的层次感，而无论在白日还是黑夜中望去，又好似五棵高直的松树，这恰好与"五棵松"地名的历史由来相呼应。由此，历史与现实通过这五棵"松树"，穿越时空连接在了一起。

■ Planning and Design Concept

Wukesong Culture and Sports Center Layout Scheme General ConceivingIt stresses the harmonious relationship among the buildings of different functions, and the functions areas are clearly and simply marked out: to organically link the basketball stadium, the baseball park, and the softball park through open space supported by steel frames and a transparent membrane top; the belt-like city residential sports, culture and leisure plaza, east of the main parks, and stadiums organically links the hotels, commerce and entertainment area with the main parks and stadiums; Wukesong Culture and Sports Meeting Plaza lying in the southeast correspond to each other in the distance, once again giving prominence to the spirit of the Olympic Sports Meeting; the outdoor sports field on the left side is integrated with the city residential leisure park among the hills, water, and the green meadows, once more forming the vigorous and open activity center on the whole left side. The whole center gives prominence to the conception of "human, the most important " and the tenet of "scientific Olympics,Human Olympics and Green Olympics"

It pays attention to the use of high technology in the construction of the Olympic Sports Meeting fields, to incarnate the spirit of the Olympic Sports Meeting with high technology. The building groups pursue the perfect combination of art, function, and environmental protection and energy saving in the city construction, organically linking the conception of :human, the most important: and the tenet of "Scientific Olympics"Human Olympics and Green Olympics"

It pays attention to ecological construction, seeking green Olympic Sports Meeting, setting up a naturally rich and colorful water sights system inside the whole Culture and Sports Center. The slopes are naturally shaped with piled earth, that is dug for water, thus effectively obstructing the outside pollution, and forming a naturally beautiful hills and water environment. Thus it enables people to integrate into the natural harmony and becomes the green ecological Center of the New Beijing, new Image slogan while challenging their own physical limitations and exerting the higher, swifter and stronger spirit of Olympic sports Meeting.

龙安建筑规划设计顾问公司（美国）
J.A.O.DESIGN INTERNATIONAL /ARCHITECTS &PLANNERS LIMITED(USA)
北京未明众志城市规划与建筑设计研究所（中国）
BEIJING WEIMINGZHONGZHI URBAN PLANNING & ARCHITECTURAL DESIGN INSTITUTE(CHINA)

Wukesong 2008

The five silver-gray round columns supporting and suspending the three main parks and stadiums Lighting roofs, towering on the cablenet structure, are the five focuses inside the culture and sport center. It enriches the sense of height of the building groups inside the Culture and Sports center. in addition, no mater viewed in the daytime or at night, they look like five high and straight pine trees, as corre-sponds to the historical origin of Wukesong. History and reality are linked together by the five pine trees across time and space.

模型 MODEL

鸟瞰图(赛时)
BIRDS-EYE VIEW (DURING GAMES)

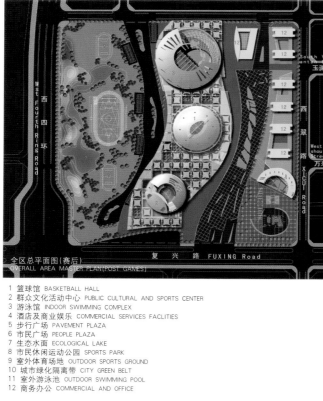

全区总平面图(赛后)
OVERALL AREA MASTER PLAN (POST GAMES)

1 篮球馆 BASKETBALL HALL
2 群众文化活动中心 PUBLIC CULTURAL AND SPORTS CENTER
3 游泳馆 INDOOR SWIMMING COMPLEX
4 酒店及商业娱乐 COMMERCIAL SERVICES FACLITIES
5 步行广场 PAVEMENT PLAZA
6 市民广场 PEOPLE PLAZA
7 生态水面 ECOLOGICAL LAKE
8 市民休闲运动公园 SPORTS PARK
9 室外体育场地 OUTDOOR SPORTS GROUND
10 城市绿化隔离带 CITY GREEN BELT
11 室外游泳池 OUTDOOR SWIMMING POOL
12 商务办公 COMMERCIAL AND OFFICE

道路交通分析图(赛时)
TRANSPORT AND TRAFFIC (DURING GAMES)

城市交通 CITY TRAFFIC
内部车流 INSIDE TRAFFIC FLOW
主要人流 POSSENGE FLOW
场馆入口 STADIO AND GYMNASIUMS ENTRANCE
公交车场 PUBLIC TRAFFIC PARKING
步行广场 PAVEMENT PLAZA
运动员停车场 PARKING FOR ATHLETES
地下车库 UNDERGROUND GARAGE
观众停车场 PARKING FOR SPECTATOR

绿化景观分析图(赛后)
GREEN NETWORK PLAN (POST GAMES)

主要景观带 MAIN LANDSCAPE BELT
主要景观点 MAIN LANDSCAPE CORE
水面景观 LANDSCAPE OF WATER
绿化景观 LANDSCAPE OF GREEN BELT

沿四环路立面图（赛后）
WEST ELEVATION ALONG THE FORTH RING ROAD (POST GAMES)

沿复兴路立面图（赛后）
SOUTH ELEVATION ALONG THE FUXING ROAD (POST GAMES)

篮球场透视（日景）
THE PERPECTIVE OF BASKEBALL HALL (DURING THE DAY)

局部透视图
THE PERSPECTIVE

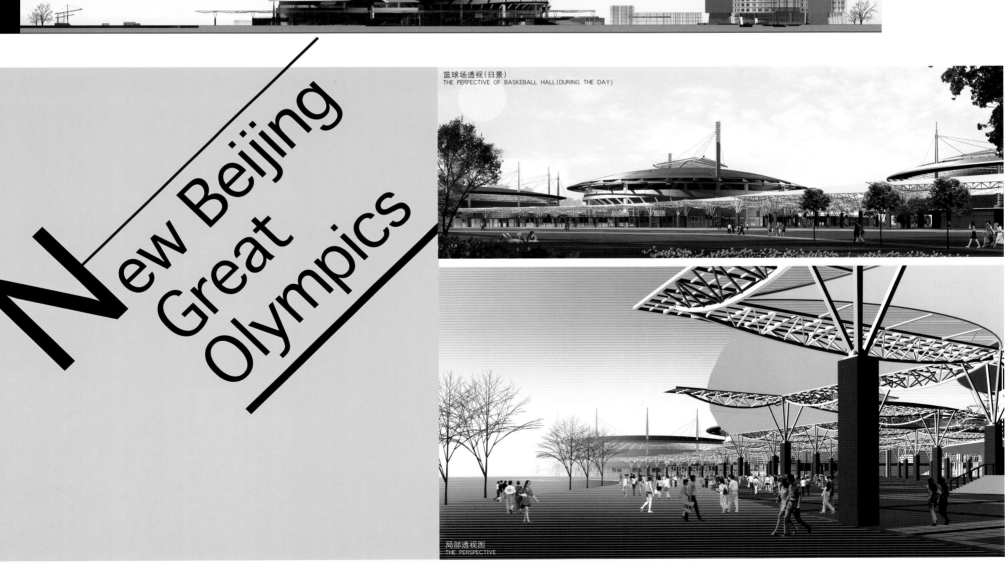

New Beijing Great Olympics

五棵松文化体育中心 Beijing wukesong Cultural and Sports Center

参赛作品　Works

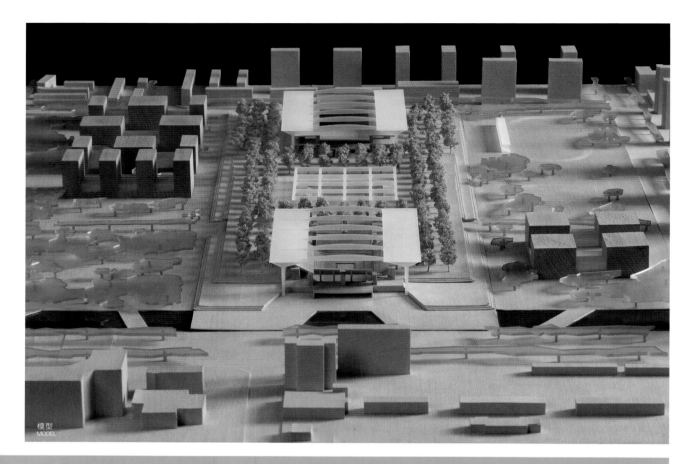

模型 MODEL

GMP 建筑师事务所（德国）
VON GERKAN, MARG U. PARTNER ARCHITECTS (GERMANY)

Wukesong 2008

■ 规划设计构思

新建五棵松篮球馆以及新建棒球场和垒球场不论是建筑上，还是造型设计上在全新而又风格迥异的建筑群体中占据突出的中心位置。本方案利用与复兴路的轴间联系而对这些体育场地的新建筑作了新的布局。结合跨越广阔水域而将复兴路和该体育运动之岛连接起来的宽大路桥，在这里形成了宽广而又富有代表性的大型体育设施平台的入口空间。宽广的阶梯和坡道如同一个舞台不仅作为体育场地的入口交通道路，而且还作为高质量的停留空间，将不同的活动场地连接在一起。

本设计方案的一个重点还表现在赛后功能的设计上。将取代棒球场和垒球场的是位于平台北部的五棵松游泳馆和位于平台中部的文化体育中心。该中心镶嵌在两个体育馆的中间。这个囊括三个建筑的中心区域可以从四面八方进入，并在西侧和东侧结合台阶设置而形成丰富多样而又引人入胜的景观和功能区域。酒店、商业娱乐和办公建筑与融入自然景观的体育场地创造了变化多端的空间。

■ Planning and Design Concept

The new Wukesong Basketball Hall together with the planned Wukesong Baseball Field will constitute the outstanding center of new, multiform architectural ensemble in both structural and architectural terms. The construction of the new sports arenas has been taken as an opportunity to create a new spatial orientation by locating them on an axis to the Fuxing Road. In conjunction with wide access brides connecting the Fuxing Road to the event facilities on the island via a generous expanse of water, well-proportioned, specious, representative access buildings on a large platform are planned. The stadia have generous stage-like entrance steps an ramps which will not only serve as access routes but also fulfil an additional role as high-quality leisure zones linking the different arenas.

This concept is underpinned by the planned post-Olympic usage. The Baseball and Softball Fields will be replaced by the Wukesong Indoor Swimming Complex in the north and the Public Cultural Sports Center in the center of the platform, which will then be flanked by both of the sports halls. To the east and west of this central area, which is framed by the three buildings and closed on all sides, flights of steps lead up to landscaped open spaces offering a wide variety of experiences and functions

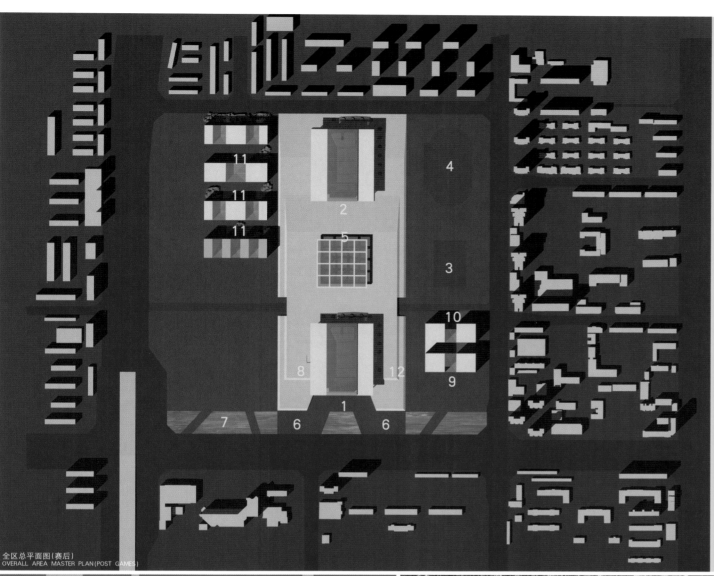

全区总平面图（赛后）
OVERALL AREA MASTER PLAN (POST GAMES)

1 篮球馆　BASKETBALL HALL
2 游泳馆　SWIMMING COMPLEX
3 足球场　FOOTBALL FIELD
4 室外体育场地　OUTDOOR SPORTS GROUND
5 群众文化活动中心　PUBLIC CULTURAL AND SPORTS CENTER
6 观众　AUDIENCE
7 水池　WATER BASIN
8 新闻媒体入口　ENTRANCE MEDIA
9 商业设施　COMMERCIAL FACILITIES
10 内部行车通道　INTERNAL PASSAGE
11 办公服务设施　BUSINESS ENTERTAINMENT FACILITIES
12 贵宾入口　VIP ENTRANCE

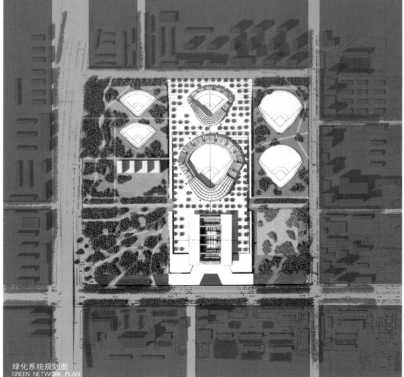

绿化系统规划图
GREEN NETWORK PLAN

篮球馆
BASKETBALL HALL

篮球馆
BASKETBALL HALL

剖面图
SECTION

沿复兴路立面图(赛时)
ELEVATION ALONG THE FUXING ROAD (DURING GAMES)

沿复兴路立面图(赛后)
ELEVATION ALONG THE FUXING ROAD (POST GAMES)

沿西四环路立面图(赛后)
ELEVATION ALONG THE WEST FOURTH RING ROAD (POST GAMES)

五棵松文化体育中心 Beijing wukesong Cultural and Sports Center

■ 规划设计构思

将其置于公园之中心位置,以人行大道穿越广阔的广场将基地一分为二,连串的曲状花园所界定出的户外休闲空间及奥运期间的临时设施将这座鲜明的建筑及行人空间围绕园景空间、路径和中央穿越道形成一个交错的路网,将此座大规模运动中心空间串联为此设计发展重心。入口两座建筑及商店街卖场,使此入口喷泉水景成为框景焦点,引导人潮进入这个贯穿整个基地的"风雨走廊"。此座壮观雄伟的廊道为视觉上的建筑节点,引导人群进入基地的心脏地带,并且表达了奥运的标志意象。

■ Planning and Design Concept

The proposal seeks to bully embrace the objectives and visions of "Green Olympics, High-Tech Olympics and People's Olympics", It integrates the permanent facilities into grand building set in the middle of the park. The complex is accessed from a pedestrian avenue which bisects the site and runs North-South through an expansive central plaza.

This dramatic building and pedestrian space is then surrounded by a series of sweeping curved gardens which define outdoor leisure spaces and the areas for temporary facilities for the Olympics. The site concept is of a grand central place of sports surrounded by an intricate web of gardens, spaces, paths and sweeping avenues.

The site is entered by a bold gateway from Fuxing Road. This entry is close to the Mass Transit Station and is therefore the primary to the site.

The Gateway has an entry fountain and is framed by two buildings and a retail mall which lead to a spectacular canopy providing weather protection across the site. The canopy is a visual, architectural ribbon which takes people along a flowing line to the center of the site. In style it is related to the Beijing Olympics Logo.

参赛作品 Works

Wukesong 2008

派森思基础设施技术公司(美国)
PARSONS INFRASTRUCTURE TECHNOLOGY GROUP INC(USA)

佩迪建筑师有限公司(英国)
PEDDLE THORPE ARCHITECTS(ENGLAND)

北京中建建筑设计院(中国)
CSCEC-BUILDING DESIGN INSTITUTE,BEIJING(CHINA)

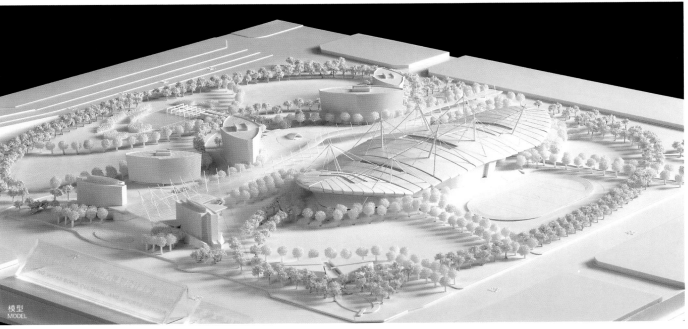

模型 MODEL

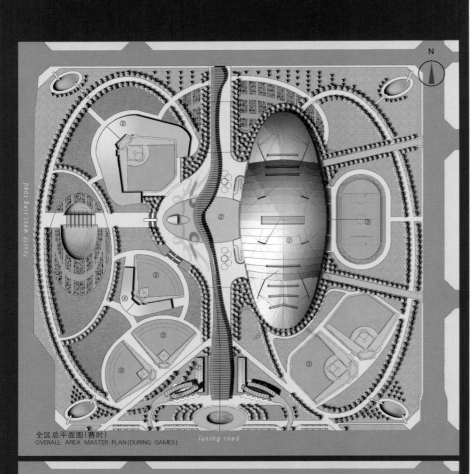

全区总平面图(赛时)
OVERALL AREA MASTER PLAN (DURING GAMES)

全区总平面图(赛后)
OVERALL AREA MASTER PLAN (POST GAMES)

1 多样性使用功能篮球馆
  MULTPURPOSE BASKETBALL HALL
2 中央广场下之 停车场及车辆出入口设施
  CARPARK, PODIUM, &SERVICE VEHICLE ENTRY UNDER CENTRAL PLAZA
3 游戏场
  PLAYING FIELDS
4 临时性看台
  SOFTBALL GRANDSTAND
5 临时性看台
  SOFTBALL GRANDSTAND
6 旅馆、公寓、房间含有会议及商业设施
  HOTEL & SERVICE APARTMENTS
7 卖场、娱乐及餐厅等设施
  RETAIL & ENTERTAINMENT
8 办公建筑
  OFFICE BUILDINGS
9 此项景观规划展现中国植物志之不同生态学
  BOTANICAL GARDENS & GLASSHOUSE FUNCTION CENTRE
10 教育性及娱乐性之综合馆 水族馆
  AQUARIUM-EDUCATION & ENTERTAINMENT COMPLEX

北——南剖面
NORTH-SOUTH SECTION

北——南立面
NORTH-SOUTH ELEVATION

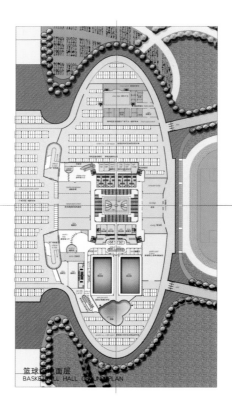

篮球馆地面层
BASKETBALL HALL GROUND PLAN

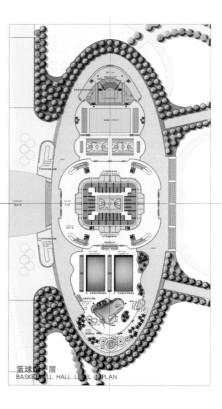

篮球馆层
BASKETBALL HALL LEVEL 1 PLAN

BASKETBALL ABENA

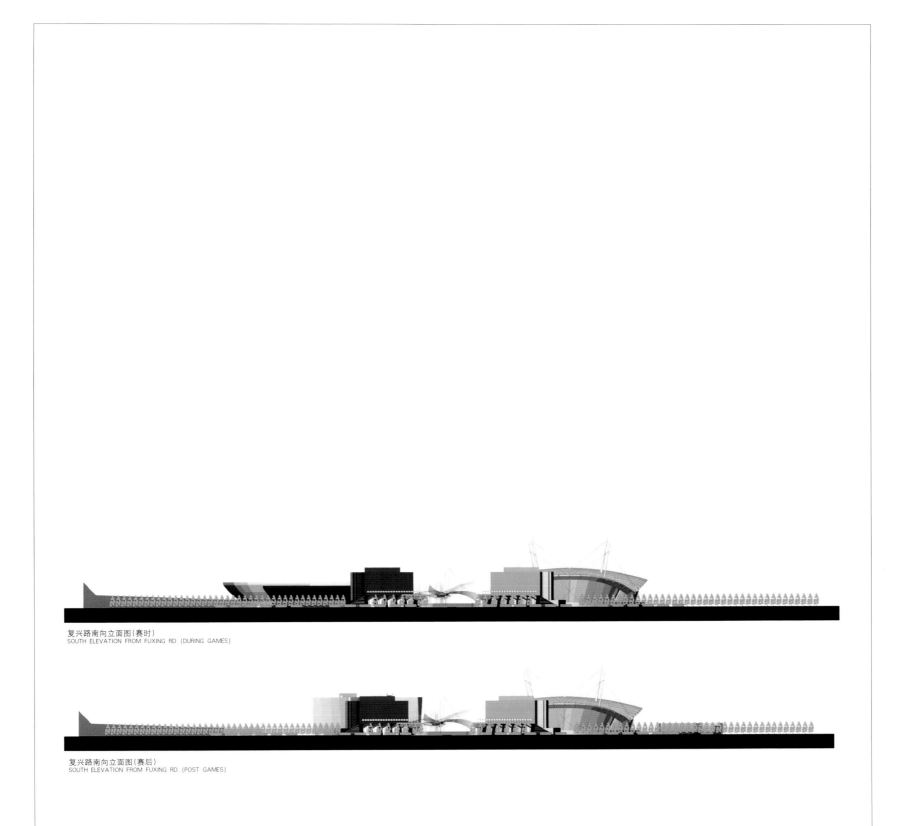

复兴路南向立面图(赛时)
SOUTH ELEVATION FROM FUXING RD. (DURING GAMES)

复兴路南向立面图(赛后)
SOUTH ELEVATION FROM FUXING RD. (POST GAMES)

五棵松文化体育中心 Beijing wukesong Cultural and Sports Center

参赛作品 Works

模型 MODEL

Wukesong 2008

万家工程有限公司（香港）
ARTORIA ROAD AND TUNNEL LIGHTING LTD.(HONGKONG)

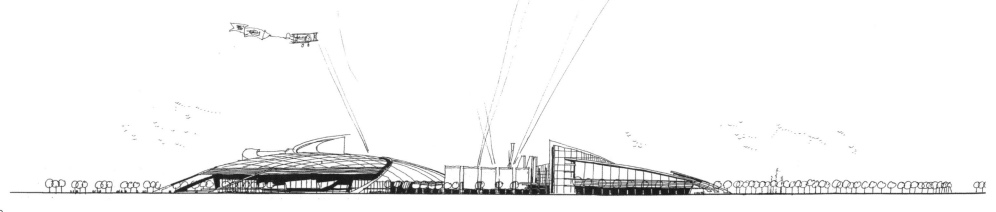

■ 规划设计构思

规划糅合了中国传统的平衡与协调原则,采用动态形式,突出运动的力量、动感和活力,形象地体现奥运会的精神。

规划以一条南北贯通的中轴线把建筑群一分为二,既将各有关活动区域结成一体,又使会场所有部分紧密相连。居住与商务,静与闹和临时与永久区域清楚地分布在中轴线的两边。宏伟而优美如画的中心广场形成了一个公众集会的地方,在广场中间有一个有盖的流线型中庭,从这里可以风雨无阻地通往文体中心的任一设施。

↓ 1 篮球场 BASKETBALL HALL
2 棒球场 BASEBALL FIELD
3 垒球场 SOFTBALL FIELD
4 游泳馆 SWIMMING COMPLEX
5 群众文化活动中心 PUBLIC CULTURAL SPORTS CENTER
6 中央广场 CENTRAL PLAZA
7 运动培训场 SPORTS TRAINNING
8 足球场 FOOTBALL

■ Planning and Design Concept

The geometrical planning approach adheres to ancient Chinese principle of balance and harmony and applies dynamic forms indicative of athletic strength, movement and energy while reflecting the official graphic symbolism of the Games itself.

The strategic planning divides the site into two via a continuous central spine to consolidate activity zones and physically connect all components of the site. Residential/commercial, quiet/noisy and temporary/permanent interfaces are clearly delineated by this spine. A magnificent central landscaped plaza provides a social meting and gathering point which connects to all parts of the site via a linear atrium with all weather protection.

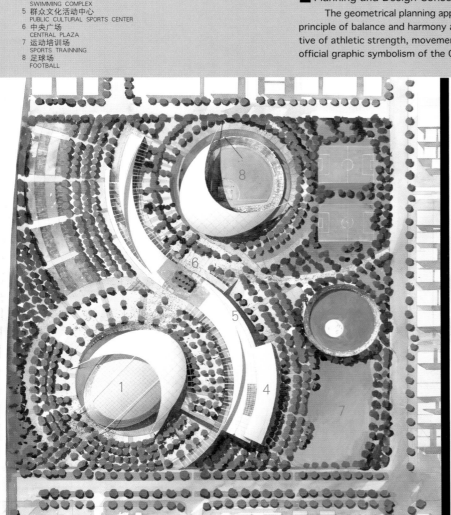

全区总平面图(赛后)
OVERALL AREA MASTER PLAN (POST GAMES)

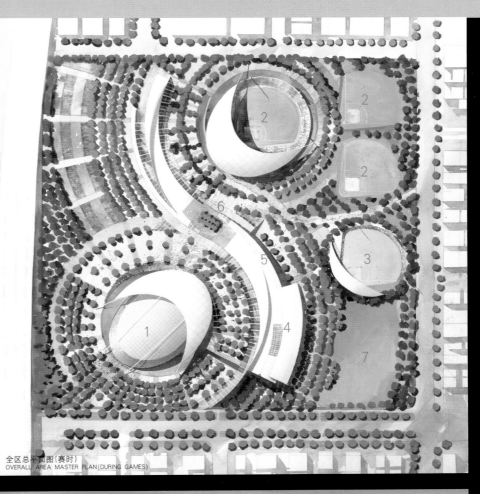

全区总平面图(赛时)
OVERALL AREA MASTER PLAN (DURING GAMES)

篮球馆
BASKETBALL HALL

■ 规划设计构思

奥运会具有丰富的历史渊源、世界性的民族参与，已经成为人类和平的象征。因此，在规划总平面设计中，我们将中心绿地以橄榄枝浮在水面上的形态呈现，配合连绵起伏的两条林带，赋予景观环境"和平"与"友谊"的象征意义。

本着以人为本的原则，我们期望规划一系列开放的空间体系，以合理的规划布局、便捷的交通流线、完善的体育设施、优美的景观环境，为运动员的出色发挥及公众的积极参与提供良好的条件。

从环境利用、保护、美化的角度出发，规划将用地范围内的绿地作为城市绿化系统的有机延伸，创造开敞的景观环境；充分利用城市中水系统为冲厕、绿地浇灌及道路喷洒、中心水体提供用水；采用环保建材，利用太阳能……体现了"生态"、"绿色"的原则。

规划在奥运会后，保留临时场馆的布局及结构形式，将大部分体育设施再利用，节约投资，并使场馆运营得以持续发展。

■ Planning and Design Concept

Having a very long history, Olympic Games, a world-class sports event, has been the embodiment of human peace. Therefore, the central green area on the site plan is designed to be the shape like an olive twig floating on the blue ripple water, along with the two waving tree belts, endowing the landscape with the symbol of peace and friendship.

Basing on the principle of "take people as the basis", we hope to plan series of open spaces, which can provide good conditions for athletes' performance and public participation. This goal can be reached through reasonable planning, convenient circulation, perfect sports facilities and beautiful landscape.

From the view of environmental utilization, protection and beautification, the green land in the area is a part of city greenbelt system so as to create a spacious landscape. Furthermore, non-potable recycled water system supplies flush water, green land irrigation, road watering and water for central water body. The plan is proposed to adopt environmental friendly materials, utilize solar energy etc., highlighting the principle of "ecological" and "green".

After the 2008 Olympic Games, the layout and structure of the temporary sports halls is planned to keep as the same and most of the sports facilities can be continually used so the investment can be saved and the sports halls can achieve sustainable development.

参赛作品 Works

广西建筑综合设计研究院（中国）
GUANGXI ARCHITECTURAL COMPREHENSIVE DESIGN & RESEARCH INSTITUTE(CHINA)

Wukesong 2008

五棵松文化体育中心 Beijing wukesong Cultural and Sports Center

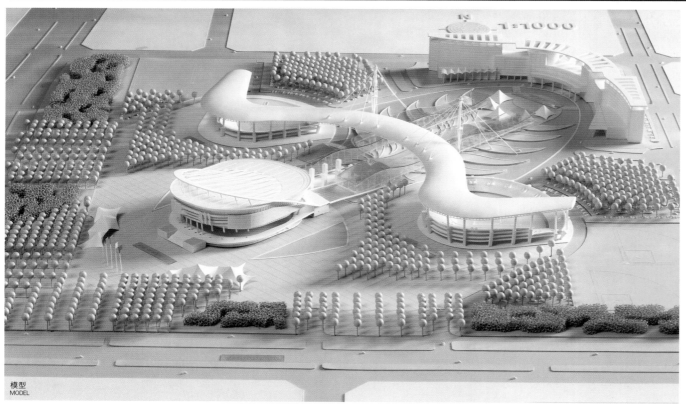

模型 MODEL

1 五棵松篮球馆 WUKESONG BASKETBALL HALL	
2 五棵松棒球场 WUKESONG BASEBALL FIELD	
3 五棵松垒球场 WUKESONG SOFTBALL FIELD	
4 棒球练习场 BASEBALL TRAINING COURT	
5 垒球练习场 SOFRBALL TRAINING COURT	
6 酒店 HOTEL	
7 商业娱乐 BUSINESS ENTERTRAINME	
8 商务办公 OFFICE	
9 中心绿地 CENTRAL GREEN AREA	
10 喷泉 FOUNTAIN	
11 地下人行出入口 UNDERGROUND PEDESTRIAN ACCESS	
12 地下车库出入口 UNDERGROUND CARPORT ACCESS	
13 永久停车场 PERMANENT PARK	
14 临时停车场 TEMPORARY PARK	

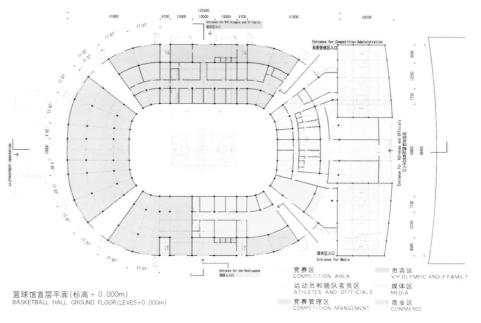

篮球馆首层平面(标高+0.000m)
BASKETBALL HALL GROUND FLOOR (LEVES+0.000m)

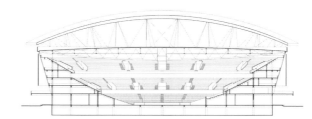

1-1 篮球馆剖面
1-1 BASKETBALL SECTION

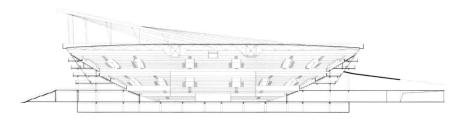

2-2 篮球馆剖面
2-2 BASKETBALL SECTION

五棵松文化体育中心
Beijing wukesong Cultural and Sports Center

鸟瞰图（赛后）
BIRDS-EYE VIEW (POST GAMES)

参赛作品 | Works

北京建筑工程学院建筑设计研究院（中国）
ARCHITECTURAL DESIGN AND RESEARCH INSTITUTE OF BEIJING INSTITUTE OF ARCHITECTURAL AND ENGINEERING (CHINA)

Wukesong 2008

■ 规划设计构思

规划设计始终贯穿"快乐参与"的宗旨。在赛时与赛后方案中，设计了大片的体育活动场所与集中绿地，展开热情洋溢的双臂，吸纳全世界各国人的参与，体现具有"中国特色"的奥运精神。

规划设计结合北京城市总体绿化规划，联通城市绿化系统，进而成为北京西部地区的公共环境、市民生活的"活力之源"。

规划设计将赛时永久建筑——篮球馆、商业服务设施等集中布置，把更多的空间预留给赛后使用。力求土地的多功能混合使用和基础设施等的赛后充分利用，提高它们的使用效率，并为赛后的操作带来了巨大的灵活性。

■ Planning and Design Concept

Keeping "pleasant participation" in mind, our design includes a large sports ground and an integrated open green space both for 2008 Games and after-Games use. The ample green space and multifunctional facilities will accommodate and attract a larger population to participate in various cultural, sports, and recreational activities.

Considering the city's layout and distribution of the green belts, this venue can be adapted to an open ground for neighboring residents' after-Games activities and an Olympic memorial park to carry forward the Olympic spirit. The sustainable development strategy will make this venue the "source of vigor and vitality" for the western Beijing.

Permanent buildings-basketball stadium, commercial and service facilities are centralized, allowing greater flexibility and efficiency for after-Games land use.

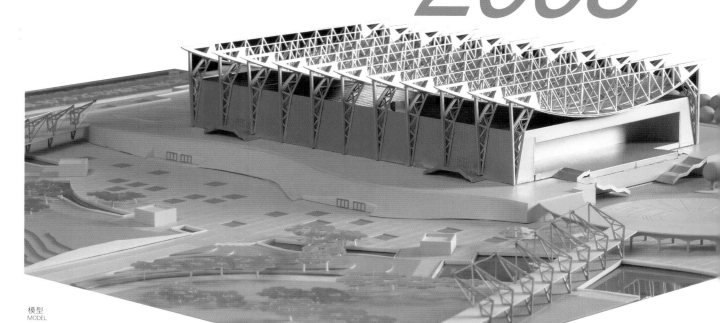

模型
MODEL

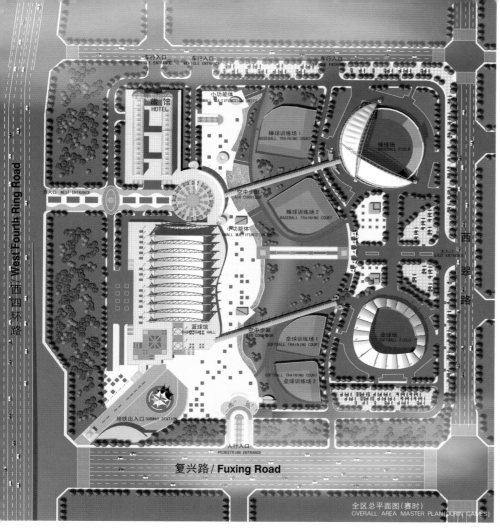

全区总平面图（赛时）
OVERALL AREA MASTER PLAN (DURIN GAMES)

篮球馆透视图
PERSPECTIVE OF BASKETBALL HALL

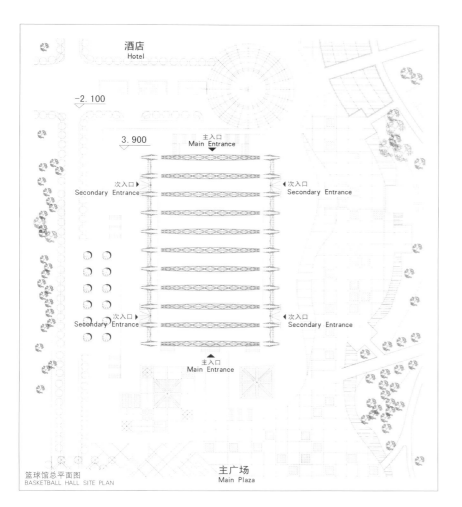

篮球馆总平面图
BASKETBALL HALL SITE PLAN

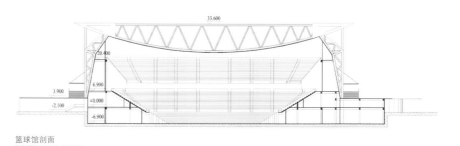

篮球馆剖面
BASKETBALL HALL SECTION

篮球馆东立面
BASKETBALL HALL EAST ELEVATION

五棵松文化体育中心 Beijing wukesong Cultural and Sports Center

篮球馆透视图
THE PERSPECTIVE OF BASKETBALL HALL

鸟瞰图（赛时）
BIRDS-EYE VIEW (DURING GAMES)

参赛作品 | Works

中元国际工程设计研究院（中国）
IPPR ENGINEERING INTERNATIONAL (CHINA)

Wukesong 2008

■ 规划设计构思

规划强调与北京古城的关系，强调与北京中轴线之间的横轴线关系。

强调建筑群体相互之间亲切宜人的尺度，庭院之间和谐的气氛，将"胡同文化"放大延伸。

强调建筑是由地下"长"起来的，跌宕起伏、绵延辗转，与大地浑然一体。

强调景观设计与周边环境的融合。

起伏的建筑采用透明的体块，尽可能暴露其内涵，展示其使用功能，让一切活动、流程、光影都暴露无遗。

强调创造生态、绿色、节能、环保、智能的室内外空间。

在公园内设置29根具有气势而造型简约的现代发光"奥林匹克"纪念柱。

■ Planning and Design Concept

It emphasizes on the connection with the ancient Beijing city and the relationship between the central axis and lateral axis of Beijing city.

The design stresses on kindness and delightful measure range between buildings, harmonious ambience between cartilages. It is an expansion and extension of "Hutong Cultre".

With an idea of integrating smoothly into ambient and growing together with the nature, it emphasizes on that buildings are "grown" up from the ground extensively and in undulate state and become one integrated mass with the mother earth.

Amalgamation of landscape design with the ambient is stressed.

Fluctuant architectures are built by using transparent blocks to expose the inside structures as far as possible to show their usable functions. Let every activity, procedure and shadow be completely exposed.

It Stresses on creating ecological, energy saving, environmental protection and intelligent inside and outside spaces.

In the park, 29# grand and simple "Olympic" columns with modern lightening are set up there as a way of memory.

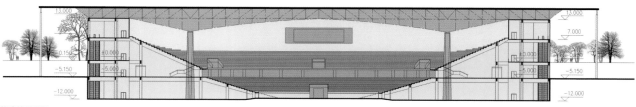

篮球馆剖面图
BASKETBALL HALL SECTION

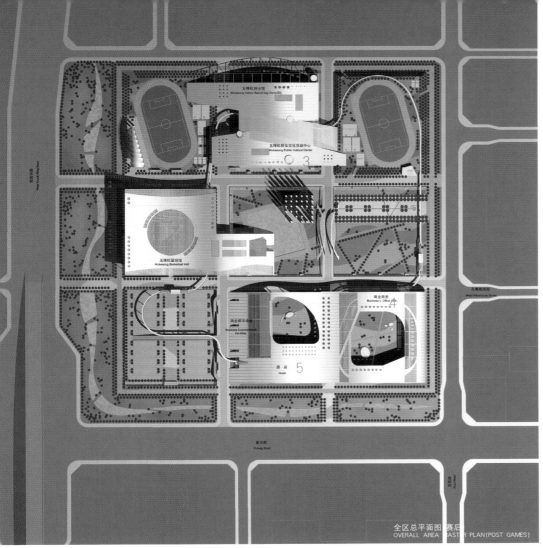

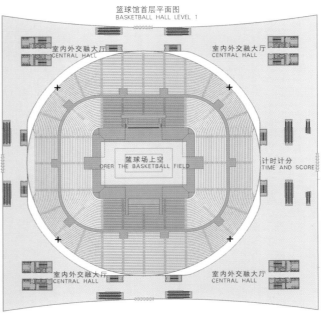

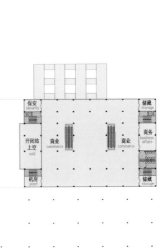

1 篮球场
  BASKETBALL HALL
2 游泳馆
  SWIMMING COMPLEX
3 群众文化活动中心
  PUBLIC CULTURAL SPORTS CENTER
4 商业服务设施
  COMMERCIAL SERVICES FACILITIES
5 酒店
  HOTEL

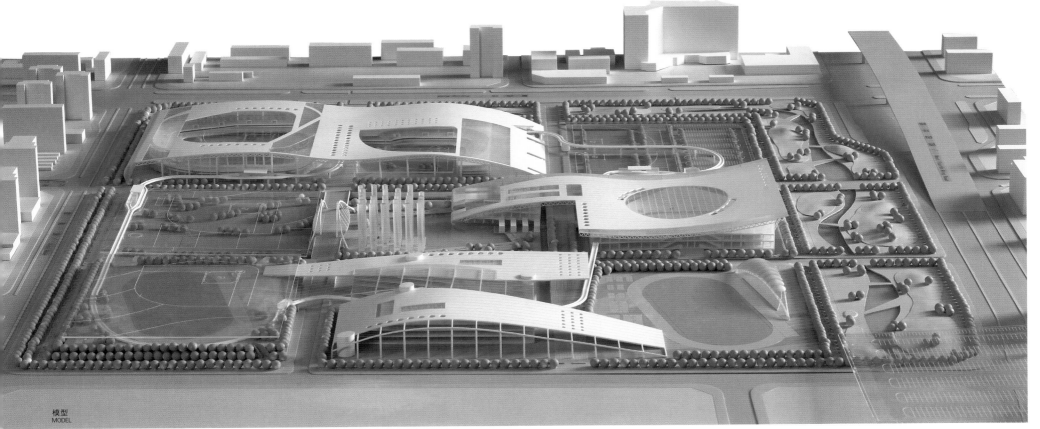

模型
MODEL

■ 规划设计构思

"九宫格"在中国古代建筑文化中有着源远流长的历史，故此在五棵松文化体育中心的总平面布局也采用此手法作构架。

每个"九宫格"面积为216m×216m，适合场地中布置种种建筑物和公共空间，由九宫格为背景再递变出适合种种建筑物比例的轴线。

216m方格可以3等分成9个72m×72m的中方格，中方格再次以3等分成9个24m×24m的小方格，将巨大空间转化成适合人类的比例。

■ Planning and Design Concept

Beijing Wukesong Cultural and Sports Center Masterplan

The site is divided into nine squares representing a long and significant history in Chinese architecture and urban planning, and it is this principle that inspires the site order for the Wukesong Olympic Precinct.

These 9 squares, measuring 216mX216m, are used to define the different uses on the site, and to order the buildings and public spaces, in the context of the required buildings on the site.

The 216m grid can be divided by 3# into another 9# squares of 72mX72m and then again into 9# squares of 24mX24m, representing the human scale in the larger context of the city.

鸟瞰图（赛后）
BIRDS-EYE VIEW (POST GAMES)

参赛作品 Works
Wukesong 2008

中国建筑科学研究院建筑设计院（中国）
THE INSTITUTE OF BUILDING DESIGN OF CHINA ACADEMY OF BUILDING RESEARCH(CHINA)

巴马丹拿国际公司（香港）
PALMER & TURNER INTERNATIONAL INC.(HONGKONG)

百和纽特公司（澳大利亚）
BLIGH VOLLER NIELD PTY.LTD.(AUSTRALIA)

模型
MODEL

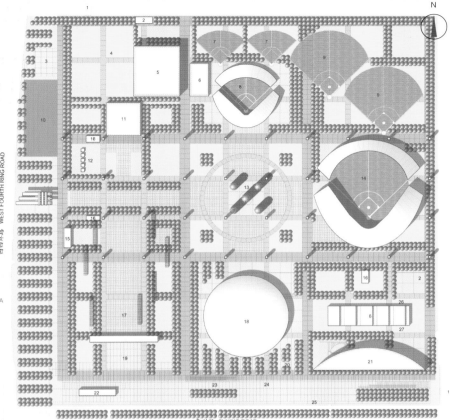

1 交通保安检测点 VEHICLE CHECK POINT	15 观众出口 SPECTATOR EXIT
2 入口处 ENTRANCE	16 卫生间 W·C
3 自行车停泊处 BICYCLE PARKING	17 奥林匹克绿化观众入口 OLYMPIC GREEN SPECTATOR ENTRY
4 特别认可专用停车场 ACCREDITED VEHICLE PARKING	18 18000座位五棵松篮球赛 18000—SEAT BASKETBALL HALL ARENA
5 物资供应站 LOGISTICS AND CATERING COMPOUND	19 保安检测点 SECURITY PLAZA
6 广播设施 BROADCAST COMPOUND	20 停车场及奥委会单元 VAHO
7 五棵松垒球练习场 SOFTBALL WARMUP	21 商业／酒店 HOTEL／COMMERCIAL
8 8500座位五棵松垒球场 8500—SEAT SOFTBALL ARENA	22 五棵松地铁站 WUKESONG STATION
9 五棵松棒球练习场 BASEBALL WARMUP	23 计程车停泊处 TAXI／WALK UP
10 小型巴士停泊处 COACH PARKING	24 巴士总站 BUS TERMINAL DROP OFF
11 奥运物品销售处 OLYMPIC MERCHANDISE MEGA STORE	25 巴士／中型巴士停泊处 BUS／COACH
12 观众食品村 SPECTATOR FOOD VILLAGE	26 传媒 PRECINCT PRESS CENTER
13 绿化区 OLYMPIC GREEN	27 职员专用休息间 STAFF CHECK-IN／BREAK
14 25000座位五棵松棒球场 25000 SEAT BASEBALL ARENA	

全区总平面图（赛时）
OVERALL AREA MASTER PLAN (DURIN GAMES)

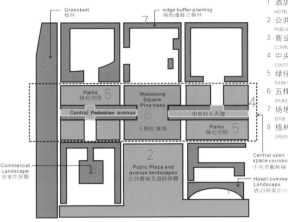

1 酒店．商业部分景观 HOTEL／COMERCIAL LANDSCAPE
2 公共广场及道路景观 PUBLIC PLAZA AND AVENUE LANDSCAPES
3 商业性景观 COMMERCIAL LANDSCAPE
4 中央景观轴线 CENTRAL OPEN SPACE CORRIDOR
5 绿化空间 PARKS
6 五棵松广场 WUKESONG SQUARE
7 场地边缘之植林 EDGE BUFFER PLANTING
8 植林 GREENBELT

绿化系统规划图（赛后）
GREEN NETWORK PLAN (POST GAMES)

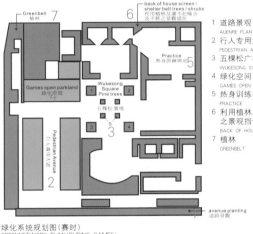

1 道路景观 AUENRE PLANTING
2 行人专用大道 PEDESTRIAN AVENUE
3 五棵松广场 WUKESONG SQUARE PINE TREES
4 绿化空间 GAMES OPEN PARKLAND
5 热身训练场地 PRACTICE
6 利用植林及灌木把噪音及不雅之景观挡住 BACK OF HOUSE SCREEN／SHELTER BELT TREES／SHRUBS
7 植林 GREENBELT

绿化系统规划图（赛时）
GREENNETWORK PLAN (DURING GAMES)

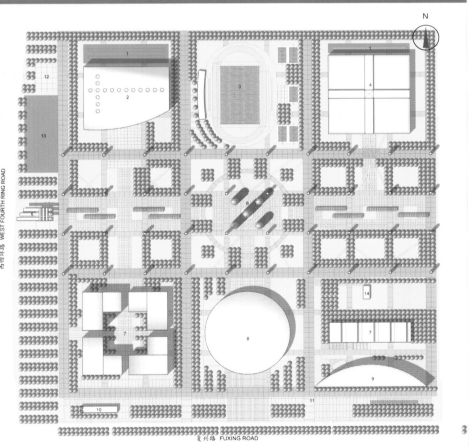

1 停车场 PARKING	8 18000座位五棵松篮球赛 18000—SEAT BASKETBALL HALL ARENA
2 五棵松游泳馆 INDOOR SWIMMING COMPLEX	9 酒店／商务中心 HOTEL／BUSS CENTRE
3 室外运动场地 OUTDOOR SPORTS GROUND	10 五棵松地铁站 WUKESONG STATION
4 群众文化活动中心 PUBLIC CULTURAL AND SPORTS CENTRE	11 巴士总站 BUS TERMINAL
5 绿化区 GREEN ZONE	12 自行车停泊处 BICYCLE PARKING
6 公众绿化区 PUBLIC GREEN AREA	13 中型巴士停车场 COACH PARKING
7 商业发展 COMMERCIAL DEVELOPMENT	14 卫生间 W·C

全区总平面图（赛后）
OVERALL AREA MASTER PLAN (POST GAMES)

MODEL

参赛作品 Works

北京清华城市规划设计研究院（中国）
URBAN PLANNING AND DESIGN INSTITUTE OF TSINGHUA UNIVERSITY(CHINA)

Wukesong 2008

■ 规划设计构思

集中绿地规划在用地西南侧，与四环路和复兴路绿化带结合。有利于最大程度地减少城市主干路与用地内部建筑的相互干扰，加强绿化的整体性。

两个永久性建筑篮球馆和四星级酒店规划靠近四环路和复兴路布置，有利于比较频繁的人流疏散，并且有助于对城市景观的长期控制，避免后期改造建设可能对城市景观产生的破坏。

由于这一地区的主要步行人流交通来自用地西南侧的公共交通和地铁交通，所以机动车交通及停车场安排在地段的东北侧，以实现人车分流。

赛后拆除垒球场建设五棵松游泳馆，结合游泳馆与四星级酒店规划群众文化活动中心。比赛时的棒球场可以部分保留，拆除东北侧看台（也可全部拆除），作为城市绿化预留地。

体育场馆与建筑呈扇形布置，与中心绿化中的弧形水面交相呼应，形成城市景观动与静的结合。

■ Planning and Design Concept

Centralized greenbelts are planned in the southwest of the center site and connected with the green buffers of the Fourth Ring Road and Fuxing Road, which will greatly decrease the inter-disturbance between the main roads and internal constructions inside the center site and intensify the whole effect.

Two perpetual buildings of a basketball court and a four-star hotel are close to the Fourth Ring Road and Fuxing Road, which will benefit the frequent evacuation and help the long-term control for the city sight, avoiding possible damages caused by late reconstruction.

Because the pedestrian traffic of the site mainly comes from southwest public transportation and subways, the vehicle traffics and parking lots are arranged in the northeast of the site to ensure the distribution of the two traffic streams.

Remove the Softball Court and construct a swimming center after the Games. Combine the swimming court with the four-star hotel to plan a public cultural activities center. Part of the Baseball court can be reserved and northeast stand will be removed. Alternatively, the whole court can be removed for the city green reserved space.

Sports center and around buildings are fan-like distributed, echoing with the arc-like water surface in the central greenbelt, and realizing the combination of moving sights and standing sights.

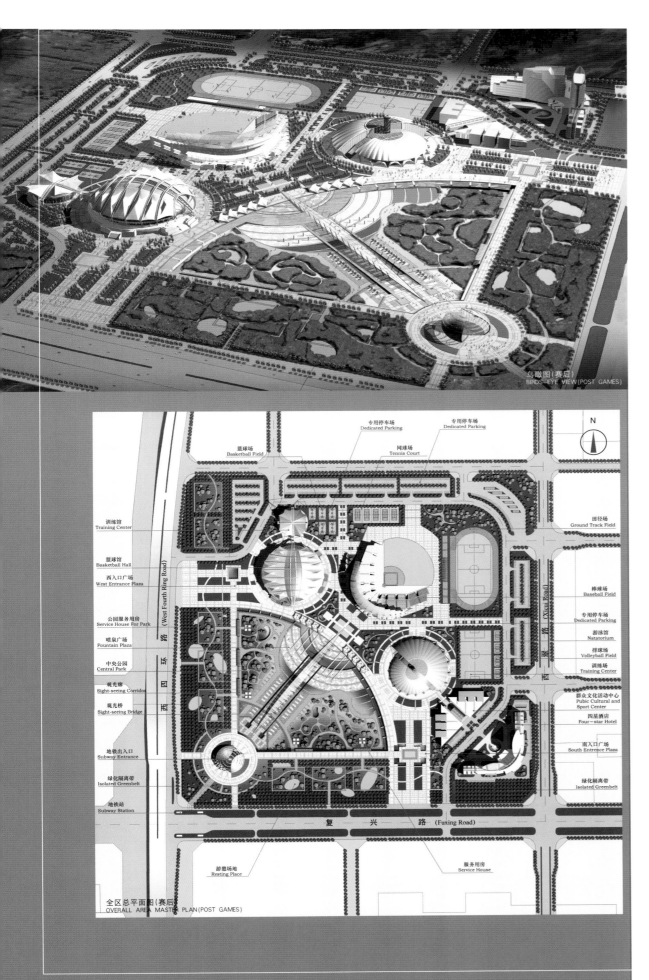

鸟瞰图(赛后)
BIRDS-EYE VIEW (POST GAMES)

全区总平面图(赛后)
OVERALL AREA MASTER PLAN (POST GAMES)

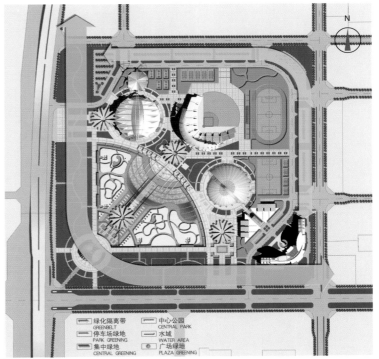

绿化系统规划图(赛后)
GREEN NETWORK (POST GAMES)

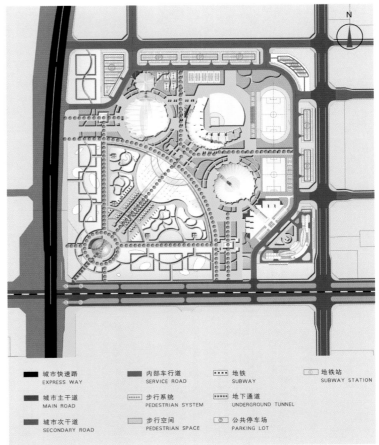

交通系统规划图(赛后)
TRANSPORT AND TRAFFIC PLAN (POST GAMES)

五棵松文化体育中心
Beijing wukesong Cultural and Sports Center

■ 规划设计构思

利用带状运动广场将文化体育中心主要场馆连接形成动感的运动轴，篮球馆、棒球场和垒球场是这条线上的三个重要节点，而这三个重要节点则与酒店、商业娱乐区域遥相呼应。运动广场东侧的奥林匹克林阴广场将酒店商业娱乐区和主要场馆有机地联系在一起，三个主要场馆和室外体育场地仅一水之隔，市民休闲运动公园与室外体育场地紧紧相邻，融于自然的和谐之中，并且成为新北京、新形象的生态绿色中心。

在整个文化体育中心建立丰富多彩的自然水系景观空间，形成优美的水景环境。

场馆建筑群追求在功能、环保节能和城市建筑艺术的结合，将"科技奥运"和"绿色奥运"有机结合起来。

不同功能的建筑群体以带状运动广场的空间变化为联系纽带，将各个功能区融为一体，体现出体育文化中心建筑群的整体严谨性。

■ Planning and Design Concept

The belt-like sports center connected to the main buildings of BWCSC will form a developmental axis: the basketball hall, baseball dome and softball field are the keys of the belt which along with the connection to the commercial services facilities will represent a five ring pattern, the symbol of the Olympic spirit. Te Olympic green that lies to the east of the sports center is fastened to the commercial services facilities and the main building in the BWCSC. The main stadiums and gymnastics harmonize with the Olympic Park so perfectly that they imply that the spirit of the Olympics goes beyond the sports meetings that are held every four years, it also contains the lively spirit of everyday sports that everyone can participate in.

The "Green Olympics" is the final goal; the completed landscape will form a beautiful natural river environment.

High technology ides will be widely employed in the Olympic gyms, the design has tried to find a balance of function, energy saving and artistic qualities of urban buildings. It will make a perfect integration of the "Scientific Olympics" and the "Green Olympics".

The sports fields within the BWCSC will contain specific modifications that will make it possible to connect all the separate functions of the different areas controlled under the strict-est organization.

参赛作品 Works

北京工业大学建筑勘察设计院（中国）
BEIJING POLYTECHNIC UNIVERSITY INSTITUTE OF ARCHITECTURAL EXPLORATION & DESIGN (CHINA)

Wukesong 2008

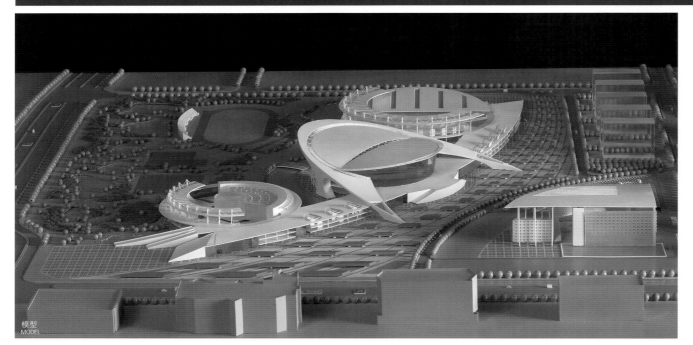

模型 MODEL

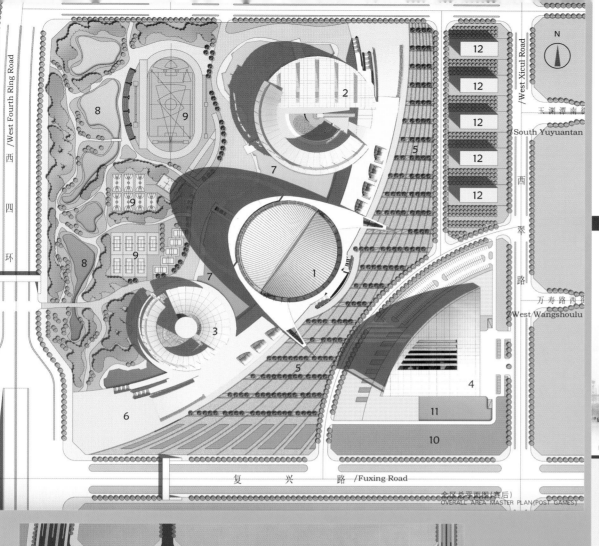

← 1 篮球馆
BASKETBALL HALL
2 群众文化活动中心
PUBLIC CULTURAL AND SPORTS CENTER
3 游泳馆
INDOOR SWIMMING COMPLEX
4 酒店及商业娱乐
COMMERCIAL SERVICES FACILITIES
5 林阴广场
SHADE AVENUE
6 文化广场
CULTURAL SQUARE
7 生态水面
ECOLOGICAL LAKE
8 市民休闲运动公园
SPORTS PARK
9 室外体育场地
OUTDOOR SPORTS GROUND
10 城市绿化隔离带
CITY GREEN BELT
11 景观水面
LAND SCAPE WATER POOL
12 商务办公
BUSIBESS OFFICE

全区总平面图（赛后）
OVERALL AREA MASTER PLAN (POST GAMES)

篮球馆透视图
BASKETBALL HALL PERSPECTIVE 1

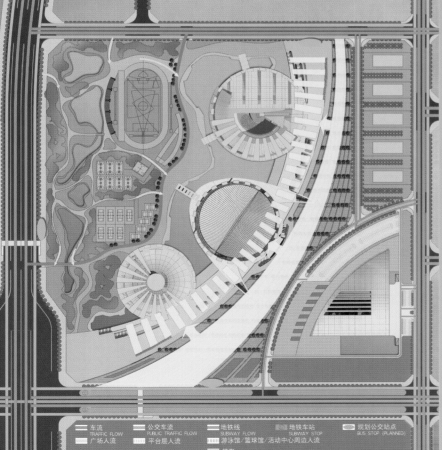

↑ 交通系统规划图（赛后）
TRANSPORT AND TRAFFIC PLANNING (POST GAMES)

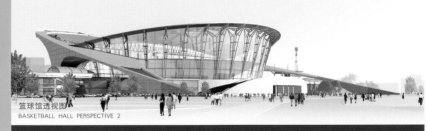

篮球馆透视图
BASKETBALL HALL PERSPECTIVE 2

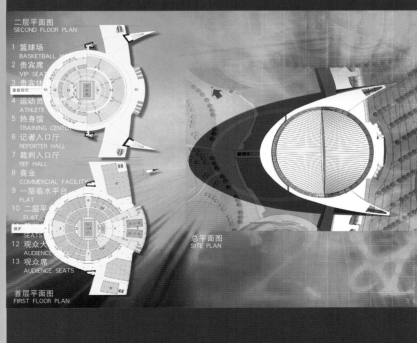

二层平面图
SECOND FLOOR PLAN

1 篮球场 BASKETBALL
2 贵宾席 VIP SEAT
3 贵宾休息
4 运动员 ATHLETE
5 热身馆 TRAINING CENTER
6 记者入口厅 REPORTER HALL
7 裁判入口厅 REF HALL
8 商业 COMMERCIAL FACILITY
9 一层临水平台 FLAT
10 二层平台 FLAT
12 观众大厅 AUDIENCE
13 观众席 AUDIENCE SEATS

总平面图
SITE PLAN

首层平面图
FIRST FLOOR PLAN

■规划设计构思

建筑沿基地东西两侧布置，顶界面缓缓坡下，与地面连为一体；六块室外运动场地下沉6m。

基地的地形起伏，与市郊的连绵西山相映成趣。

下沉的室外运动场，以其完整的几何形态、庞大的尺度、自如的分布，隐喻月球表面陨石撞击的痕迹，暗示自然不可知的神秘力量。

东西两侧的建筑界面打破垂直线条，随机地对立柱和幕墙竖向分割线加以倾斜，同时追求材质对比与光影变化，再现日月光照下纷纭错落的林间景象。

冰凌般的玻璃通廊以坚硬的棱角、刚直的轮廓，象征着冰山雪峰，凸现生命意志的顽强与坚韧。

露珠般的建筑采光天窗和出入口散布在绿草覆盖的"山坡"上，阳光下散发出自然灵性的光芒。

曲折蜿蜒的道路象征着承载生命的水源、河流。

这些超常规的抬升、下沉、变形以及建筑与土地的穿插交错，重新建构了整个街区的地质机理，为城市找到了一种趋向自然的空间层次，从而实现了：

土地的多功能高效混合使用，完备的设施及充足的公共绿化开放空间的同时获得；

机动车、人行的分层组织，地面人行步道系统的形成；

景观的整体性及景观与功能的高度统一；

近期、远期功能的顺利转换。

■ Planning and Design Concept

The project fulfils a serious of requirements given by the Commission sufficiently with a plan which focuses on heaving and sinking distinctively and enriched creativity. Located along the east and west edge, the building's roofs slope down slowly and dissolve with the ground as a whole. Six air-open sports are lowered down 6 meters from the ground.

The undulating terrain of the site forms a delightful contrast with the lush West Mountain in the distance.

The recessed outdoor sport fields, bring to mind lunar craters. Their perfect circular shapes, grand scale and flexible distribution hint at Nature's mysterious forces.

Randomly angled columns and vertical glass panes, explore contrasting materials and changes of shadows, revealing a diverse woodland scene bathed in the sunshine and moonlight.

The icicle-like glass gallery shows the resolute and dauntless will of life with its solid edges, corners and strong profile.

The floe-like skylights provide natural architectural lighting. Scattered across the grassy "hillside" with entrances and exits, they glitter with natural and intelligent brilliance.

The twisting and turning roads symbolize headwaters and rivers which bear and support life.

All these heavings, distortions and mutual penetrations of buildings and ground would make up the geological profiles of this new district and would contribute to the natural stratification of the city, realizing:

Efficiently intermix use of land obtaining maturity facilities and ample green public open space at the same time.

Organizing people and vehicle transportation separately; establishing ground pedestrian system;

Integration of landscape, while unifying landscape with function;

Transforming the functions of the buildings successfully from near future to far future.

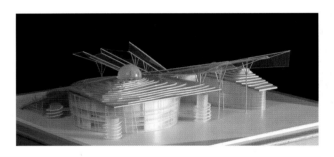

Wukesong
2008

哈尔滨工业大学建筑设计研究院（中国）
ARCHITECTURAL DESIGN AND RESERARCH INSTITUTE OF HIT(CHINA)

莫斯科第四国立设计院（俄罗斯）
GUE MNIIP "MOSPROECT-4"(RUSSIA)

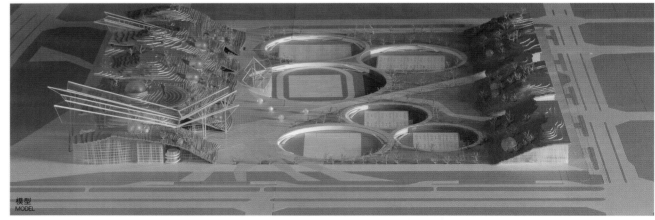

模型
MODEL

全区总平面图(赛时)
OVERALL AREA MASTER PLAN (DURING GAMES)

1 篮球馆(18200座)
  BASKETBALL HALL (18,200 SEATS)
2 主入口
  MAIN ENTRANCE
3 垒球场(8500座)
  SOFTBALL (8500 SEATS)
4 棒球场(25000座)
  BASEBALL COURT (25,000 SEATS)
5 垒球训练场地
  SOFTBALL TRAINING CENTER
6 棒球训练场地
  BASEBALL TRAINING CENTER
7 观众停车场(1270车位)
  PARKING FOR SPECTATORS (1,270 CARS)
8 贵宾、运动员、媒体停车场(940车位)
  PARKING FOR VIP, SPORTSMEN, MEDIA (940 CARS)
9 活动场地
  ACTIVITY AREA
10 酒店
   HOTEL
11 商务、商业娱乐设施
   ANCILLARY BUSIBESS AND ENTERTAINMENT FACILITIES

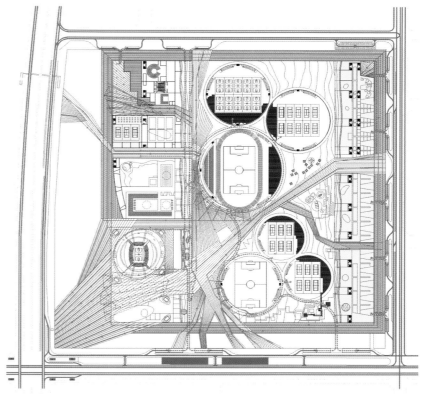

平面图±0.0米(赛时)
GROUND FLOOR LEVALE ± 0.0m (DURING GAMES)

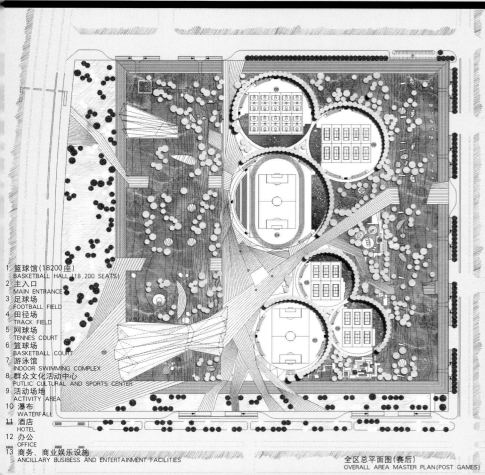

1 篮球馆(18200座)
  BASKETBALL HALL (18,200 SEATS)
2 主入口
  MAIN ENTRANCE
3 足球场
  FOOTBALL FIELD
4 田径场
  TRACK FIELD
5 网球场
  TENNES COURT
6 篮球场
  BASKETBALL COURT
7 游泳馆
  INDOOR SWIMMING COMPLEX
8 群众文化活动中心
  PUTLIC CULTURAL AND SPORTS CENTER
9 活动场地
  ACTIVITY AREA
10 瀑布
   WATERFALL
11 酒店
   HOTEL
12 办公
   OFFICE
13 商务、商业娱乐设施
   ANCILLARY BUSIBESS AND ENTERTAINMENT FACILITIES

全区总平面图(赛后)
OVERALL AREA MASTER PLAN (POST GAMES)

鸟瞰图(赛后)
BIRDS-EYE VIEW (POST GAMES)

五棵松文化体育中心 Beijing wukesong Cultural and Sports Center

参赛作品 Works

模型 MODEL

Wukesong 2008

北方工业大学建筑学院（中国）
COLLEGE OF ARCHITECTURE OF NORTH CHINA UNIRERSITY OF TECHNOLOGY(CHINA)

北京中色北方建筑设计院（中国）
BEIJING ZHONGSE NORTH ARCHITEETURAL DESIGN INSTITUTE(CHINA)

■ 规划设计构思

本方案规划布局采用绿地与建筑相互环绕的方式。以10ha大片绿地公园为规划区中心，以建筑环绕。同时，以篮球馆为规划区的核心建筑，周围环绕绿地水面。

在规划区内，按季节要求、时间要求、年龄要求交叉安排市民喜爱的多功能室内外空间，实现北京西部地区市民文化体育中心的功能定位，填补该地区目前的功能空白。通过人行通路与"视线走廊"的多方位穿透，实现大众性、欢快性、亲切性与开放性气氛的营造。

本方案通过地下空间与地上空间、绿地与广场、水面与建筑有机结合，体现出合理利用土地的开发观念。充分结合自然气候条件、合理有效地进行雨水收集与污水再利用，实现节水节能，充分体现可持续性发展的开发观念。

■ Planning and Design Concept

The layout of this plan is to adopt a designing mode in which the green space and buildings cincture each other. With a vast green space and a park of 10 hectares at its center, the planned district will be encircled by a variety of buildings, among which the Basketball Hall will be the core building encircled by a green space and water.

In the planned district, a multifunctional outdoor activity space will be arranged to offer different intersecting programs for the residents according to seasons, time of a day, and the age of the residents, thus rendering the functional positioning of the cultural and sports center of the residents in the western district of Beijing, filling up the existing gaps in functions of this district, building an atmosphere of popularity, affability, cheerfulness, and openess by arranging for the passageways and the corridor of the line of vision to penetrate each other in a multiple way so that they will represent Beijing residents' persistent pursuit of unity, progress, better life and more beautiful future by diversifying the images of buildings that will look as natural as flowing water and floating clouds.

Through an organic combination between the underground space and ground space, between the green space and square, water and buildings, this plan will best embody the developmental concept of rational use of land. In terms of environmental technology, this plan will make full use of the natural and climate conditions, and rationally and effectively collect rainwater and reuse sewage to save water and energy resources, thus fully materializing the developmental concept of sustainable development of the city.

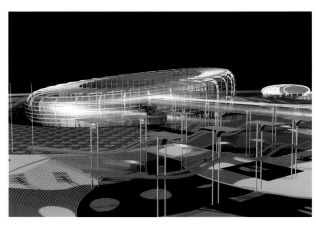

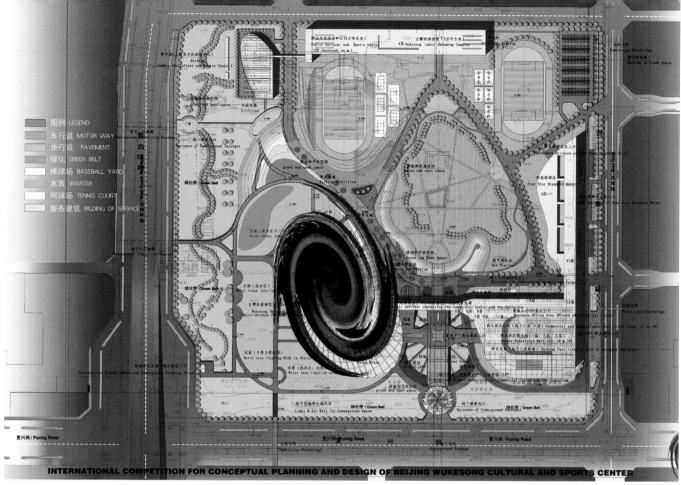

←全区总平面图(赛后)
OVERALL AREA MASTER PLAN (POST GAMES)

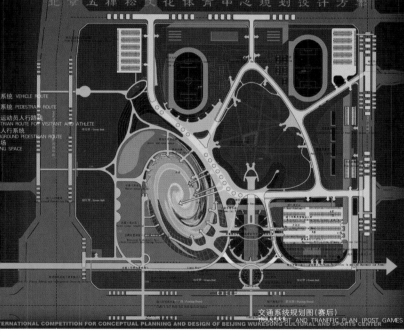

交通系统规划图(赛后)
TRANSIT AND TRAFFIC PLAN (POST GAMES)

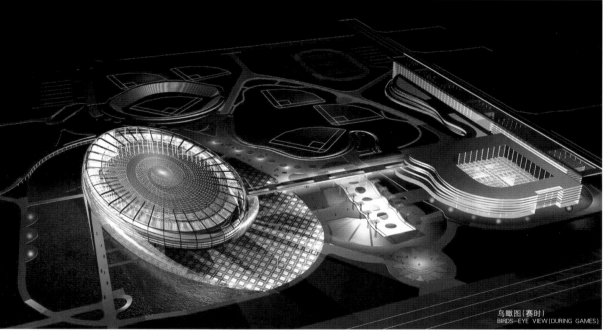

鸟瞰图(赛时)
BIRDS-EYE VIEW (DURING GAMES)

五棵松文化体育中心 Beijing wukesong Cultural and Sports Center

参赛作品 Works

模型
MODEL

Wukesong 2008

DP建筑设计有限责任公司（新加坡）
DP ARCHITECTS PTE LTD.(SINGAPORE)

■ 规划设计构思

中央广场

整个建筑布局可清楚地看到三座巨型文化体育设施耸立在浮离地面6.0m的广场上。广场与地面隔离将起着分隔行人与机动交通的作用。广场层形成了主要行人交通的步行区，并以斜坡道把地面行人从场地四个角落的集散点索引至广场。

商业区

酒店及商业娱乐设施主要聚集在整个规划区以北的区域。商业区与广场上的零星商业设施相连，无形中使各方的商业活动受到互动，形成了商业热带。

适应性再使用

方案将远期的建筑设施建设在近期的临时建筑上。在赛后，棒球场和垒球场将分别改装成室内游泳中心和群众文化活动中心。此外，棒球和垒球的训练场地则在赛后成为室外体育场地。

■ Planning and Design Concept

The Central Plaza

The plaza spaces between the stadiums connect to the corners of the site, most importantly the subway station on the southwest. Towards the west and south of the site, the park is lined with series of Olympics and sports sculptures.

Commercial Belt

The commercial zone is located along the north boundary, fronting the commercial and residential neighbourhood. The commercial activity is connected & flows to the entire plaza level, allowing the plaza area to be more lively and vibrant.

Adaptive Reuse

After the Olympic Games, with some alteration works, the Baseball Stadium will be converted into the Indoor Swimming Complex and the Softball stadium will be converted into the Sports & Cultural Complex with strong emphasis on environmental issues such as natural lighting and landscape for the enjoyment of the internal and external spaces.

1 篮球馆 BASKETBALL HALL
2 垒球场 SOFTBALL
3 棒球场 BASEBALL COURT
4 垒球训练场地 SOFTBALL TRAINING CENTER
5 棒球训练场地 BASEBALL TRAINING CENTER
6 中央广场 CENTRAL PLAZA

全区总平面图（赛时）
OVERALL AREA MASTER PLAN (DURING GAMES)

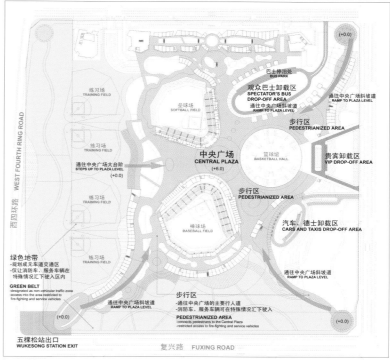

交通系统规划图
TRANSPORT AND TRAFFIC PLAN

鸟瞰图（赛后）
BIRDS-EYE VIEW (POST GAMES)

篮球馆内部透视图
INTERNAL PERSPECTIVE

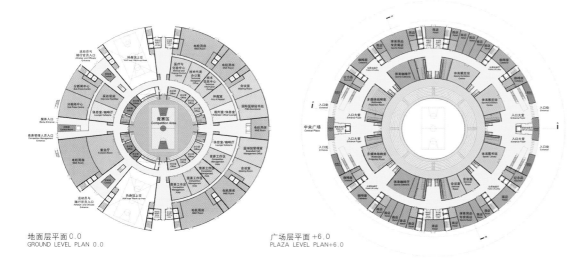

地面层平面 0.0
GROUND LEVEL PLAN 0.0

广场层平面 +6.0
PLAZA LEVEL PLAN +6.0

前立面（沿中央广场）
FRONT ELEVATION (ALONG CENTRAL PLAZA)

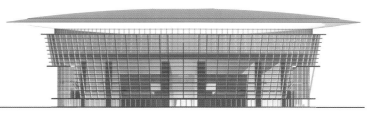

侧立面（沿卸载区）
SIDE ELEVATION (ALONG DROP OFF AREA)

■ 规划设计构思

规划方案通过主要干路的划分，将每个建设用地划分成三部分。

第一部分是运动场馆，它位于用地东南至西北的中间地。

第二部分是商业服务设施，它需要相对独立的出入口，远离人流频繁密集的体育场馆。

第三部分是绿地和开放空间，它始于宾馆与篮球馆之间的广场绿地直至西南部广阔的绿化空间，与其西南两侧的城市公共绿化带连成一体，如绿色的织锦。

交通流线，简捷顺畅。

所有的建筑与道路都锁定在以篮球馆为中心的南北东西的轴线和其旋转45°的两组轴线网上。

布局紧凑，节省用地，并留有发展余地。如赛时北面沿街停车场，在赛后可改建成商业用房，车库改设于地下。绿化用地约50%，北入口东侧绿地赛后可作为建筑发展用地。

景观设计将规划区分为五大区域 一为中心主场馆西南边的林阴休闲广场；二为水池前的下沉台地广场；三为主场馆北边喷泉水池广场；四为健身区；五为儿童乐园区。

■ Planning and Design Concept

In this scheme, the whole construction land can be divided into three portions based upon the division of the main trunk road.

This first portion is the sports venue, which is located in the central part of the construction land from the southeast to northwest.

The second portion is commercial services facilities.

The third portion is green land and open space.

Traffic streamlines are simple and smooth

All the buildings and roads are locked on the axle of south, north, east and west directions of the centered basketball hall and on the net of the two groups of axles after the former axle circumvolve about 45 degree.

Layout is closely compacted, land is accordingly saved, and space is left out for further development. Take an example, the parking lot along the street on the north during the Games will be reconstructed into commercial buildings after the Games, and the garages will be set up under the ground. The green coverage is about 50%. The green land at the east side of the north entrance will be land reserved for further development the architectural buildings.

Our design divides this land into five large areas: the first area is the green shades lounging plaza to the southwest of the central main sports venues; the second area is the sinking mesa plaza in front of the water pool; the third area is fountain pool plaza to the north of the main sports venues; the fourth is body-building zone; and the fifth is the play garden for the children.

总参工程兵第四设计研究院(中国)
FOURTH DESIGN INSTITUTE, CORPS OF ENGINEER, GS(CHINA)

Wukesong 2008

模型 MODEL

鸟瞰图(赛后)
BIRDS-EYE VIEW (POST GAMES)

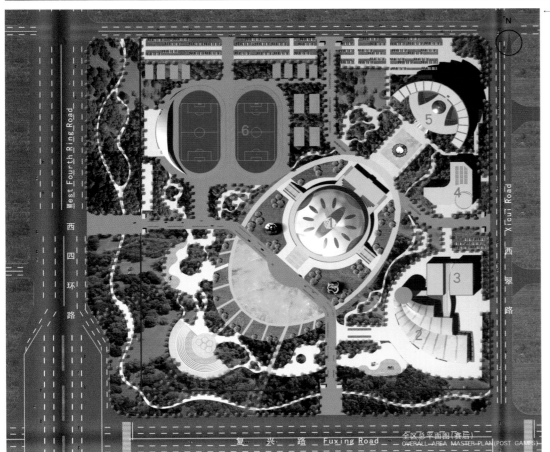

全区总平面图(赛后)
OVERALL-AREA MASTER-PLAN (POST GAMES)

1. 篮球场 BASKETBALL HALL
2. 群众文化活动中心 PUBLIC CULTURAL SPORTS CENTER
3. 游泳馆 SWIMMING COMPLEX
4. 商业服务设施 COMMERCIAL SERVICES FACILITIES
5. 酒店 HOTEL
6. 室外运动场 OUTDOOR SPORTS GROUND

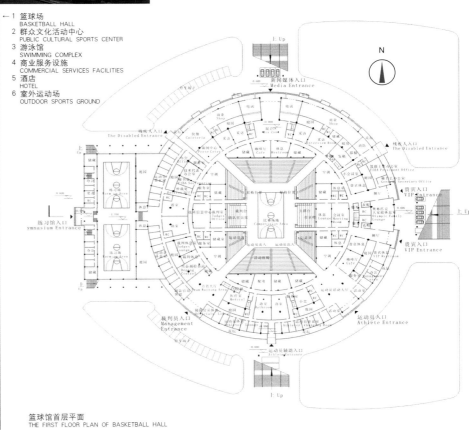

篮球馆首层平面
THE FIRST FLOOR PLAN OF BASKETBALL HALL

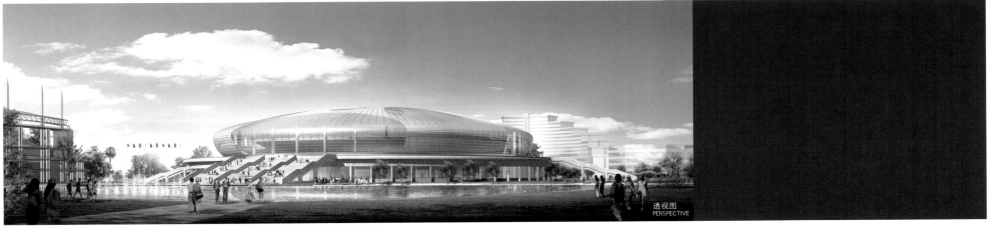

透视图
PERSPECTIVE

■ 规划设计构思

以两个广场和两条空间轴线，定位主体建筑。

两个广场：南端文化广场为下沉式音乐旱喷泉广场（露天剧场）是整个规划区的形象中心。北端火炬广场是核心空间也是整个规划的交通中心。

两条主轴线：文化广场和火炬广场形成南北向空间主轴线。

奥运之门与北端火炬广场，以及东西向主要道路形成东西向空间主轴线。

东西向主轴线以火炬广场为中心设主要道路与通向奥体中心主会场道路——四环路相接。

两条轴线贯通南北与东西两个空间，统领三个主要比赛场，突出构图的均衡性和协调性。

开放空间的形态：整个文化体育中心两条空间轴线，四角点状布置主体建筑。文化中心为下沉式广场，降低中轴上文化活动中心的高度，形成东西和南北两条宽阔的空间视线走廊，使半封闭空间具有全开敞的空间效果，不论在任何位置都可欣赏京西西山的绿色风光和翠微山的优美天际线。

■ Planning and Design Concept

The orientation of principal architectures will be done with 2 squares and 2 space axes.

squares: The Cultural Square in the south, a sunken music spring square (open-air theatre) will be the center of the entire planning zone.

The Torch Square in the north will be not only the core space, but also the traffic center of the whole plan.

major axes: The side both Cultural Square and Torch Square will form the major axis of the space in south and north directions.

A basketball gymnasium and a hotel will be arranged on 2 sides of the Cultural Square, thus, the 2 emblematic principal buildings with different types will be symmetric to each other.

The side 2 axes will pass through the spaces in the directions of south/north and east/west and lead to the 3 major competition sites. So, the proportionality and harmony of the composition will be prominent.

Pattern of open spaces: The principal architectures on the 2 space axes of the entire Cultural and Sports Center will be arranged in quadrangular spot form. The Cultural Center will be of sunken square and the height of cultural activity center on the central axis will be reduced. Therefore, 2 wide space sightseeing corridors will be formed in the directions of east/west and south/north, thus making the semi-closed spaces have a full-open space effect. No matter where a man stands, he can enjoy the green scenery of the West Hills and the exquisite skyline of Suiwei Hill in west part of Beijing.

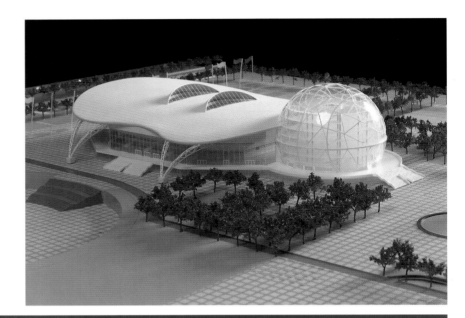

Wukesong 2008

北京首钢设计院（中国）
BEIJING SHOUGANG DESIGN INSTITUTE (CHINA)

鸟瞰图（赛后）
BIRDS-EYE VIEW (POST GAMES)

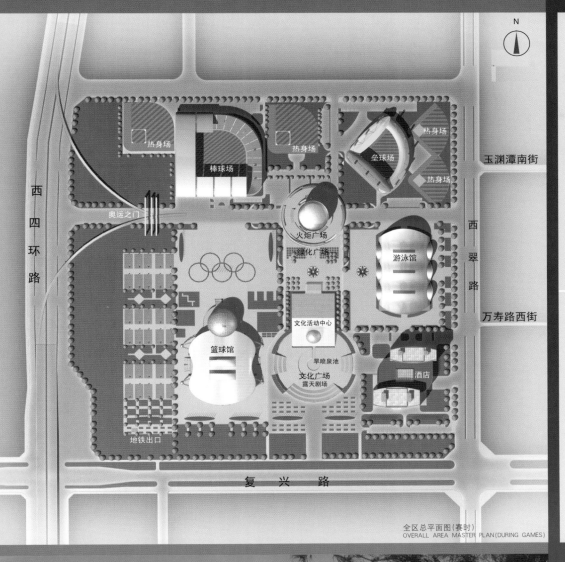

全区总平面图（赛时）
OVERALL AREA MASTER PLAN (DURING GAMES)

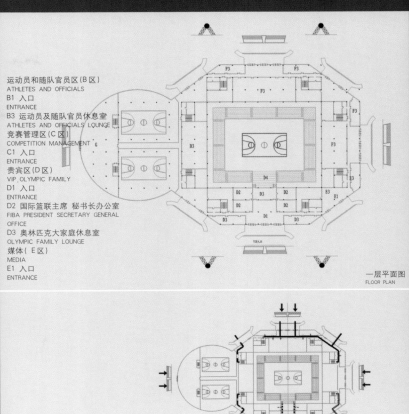

运动员和随队官员区(B区)
ATHLETES AND OFFICIALS
B1 入口
ENTRANCE
B3 运动员及随队官员休息室
ATHLETES AND OFFICIALS LOUNGE
竞赛管理区(C区)
COMPETITION MANAGEMENT
C1 入口
ENTRANCE
贵宾区(D区)
VIP, OLYMPIC FAMILY
D1 入口
ENTRANCE
D2 国际篮联主席 秘书长办公室
FIBA PRESIDENT SECRETARY GENERAL OFFICE
D3 奥林匹克大家庭休息室
OLYMPIC FAMILY LOUNGE
媒体(E区)
MEDIA
E1 入口
ENTRANCE

一层平面图
FLOOR PLAN

运动员人流　ATHLETE FLOW
竞赛管理人流　COMPETITION MANAGEMENT PERSONNEL FLOW
贵宾人流　DISTINGUISHED GUEST FLOW
媒体人流　MEDIA FLOW
观众人流　AUDIENCE FLOW

交通流线分析图
COMMUNICATION LINE CHART

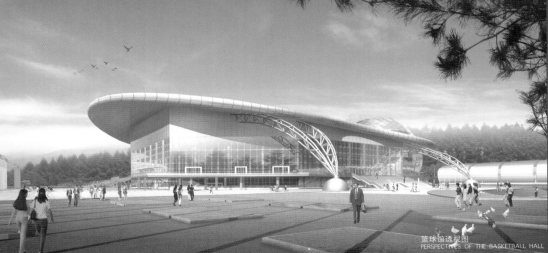

篮球馆透视图
PERSPECTIVES OF THE BASKETBALL HALL

273

Wukesong 2008

浙江省建筑设计研究院(中国)
ZHEJIANG PROVINCE ARCHITECTURAL DESIGN AND RESEARCH INSTITUTE(CHINA)

■ 规划设计构思

将篮球馆放在地块西南角,对城市景观及人流组织都非常有利。

将酒店设在地块的西北角,主要是考虑到城市形象在一期便能比较完整。

在地块设计中着眼于城市空间的角度,在两条主要马路前均保留足够的空间,让城市渗透到本地块之中,尤其是在靠复兴路一侧,设有大型广场,既可作为场馆大型活动时的集散场地,又可作为平时市民休憩、锻炼之用。

将地铁人流通过地下通道引入地块内,结合下沉广场空间布置商业街,既活跃了本地块的商业气氛,又避免城市景观过于杂乱。

环境设计强调建筑和景观一体化,使城市环境质量得到改善。

在建筑单体设计中,充分考虑广场路自身"造血"的机能,充分运用最新科技手段和建筑材料,以便于节约能耗,保护环境。

■ Planning and Design Concept

Basketball hall is located on the south-west corner of the site which benefits the urban landscape and pedestrian circulation system.

Hotel is served on the northwest corner of the site mainly considering the integrity of the urban identity at the first phase.

Respectful of the urban planning criteria, enough space is reserved along the two driveways. Master planning of the site will be designed and articulated into the urban landscape, especially provided with a grandiose plaza near Fuxing road acting as the concourse area for large events and the destination to relax and exercise for the citizen.

Pedestrian circulation from the subway station will be directed to the site. Incorporated into the sunken plaza, commercial festive streets will help boost the business around the site and avoid the urban landscape to be overloaded out of order.

Landscaping design focuses on the integration of the architecture and the immediate context which improves the urban environmental quality.

In designing the components within the site, venues are conceived to sustain themselves by business activities. Focusing on the efficient and effective use of new technology and material system, consumption of the energy source will be minimized and the environment will be protected.

MODEL

鸟瞰图（赛后）
BIRDS-EYE VIEW (POST GAMES)

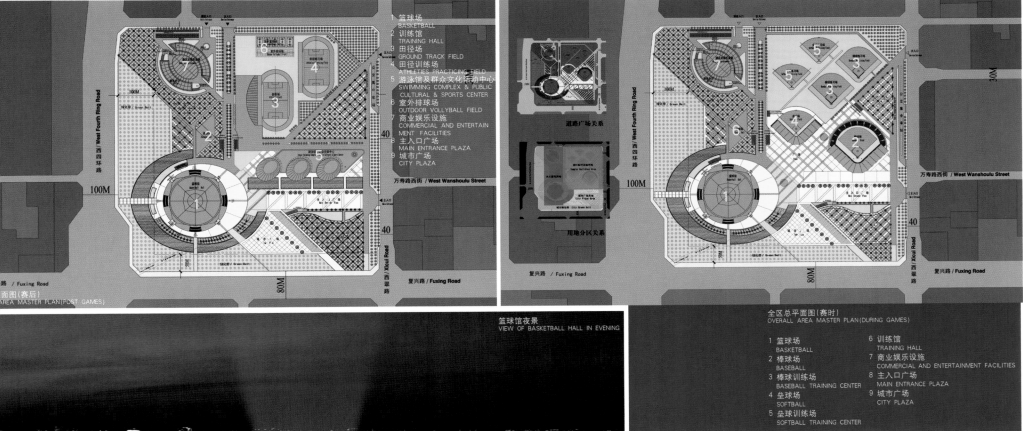

篮球馆夜景
VIEW OF BASKETBALL HALL IN EVENING

全区总平面图（赛时）
OVERALL AREA MASTER PLAN (DURING GAMES)

1 篮球场 BASKETBALL
2 棒球场 BASEBALL
3 棒球训练场 BASEBALL TRAINING CENTER
4 垒球场 SOFTBALL
5 垒球训练场 SOFTBALL TRAINING CENTER
6 训练馆 TRAINING HALL
7 商业娱乐设施 COMMERCIAL AND ENTERTAINMENT FACILITIES
8 主入口广场 MAIN ENTRANCE PLAZA
9 城市广场 CITY PLAZA

Wukesong 2008

考克斯集团（澳大利亚）
COX GROUP (AUSTRALIA)

五棵松文化体育中心
Beijing Wukesong Cultural and Sports Center

参赛作品 | Works

■ 规划设计构思

方案的主要设计战略，是在现有街面之上开发出一个全新的行人广场。访客可以走上楼梯斜坡，进入焦点和活动的中心广场。到达五棵松，人们就是进入期望、预料之地。广场的设计，能让公共空间下面出现层次完全不同的人流现象。体育文化建筑物的下层经过设计，可以让官员和运动员无需通过公共区域，就能进入运动场层面。

■ Planning and Design Concept

The key design strategy is to develop a new pedestrian plaza one level above existing street level. This achieves a number of objectives: visitors will ascend stairs and ramps to the plaza which focus attention and activity. The plaza allows the development of a completely separate level of circulation below the public space. The lower level of the sporting and cultural buildings is positioned in a way to give officials and athletes access to the playing level without entering the public domain.

模型 MODEL
北京五棵松文化体育中心规划（赛后）
BEIJING WUKESONG CULTURAL AND SPORTS CENTRE PLANNING (POST-GAMES) 1:1000

鸟瞰图(赛后)
BIRDS-EYE VIEW (POST GAMES)

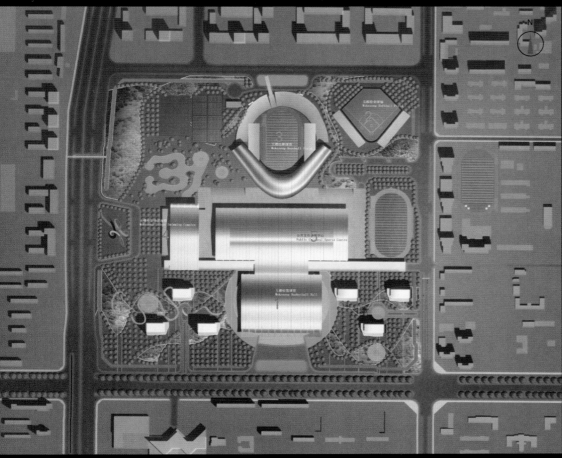

全区总平面图(赛后)
OVERALL AREA MASTER PLAN (POST GAMES)

1 篮球馆 BASKETBALL HALL
2 垒球场 SOFTBALL
3 棒球场 BASEBALL COURT
4 室内游泳馆 INDOOR SWIMMING COMPLEX
5 公共文化体育中心 PUBLIC CULTURAL SPORTS CENTER

模型
MODEL

五棵松文化体育中心 Beijing wukesong Cultural and Sports Center

参赛作品 | Works

■ 规划设计构思

规划用一个45°扇形布局,将西侧100m及南侧50m的城市绿化带连成一整片,使本规划得到总共近20ha绿地。

用一条50m宽、包含着12m宽绿化带的"道路——广场",将10ha的集中绿地和五棵松建筑场馆东西连接起来。10ha的绿地,50m的大道加上5栋场馆,形成了一个城市大道空间的延伸。规划主轴如"直箭",副轴如"弓弧",似"孔雀开屏"。

四星级以上的饭店,设置在西南角地铁出口。这个位置将为饭店带来很好的效益,使城市景观更趋丰满。带来生活气息和繁荣商机。

市民文化活动中心和市民用的运动场,设置在东北角。这是因为未来住宅开发区及现有居民区集中住在北面和东面。

■ Planning and Design Concept

We consider that the only way that could connect the 100m and 50m city green belts to West and South side is to plan a "45°„Fan Shape Plan".

There is a 50m wide radius main "Avenue-plaza", which includs a tree/green belt of 12m wide. This "Avenue plaza" fulfills the function, not only for the main traffic flow, but also for connecting the 10 hectare green/open space and the 5 sports buildings complex.

The hotel of four star class is located at the South West corner of the site where is also the exit of subway. This will bring well profit to the hotel and enrich the urban life.

The Community Cultural Center and the track field are located at the North East side of the site. This is because most of the residential area is to the North and East of the site.

大地建筑事务所(国际)(中国)
GREAT EARTH ARCHITECTS & ENGINEERS,INTERNATIONAL(CHINA)

异空建筑师事务所(韩国)
BEYOND SPACE ARCHITECTS(KOREA)

沈祖海建筑师事务所(中国台湾)
HAIGO SHEN & PARTNERS,ARCHITECTS AND ENGINEERS(TAIWAN CHINA)

核工业第二研究设计院(中国)
BEIJING INSTITUTE OF NUCLEAR ENGINEERING(CHINA)

Wukesong 2008

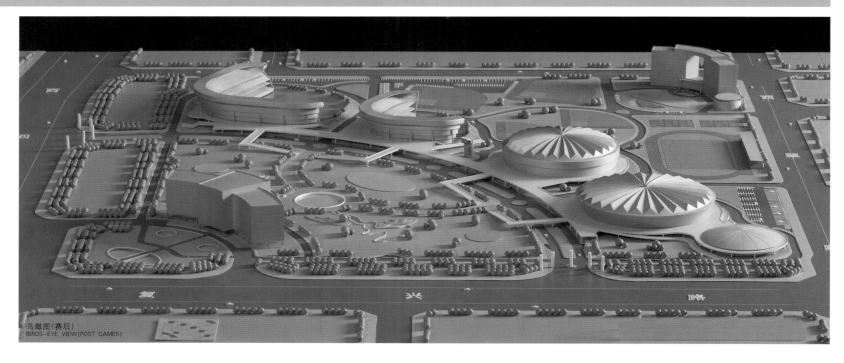

鸟瞰图(赛后)
BIRDS-EYE VIEW(POST GAMES)

← 鸟瞰图（赛时）
BIRDS-EYE VIEW (DURING GAMES)

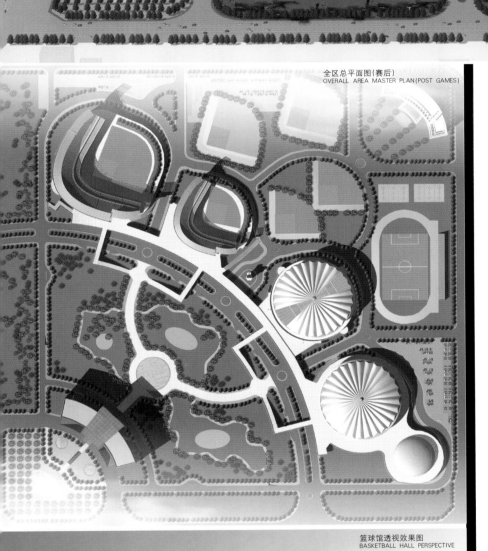

全区总平面图（赛后）
OVERALL AREA MASTER PLAN (POST GAMES)

篮球馆透视效果图
BASKETBALL HALL PERSPECTIVE

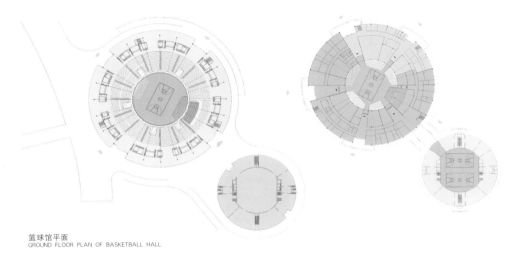

篮球馆平面
GROUND FLOOR PLAN OF BASKETBALL HALL

篮球馆西立面
BASKETBALL HALL WEST ELEVATION

篮球馆立剖面
BASKETBALL HALL VERTICAL

五棵松文化体育中心
Beijing Wukesong Cultural and Sports Center

参赛作品 Works

■ 规划设计构思

用地的功能布局可以用一条南北主轴线，两条斜交叉轴线，以及内外围两个环的建筑场地设施来概括。

五棵松体育中心的主要功能分为三方面：奥运会场馆设施区、群众性文化体育设施区、商业设施区，三者交叉布置于规划用地上。

鉴于复兴路与西四环路均为城市干道，不宜有过多的出入口，故在场地东侧及北侧设置车辆地面及地下出入口，在出入口附近设置大型车辆地面停车场及地下停车库。

在生态环境规划中采用热循环系统、雨水循环系统。

■ Planning and Design Concept

One main axis from north and south, two intersected oblique axis and two ring fields with installations inside and outside.

The main function of Wu Ke Song Sports Center can be divided into three aspects: Olympics assembly hall building facilities area, crowd culture athletics facilities area and business facilities area. The three areas are crisscross arranged in the programming land.

Concerning the Fuxing Street and the 4th West road circling the city are the city's arterial roads, not many entrances and exits are probably to be built there. So presently the ground and underground exits and entrances are set up at the east and north side of the field. Large parking lots on the ground and underground will be established nearby.

Ecology environment planning: Use Of Heat circulation system and rainwater circulation system.

北京大学城市规划设计研究中心（中国）
THE CENTER FOR URBAN PLANNING AND DESIGN (CUPD), PEKING UNIVERSITY (CHINA)

Wukesong 2008

MODEL

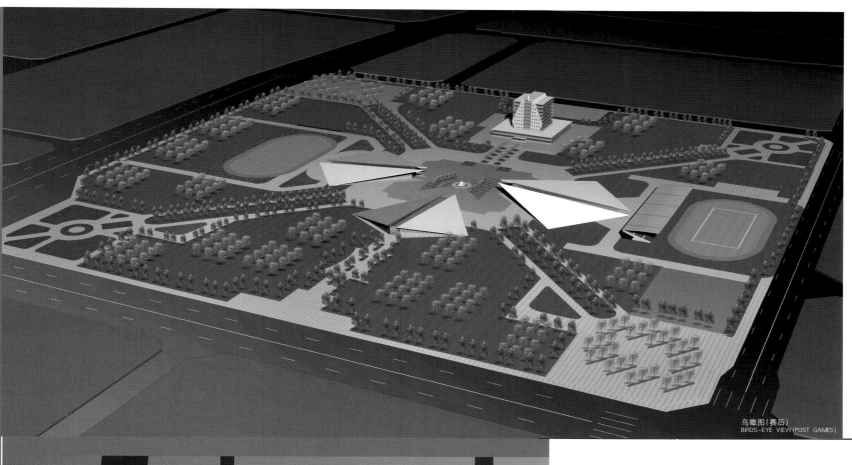
鸟瞰图(赛后)
BIRDS-EYE VIEW (POST GAMES)

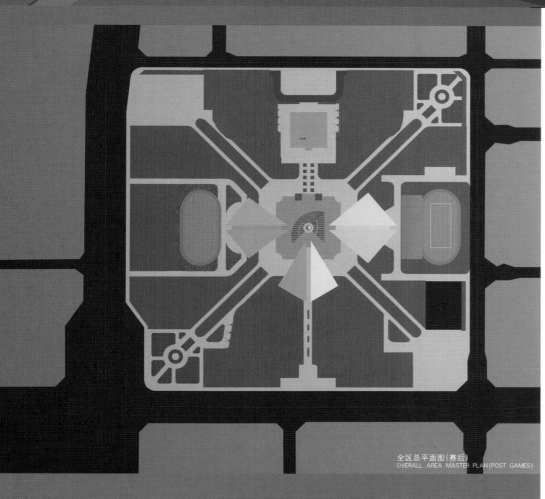
全区总平面图(赛后)
OVERALL AREA MASTER PLAN (POST GAMES)

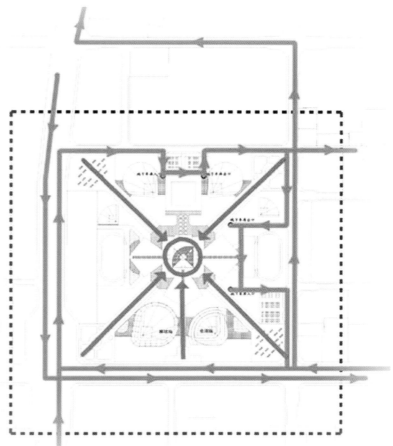
交通流线图(赛后)
TRAFFIC STREAMLINE (POST GAMES)

五棵松文化体育中心 Beijing Wukesong Cultural and Sports Center

参赛作品 Works

■ 规划设计构思

北京的城市肌理源于中国传统的规划手法及精神理念,是一个有着严格秩序及观念的城市脉络,其规划理念几千年来始终沿袭,这一特征在每个城市风貌中都得到过体现。作为地处这一特定背景下的规划项目,本方案设计融于城市肌理中,使之符合传统的规划手法,同时又引入新概念、新手法,彰显自我。规划方案依据轴线组织各大场馆,并将奥运场馆置于中轴线上,在身份及性质上与周边建筑及其他功能建筑区分开,同时又便于组织集中的人流。本规划方案在总体形态上以一个外圆联系各不同分区,隐喻2008年全世界将团聚于此,展开公平竞赛。

■ Planning and Design Concept

The center is situated in the city context of rigid order which originated from tradition Chinese planning ideas. This proposal is intended to express it self with modern design ideas and also harmonize with the city fabric, arranging the courts of basketball, baseball and softball at the axis. This scheme connects the different areas with an external to symbol the gathering of the peoples all over the world and the competition justice.

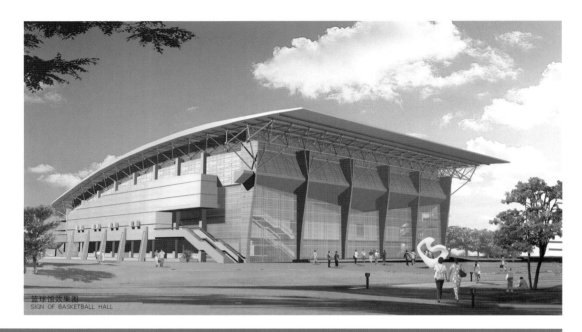
篮球馆效果图
SIGN OF BASKETBALL HALL

Wukesong 2008

北京中标勘察设计咨询公司(中国)
BEIJING ZHONGBIAO RECONNAISSANCE DESIGN AND CONSULTATION COMPANY (CHINA)

模型
MODEL

鸟瞰图（赛后）
BIRDS-EYE VIEW (POST GAMES)

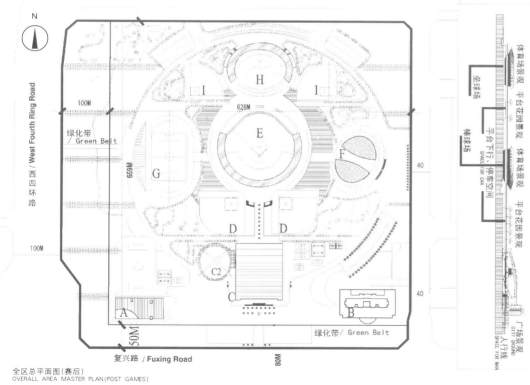

全区总平面图（赛后）
OVERALL AREA MASTER PLAN (POST GAMES)

A 商场 SHOPPING CENTER	C2 篮球训练馆 BASKETBALL TRAINING HALL	F 游泳馆 SWIMMING HALL	I 垒球训练场 TRAINING CENTER	
B 酒店 HOTEL	D 棒球训练场 BASEBALL TRAINING FIELD	G 田径场 SPORTS GROUND		
C 篮球馆 BASKETBALL HALL	E 棒球场 BASEBALL FIELD	H 垒球场 SOFTBALL FIELD		

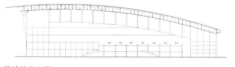

篮球馆东立面
BASKETBALL EAST ELEVATION

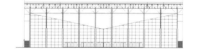

篮球馆南立面
BASKETBALL SOUTH ELEVATION

1-1 剖面 1:500
(Sections)

篮球馆剖面
BASKETBALL SECTIONS

283

五棵松文化体育中心 Beijing wukesong Cultural and Sports Center

■ 规划设计构思：

篮球馆位于场地的西南角,使其成为复兴路及西四环路上的标志建筑。用旅馆和文化中心建筑来共同强调篮球馆在复兴路上的重要位置。地铁线和五棵松地铁站也与篮球馆有直接的交通联系。

该设计将室外体育设施集中在场地中心。这有利于对比赛期间的功能组织，同时在奥运会后给周围居民区保留一个开放的绿色空间。

利用景观软环境和水体在建筑物和开放空间之间达成一种平衡。并且，景观环境设计可延续至奥运公园之外。

沿复兴路及西四环路上的红线内绿地也结合到总的景观设计中。

本设计对奥运期间和奥运会后的公众交通流线做了明确的界定。

本设计对商业和零售业的分期建设做了灵活的安排以满足社区和市场的需求。

本总体规划一开始就围绕着多功能组织分析的要求来进行设计的。

■ Planning and Design Concept

An Arena building which anchors the south-west corner of the site.providing a landmark building with a clear visual connection to both Fuxing Road and the West Fourth Road .

Placement of the Hotes Building and the Cultural Centre to reinforce the Fuxing Road address.This was to utilize a simple connection to the existing subway line and Wukesong Station.

The plans allow for the outdoor sports facilities to be centrally located on the site.This allows for practical"overlay"considerations during the Games mode and a generous green space plan for the surrounding residential neighborhoods in the post-Games mode.

It uses soft landscape and water features to create a balance between built forms and open space.The landscape is further designed to extend beyond the site as surrounding precincts are redeveloped in the future. The green reserved zones along Fuxing Road and the West Fourth Ring Road are also incorporated into the overall landscape concept.

Public circulation areas are clearly defined in both the Games mode and Post Games mode.

The plan adopts a flexible approach to the construction phasing of the commercial and retail spaces to meet demands by the community and the market place.

The master plan is designed around practical overlay requirements at the beginning of the design phase,rather than a complicated retrospective design at later stages of the project.

参赛作品 Works

Wukesong

HOK 体育建筑设计公司（澳大利亚）
HOK SPORT+VENUE+EVENT(AUSTRALIA)

2008

模型 MODEL

五棵松文化体育中心
Beijing wukesong Cultural and Sports Center

参赛作品 Works

■ 规划方案构思

本设计为整个用地提供一个人们能在其中自由行走和从事各种活动的森林。

本方案将整个用地看作是一个均等的界面,采用多种质感的、人们可在其上活动和行走的地面,并将其进行随机安排,最大效率地运用土地,并为赛时人员疏散提供最大的空间。在远期规划中,整个用地将没有阻挡人们行走的边界,它将成为一个与周围城市空间全面融合的市民活动中心。

本方案将用地内的各类设施设计为一个个在绿色海洋中的岛屿,这些岛屿在设计阶段是浮动的岛,所有的人都可以改动岛的位置,以便使其更合理,而这些改动将不会影响规划的构思。本方案是一个最易修改的方案。

本方案将树按 8m × 8m 网格布置,并全部采用可行走和使用的地面铺装,同时四周全部开敞,为场地的使用提供了巨大的灵活性。

在布局上,"岛屿"和建筑均南北向布局,以便和传统北京城市格局相一致,并且能达到节约使用土地的效果。

■ Planning and Design Concept

Our conception for this Cultural and Sports Center is a place in the city that will offer unique conveniences and multiple choices in a "green forest" square where people can leisurely stroll and participate in a multitude of activities.

This design envisions the whole site as a homogeneous interface with surfaces of diverse textures on which people could walk and participate in various activities. The random arrangement maximizes the efficiency of land usage and provides the extensive space requirements for people dispersing during the games. In long term planning, the whole site without any border obstructing pedestrians will be a citizens', activity center merging with the urban space as a whole.

In this design, the various facilities in the site are grouped and conceived as "islands" in a sea of green. These islands "float" in the design, and can be rearranged when required. Any changes will not affect the planning concept, making this planning a most revisable one.

In this design, all trees are arranged in a 8mX8m grid, and all ground pavements are for pedestrian and usable. The borders are open to provide great flexibility in use.

"Islands" and buildings are in a north-south orientation in accordance with the traditional city pattern of Beijing, and resulting in economy and efficiency in land usage.

Wukesong
2008

北京市建筑设计研究院(中国)
BEIJING ARCHITECTURAL DESIGN AND RESEARCH INSTITUTE(CHINA)

交通部规划研究院(中国)
PLANNING INSTITUTE OF MINISTRY OF COMMUNICATIONS(CHINA)

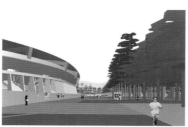   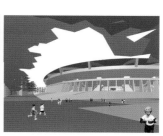

模型
MODEL

1 篮球馆 BASKETBALL HALL
2 游泳馆 INDOOR SWIMMING COMPLEX
3 文化活动中心 PUBLIC CULTURAL CENTER
4 休闲娱乐中心 CETERTAINMENT CENTER
5 商业 COMMERCIAL BUILDING
6 商业服务设施 COMMERCIAL SERVICES FACILITIES
7 酒店 HOTEL

节点索引图（赛时）
NODS INDEX (DURING GAMES)

节点索引图（赛后）
NODS INDEX (POST GAMES)　节点索引图/赛后　Nodes Index/After the Games

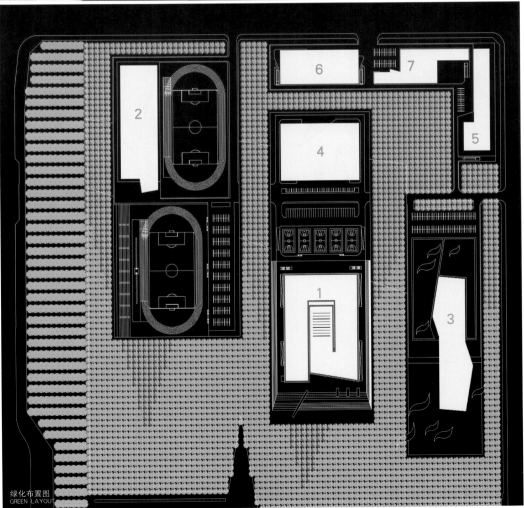

绿化布置图
GREEN LAYOUT

■规划设计构思

三角舟水晶篮球馆和绿孔雀开屏的中国扇酒店,都架空于正负零地面而形成100%绿覆盖新概念生态社区。由西北向东南被分割成主大街与次大街两大板块。三大赛事布置于西南地块中,形成城市奥运氛围。月亮湖别墅及生态酒店架空于东北三角块里。闹中取静中又多了片水晶游泳馆。在中心广场中布置约10 000m² 的移动式太空仓客栈,用于赛事之需,20 000m²的群众文化活动中心布置于有充分自然采光的篮球馆水晶湖负一层的地下花园中。

■ Planning and Design Concept

The triangle sailboat-shaped crystal basketball hall and the Chinese fan-shaped hotel looking like a peacock in his pride will be constructed on a positive & negative zero stilted level. That will create a new concept ecological community with 100% green coverage. From the northwest to the southeast, the Cultural & Sports Center will be divided into two major blocks. Three major competition events will be arranged in the southwest part of the center, thus adding an Olympic air to the city. The stilted Moon Lake villas and ecological hotel will be erected in a triangle region in the northeast. A Planned crystal swimming hall will create a quiet place in the busy mid-town. A mobile space module-like inn of about 10000m² at the central square is designed with competition events in mind. A mass cultural activity center of 20000m² is planned at level 1 of the basement garden of the Crystal Lake of the basketball hall. With such design, the mass cultural activity center will obtain full natural lighting.

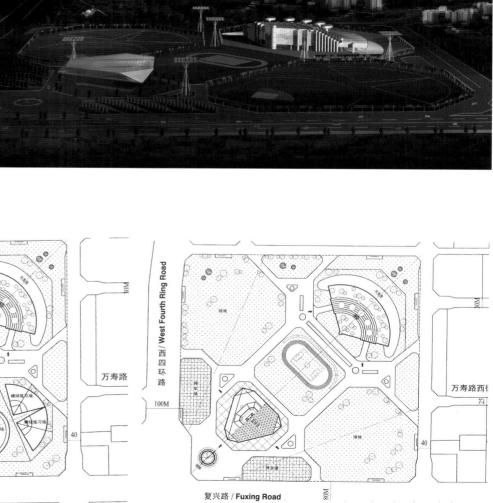

模型
MODEL

Wukesong 2008

核工业部第四研究设计院深圳设计院(中国)
THE FOURTH DESIGN INSTITUTE OF THE MINISTRY OF NUCLEAR INDUSTRY(CHINA)

伍凸设计师事务所(深圳有限公司)(中国)
WUTU ARCHITECTS ASSOCIATES(SHENZHEN LIMITED)(CHINA)

全区总平面图(赛时)
OVERALL AREA MASTER PLAN (DURING GAMES)

全区总平面图(赛后)
OVERALL AREA MASTER PLAN (POST GAMES)

■规划设计构思

基地指向奥林匹克公园的轴线作为方案的主轴线,使得本基地在更大的空间尺度上,同北京,同奥林匹克公园联系在一起,将它们之间精神上的对应关系直接展现出来。

依据垒球场和棒球场在北京的纬度上的最佳朝向布置赛场,是方案的基点所在。

通过一个向五个方向放射的公共空间体系将上述的各个功能区联接成一个整体,形成一个功能复合,交通换乘便捷的地区文化商业中心,利于政府和社会共同开发局面的形成。

■ Planning and Design Concept

The axis from the base to the Olympic Park will be set at the main axis. The use of this linkage as the main axis, the base is, in terms of its spatial measurements, more closely linked with Beijing and the Olympic Park and it also exhibits a spiritual relevance between them.

Sports venues will be arranged according to the best orientation of softball and baseball rings in terms of their latitude in Beijing. This is the base point for this design plan.

The functional areas mentioned above are integrated by a radiant pentangular public space, resulting in a local cultural and commercial center with multi-functions, which makes easy transit between bus lines and subway lines and will help to promote joint development of this area by government and general public.

上海同济城市规划设计研究院(中国)
URBAN PLANNING & DESIGN INSTITUTE SHANGHAI TONGJI UNIVERSITY(CHINA)

高柏伙伴规划园林和建筑顾问公司(荷兰)
KUIPERCOMPAGNONS, OFFICE FOR URBAN PLANNING, ARCHITECTURE AND LANDSCAPE CONSULTANCY (HOLAND)

模型
MODEL

五棵松文化体育中心
Beijing wukesong Cultural and Sports Center

■ 规划设计构思：

体现奥运精神，满足功能需求。折板延绵如翻腾的巨龙，以其舒展、开放、包容的形态，体现泱泱大国的豪气。在该主题的引领下，各部分功能随这一变化，形成一个有序的、不可分割的整体。

绿色的概念——土地赛后重复利用

功能区域划分——按功能分为东西两大部分，中间为开放的绿化空间。

交通流线的组织——划分为地面步行系统、地下交通系统和空中交通系统三个层面。

中心设计纳入城市安全防灾系统考虑，弥补目前公园作为城市防灾、救灾空间的设计缺憾。

中心预留未来发展空间用地。

■ Planning and Design Concept

Embody Olympics spirit, to satisfy the request of functions. The table -flap as a huge seething dragon, to embody the pride of the great nation with its unfolded, opened, tolerated form. Under guiding of that topic, all parts of functions become aligned, indivisible whole along with this continuously changing space.

Make use of the land, function after the games.

Division of the function parts-The center is divided into two big parts by functions, and all used grounds have located either side of the greenbelts.

Transportation organization - divided into three levels as foot systems, underground transportation systems and the air transportation systems.

Consideration of the city security systems, and design that center into the disaster prevention systems of city, complement the park to prevent disaster.

Leave spaces for use in the future.

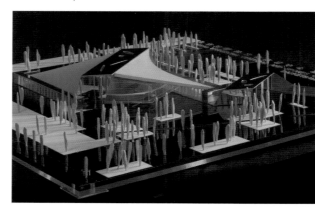

北京凯帝克建筑设计有限公司（中国）
BEIJING CADTIC ARCHITECTURE DESIGN CO,. LTD(CHINA)

参赛作品 Works

模型
MODEL

ANTHONY BECHU建筑设计公司(法国)
AGENCE D'ARCHITECTURE ANTHONY BECHU(FRANCE)

模型 MODEL

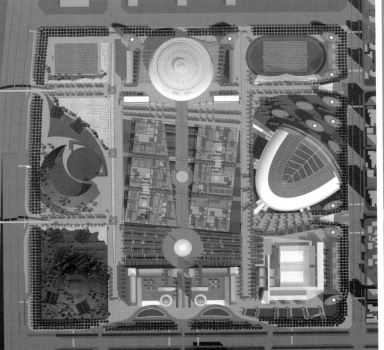
全区总平面图(赛后)
OVERALL AREA MASTER PLAN(POST GAMES)

交通系统规划图(赛后)
TRANSPORT AND TRAFFIC PLAN(POST GAMES)

五棵松文化体育中心 Beijing Wukesong Cultural and Sports Center

参赛作品　Works

Wukesong 2008

POSTSCRIPT

后记

  此次《北京奥林匹克公园和五棵松文化体育中心规划设计方案》征集活动，得到国内外众多单位和个人的热情支持和关注。从2001年底开始筹备，2002年7月份完成方案评审和公开展览。此后，两个地区的规划均在获奖方案中选择确定了实施方案的蓝本，在此基础上进行总体规划的调整深化和详细规划。这个过程，凝聚着无数参与此项工作的单位和个人的辛勤劳动和智慧结晶。在此，我们再次向他们表示最衷心的感谢！

2002 年 11 月

Postscript

  The Planning and Design Competition for the Beijing Olympic Park and Wukesong Culture and Sports Center has enjoyed the enthusiastic support both from people of China and abroad. The preparation of the event started at the end of 2001 till July of 2002 when the appraisal was finished and submitted proposals were on display in the public. Afterwards, the blueprints for implementation were selected out of the prize winners on the basis of which general and meticulous planning will be conducted. The whole process is the result of strenuous efforts and wisdom of many individuals and units. Therefore, we would like to express our heartfelt thanks to them all!

November 2002